Introduction to Algebraic and Constructive Quantum Field Theory

Princeton Series in Physics

Edited by Philip W. Anderson, Arthur S. Wightman, and Sam B. Treiman

Quantum Mechanics for Hamiltonians Defined as Quadratic Forms *by Barry Simon*

Lectures on Current Algebra and Its Applications *by Sam B Treiman, Roman Jackiw, and David J Gross*

Physical Cosmology *by P J E Peebles*

The Many-Worlds Interpretation of Quantum Mechanics *edited by B S DeWitt and N Graham*

Homogeneous Relativistic Cosmologies *by Michael P Ryan, Jr , and Lawrence C Shepley*

The $P(\phi)_2$ Euclidean (Quantum) Field Theory *by Barry Simon*

Studies in Mathematical Physics Essays in Honor of Valentine Bargmann *edited by Elliott H Lieb, B Simon, and A S Wightman*

Convexity in the Theory of Lattice Gases *by Robert B Israel*

Works on the Foundations of Statistical Physics *by N S Krylov*

Surprises in Theoretical Physics *by Rudolf Peierls*

The Large-Scale Structure of the Universe *by P J E Peebles*

Statistical Physics and the Atomic Theory of Matter, From Boyle and Newton to Landau and Onsager *by Stephen G Brush*

Quantum Theory and Measurement *edited by John Archibald Wheeler and Wojciech Hubert Zurek*

Current Algebra and Anomalies *by Sam B Treiman, Roman Jackiw, Bruno Zumino, and Edward Witten*

Quantum Fluctuations *by E Nelson*

Spin Glasses and Other Frustrated Systems *by Debashish Chowdhury*
(*Spin Glasses and Other Frustrated Systems* is published in co-operation with World Scientific Publishing Co Pte Ltd , Singapore)

Large-Scale Motions in the Universe A Vatican Study Week *edited by Vera C Rubin and George V Coyne, S J*

Instabilities and Fronts in Extended Systems *by Pierre Collet and Jean-Pierre Eckmann*

From Perturbative to Constructive Renormalization *by Vincent Rivasseau*

Maxwell's Demon Entropy, Information, Computing *edited by Harvey S Leff and Andrew F Rex*

More Surprises in Theoretical Physics *by Rudolf Peierls*

Supersymmetry and Supergravity Second Edition, Revised and Expanded *by Julius Wess and Jonathan Bagger*

Introduction to Algebraic and Constructive Quantum Field Theory *by John C Baez, Irving E Segal, and Zhengfang Zhou*

Introduction to Algebraic and Constructive Quantum Field Theory

JOHN C. BAEZ

IRVING E. SEGAL

ZHENGFANG ZHOU

PRINCETON UNIVERSITY PRESS

PRINCETON, NEW JERSEY

Copyright © 1992 by Princeton University Press
Published by Princeton University Press, 41 William Street,
Princeton, New Jersey 08540
In the United Kingdom: Princeton University Press, Oxford

All Rights Reserved

Library of Congress Cataloging-in-Publication Data

Baez, John C., 1961-
Introduction to algebraic and constructive quantum field theory /
by John C. Baez, Irving E. Segal, Zhengfang Zhou.
p. cm. — (Princeton series in physics)
Includes bibliographical references and index.
ISBN 0-691-08546-3
1. Quantum field theory. 2. C*-algebras. I. Segal, Irving Ezra.
II. Zhou, Zhengfang, 1959- . III. Title. IV. Series.
QC174.45.B29 1991
530.1'43—dc20 91-16504 CIP

This book has been composed in Linotron Times Roman

Princeton University Press books are printed
on acid-free paper, and meet the guidelines
for permanence and durability of the Committee
on Production Guidelines for Book Longevity
of the Council on Library Resources

Printed in the United States of America by
Princeton University Press, Princeton, New Jersey

10 9 8 7 6 5 4 3 2 1

Designed by Laury A. Egan

Contents

Preface

This book is an amplified and updated version of a graduate course given by one of us (IES) at various times during the past two decades. It is intended as a rigorous treatment from first principles of the algebraic and analytic core of general quantum field theory. The first half of the book develops the algebraic theory centering around boson and fermion fields, their particle and wave representations, considerations of unitary implementability, and the representation-independent C^*-algebraic formalism. This is tied in with the infinite-dimensional unitary, symplectic, and orthogonal groups, and their actions as canonical transformations on corresponding types of fields. In part, this represents a natural extension of harmonic analysis and aspects of the theory of classical Lie groups from the finite to the infinite-dimensional case. But certain features that are crucial in the physical context—*stability* (positivity of the energy or of particle numbers) and *causality* (finiteness of propagation velocity), both of which play essential roles in the quantization of wave equations—are also central in our treatment. These features are integrated with *symmetry* considerations in a coherent way that defines the essence of algebraic quantum field theory and subsumes the quantization of linear wave equations of a quite general type.

But the key process that distinguishes *quantum* from *classical* physics is that of *particle production*. The mathematical description of this phenomenon in accordance with the physical desiderata underlined above is, at the fundamental level, by quantized *nonlinear* wave equations. The algebraic theory developed in the first part of the book is used extensively in the second part in developing the mathematical interpretation, and solution in a prototypical context, of such equations. New analytic, rather than algebraic, issues of renormalization and singular perturbation arise in this connection. In particular, nonlinear local functions of quantum fields—such as fields that are not necessarily free, or defined over arbitrary Riemannian manifolds—are treated in general contexts. These are in the direction of ultimate extension to physically more realistic cases than the model treated here. This consists of the formulation of the concept of solution of a quantized local nonlinear scalar wave equation in a sense that is independent of the existence of an associated free field, and of the establishment of such solutions in two space-time dimensions (albeit by a technique that starts from a free field).

This book is intended as a self-contained introduction, and not as a treatise or compendium of results. A logically coherent and detailed treatment of a subject as indivisible and multifaceted as quantum field theory must be highly selective. We eschew both "practical" and the now classical "axiomatic" quantum field theory, on which there are excellent textbooks. Those in axiomatic theory include the authoritative texts by Bogolioubov, Logunov, and Todorov (1975); Jost (1965); and Streater and Wightman (1964). The plenitude of texts on practical theory includes the lucid treatments, in order of increasing scope, of Mandl (1959); Bjorken and Drell (1965); and Cheng and Li (1984).

We also fail to do justice to the geometrical side of the theory, partly because to do so would require another volume of the present size, but also because our approach is designed to apply in essence to virtually arbitrary geometries, the particularities of which need not be specified. Applications and illustrations are usually given in terms of Minkowski space, because this is the simplest and the most familiar. The key developments would, however, apply to general classes of space-times endowed with causal structures and corresponding transformation groups, including the context associated with the conformal group.

We include brief historical comments, but refer to the classics by Dirac (1958), Heisenberg (1930), and Pauli (1980) for the nascent formalism and spirit of heuristic quantum field theory.

We assume the reader has some background in modern analysis and exposure to the ideas of quantum theory, because it would be impractical to do otherwise in a volume of this size. We have, however, appended a glossary to clarify terminology and to help tide the reader over until he has a chance to consult a source of technical detail, in case he wishes to do so; it may also serve as a review of the technical background. A series of Lexicons in the main text serves to correlate the mathematical formulation with the physical interpretation. The Problems and Appendix B further round out the treatment and provide contact with the physical roots and interesting, if peripheral, mathematical issues.

The first half of the book should provide a basis for a one semester, first- or second-year graduate course on group, operator, and abstract probability theory, as well as quantum field theory per se. The full book should do the same for a yearlong course in its primary topic of quantum field theory. It can easily be enhanced with specific mathematical or physical applications, from number theory and theta functions to path integrals and many-body formalism, depending on the direction of the class or reader. Our purpose here has simply been to provide a coherent and succinct introduction to a general area and approach that connects, and in part underlies, all of those fields.

This book is primarily a synthesis, and only secondarily a survey. Detailed reference could not have been made briefly in an accurate and balanced way, and the alternative—lengthy accounts of the development of the subject—

would have distracted from the logical development and added much to the bulk of the book. We have followed a middle course of providing each chapter with bibliographic notes on key publications related to our presentation. A more general list of references enlarging on our themes, directions, and techniques is given at the end of the book. The literature is now so vast that such a list cannot be complete, and we have given priority to sources of special relevance to or coherence with the rigorous treatment here.

We thank students and colleagues who have made corrections or suggestions regarding notes forming a preliminary version of this book, and Jan Pedersen for a final reading.

Introduction

The logical structure of quantum field theory

Quantum field theory is quintessentially the algebra and analysis of infinite-dimensional dynamical systems, as constrained by quantum phenomenology, causality, and symmetry. Although it has a clear-cut central goal, that of the realistic description of particle production and annihilation in terms of the localized interactions of fields in space-time, it is clear from this description that it is a multifaceted subject. Indeed, both the relevant mathematical technology and the varieties of physical applications are extremely diverse. In consequence, any linear (i.e., sequential) presentation of the subject inevitably greatly oversimplifies the interaction between different parts of the subject.

At the level of mathematical technology, the subject involves Hilbert space and geometry, wave equations and group representations, operator algebra and functional integration, to mention only the most basic components. At the level of practical applications, much of physical quantum field theory is heuristic and somewhat opportunistic, even if it is strongly suggestive of an underlying distinctive and coherent mathematical structure. At the overall foundational level, the logical basis of the subject remains unsettled, more than six decades after its heuristic origin. There is not even a general agreement as to what constitutes a quantum field theory, in precise terms.

In this context, logical clarity and applicable generality take on particularly high priority, and have been emphasized here. At the same time, there has been outstanding progress in the mathematical theory of quantum fields, although the ultimate goal of an effective theory of relevant equations in four-dimensional space-time has by no means been achieved. To present the established developments in a lucid and succinct way, while treating the issues involved in the ultimate goal of the theory, it is unfortunately necessary to be somewhat abstract and to focus on the algebraic and constructive core of the theory. This results in an approach that may, at first glance, seem unfamiliar to those introduced to the subject in the conventional manner.

The traditional approach commonly involves presenting a prototypical, classical "free" equation, often the Klein-Gordon equation

$$\Box\varphi + m^2\varphi = 0,$$

and explaining that in view of quantum mechanics, what is wanted is rather an operator-valued function φ that satisfies the same equation, together with the nontrivial commutation relations at an arbitrary fixed time t_0.

$$[\varphi(t_0, x), \varphi(t_0, y)] = 0 = [\partial_t\varphi(t_0, x), \partial_t\varphi(t_0, y)];$$
$$[\varphi(t_0, x), \partial_t\varphi(t_0, y)] = -\, i\, \delta(x - y),$$

which are analogous to the Heisenberg commutation relations. Having explained heuristically how the appropriate "function" φ may be obtained, the text goes on to treat vector and other higher spin fields, the quantization of wave equations on other space-times, and so on.

This is a natural and sensible approach, but it suffers from some scientific limitations, and there are practical advantages to a more abstract treatment. More specifically, the value of a quantized field at a point is a highly theoretical entity, which is relatively singular mathematically, and not really conceptually measurable from a foundational physical standpoint. It is now widely realized that it is mathematically more efficient and physically more appropriate to treat, instead of the value of the quantized field at a point, its average with respect to a smooth function f supported in a neighborhood of the point, such as $\phi(f) = \int\varphi(X)f(X)dX$, where $X = (t,x)$. What is less widely realized is that the test functions f are *effectively* in an invariant Hilbert space, the so-called "single-particle space." From the fact that all Hilbert spaces of the same dimension are unitarily equivalent, it follows that the underlying quantum field theory does not depend, for the most part, on the particularities of the wave equation or space-time under consideration. The point is that there are universal boson and fermion quantizations applicable to an arbitrary Hilbert space.

Moreover, the use of these universal fields for the quantization of given equations is not only a major theoretical economy, but definitively simplifies the treatment of many important topics. Thus, there is a simple invariant condition for the unitary implementability of canonical transformations on a free quantum field in Hilbert space terms that is necessary and sufficient, in contrast to the complex and unnecessarily stringent sufficient conditions derivable from analysis in a more specialized format.

More generally, instead of starting out with a given Hilbert space, it is necessary in more complicated cases to begin with a space with less structure, a real orthogonal or symplectic space. The introduction of an appropriate complex structure in such a space, on the basis of symmetry and stability (positive energy) considerations, underlies the determination of the "vacuum" and, indeed, the use of complex numbers generally in quantum mechanics. At the same time, these real variants of a Hilbert space provide the essential basis for the C^*-algebraic formalism, which uses an infinite-dimensional version of the Clifford algebra in the orthogonal case and of Weyl algebra in the symplectic case. These C^*-algebras are representation-independent (i.e., do not depend on a particular choice of a representation of the canonical or anticanonical commutation relations). The C^*-algebraic formalism is a natural extension of the Hilbert space line and at the same time permits the quantization of fields that do not necessarily possess a vacuum, such as those of "tachyons" (represented, e.g., by the Klein-Gordon equation with imaginary mass).

Last, but not at all least, the use of a succinct and invariant algebraic formalism facilitates the overall view of the forest in a subject in which this view is obscured by a very large number of trees. The development of a foundation for four-dimensional quantum field theory that is consistent with the simple and natural ideas that motivated the subject—and there is no compelling reason to doubt that one exists—seems likely to be expedited thereby, even if it may ultimately be describable in other terms.

For these reasons, the first part of the book, consisting of Chapters 1–5, emphasizes the algebra and treats the Klein-Gordon and Dirac equations in Minkowski space only as examples. The second part, or Chapters 6–8, treats differential equations and nonlinear issues, in which the notion of the quantized field as a point function becomes quite pertinent. The basic principle of the locality of causal interactions is naturally formulated in terms of point values of fields, and the constructive theory is more sensitive than the algebraic to the underlying geometry and wave equation.

The connection between the global Hilbert space standpoint and the intuitive idea of fields as localized entities will emerge in opportune places throughout the book and will be crucial to Chapter 8. This chapter will develop in particular the concept of a nonlinear local function of an interacting quantum field. The reader who would like early on to see an example of the connection may find it in Appendix B.

Organization of the book

The main technical prerequisite for this presentation of quantum field theory is basic functional analysis and operator theory. No familiarity with heuristic quantum field theory is assumed, but it should be helpful. Correlation of the rigorous theory with the conventional looser and more specialized physical formalism is made in the Examples, Problems, and Lexicons. The Problems are also used to make contact with some interesting but seemingly peripheral issues in quantum field theory.

Chapter 1 treats the universal free boson field. The unboundedness of the field operators and the incorporation of functional integration theory into this chapter make it relatively lengthy. Chapter 2 treats the universal fermion field along parallel lines, more briefly because of the boundedness of the field operators and the subsumption of the counterpart to functional integration theory under operator algebra. Each of these chapters details the three basic unitarily and physically equivalent representations of the universal fields: the particle representation, which diagonalizes the "occupation numbers" of states; the "real" wave representation, which essentially diagonalizes the hermitian field operators at a fixed time; and the "complex" wave representation, which pseudo-diagonalizes the creation operators.

Chapter 3 establishes general properties of the universal fields, such as the relation of the field over a direct sum to the fields over the constituents, and their unicity subject only to the constraint of positive energy. The remarkable parallelism between the structure and representations of the boson and fermion fields, notwithstanding their striking differences, comes through strongly in this chapter. At the same time, the assumption of an underlying (single-particle) Hilbert space is relaxed to treat boson fields over symplectic spaces and fermion fields over orthogonal spaces.

Chapter 4 deals with absolute continuity of distributions in function spaces and the related question of the unitary implementability on the universal fields of given canonical transformations on the underlying single-particle space (or classical field). The result, that the Hilbert-Schmidt condition on the commutator of the given transformation with the complex unit i is necessary and sufficient for unitary implementability, confirms the relevance and essentiality of the Hilbert space format (for the underlying classical field or single-particle space) used in the earlier chapters. At the same time, this condition shows a need for a broader formalism that will encompass the actions on the quantized field of general canonical transformations. In Chapter 5 it is shown that these transformations act naturally, essentially in extension of their Hilbert space actions when these exist, on C^*-algebras associated with the underlying symplectic or orthogonal structures.

Chapter 6 applies the algebraic theory to the quantization of wave equations (using the Schrödinger, Klein-Gordon, and Dirac as prototypes) and marks the transition to constructive quantum field theory. Chapter 7 develops in a quite general format the theory of renormalized local products of boson fields. These are shown to exist as generalized operators—more specifically, continuous sesquilinear forms on the domain **D** of infinitely-differentiable vectors for the field Hamiltonian—without any special assumption as to the character of space or an underlying wave equation. The last section details the specialization to the formulation of a quantum field as an essentially self-adjoint operator-valued distribution on Minkowski space and establishes the existence of Wick powers of this field as a Lorentz-covariant operator-valued distribution.

Chapter 8 focuses on the case of a nonlinear scalar field in two space-time dimensions as a vehicle for illustrating some of the basic ideas and methods in constructive quantum field theory. A nonlinear variant of the Weyl relations is used to give effective meaning to polynomials in a quantum field that is not necessarily free, at a fixed time, as required for the fundamental partial differential equations of interacting quantum field theory to make mathematical sense as such. The existence of quantized solutions to local nonlinear scalar wave equations, in which the nonlinearities are renormalized relative to the "physical" rather than a somewhat ambiguous and hypothetical "free" vacuum, is shown in detail. The real-time (Lorentzian) approach of the original rigorous treatment seems physically more natural and generally adaptable than the later-developed imaginary time (Euclidean) treatment, and is adopted here.

The numbering system is as follows. Theorems are numbered seriatim in each chapter: e.g., Theorem $a.b$ denotes the bth theorem in Chapter a. Lemmas and corollaries are numbered serially following the underlying theorem number. Thus, Lemma $a.b.c$ denotes the cth lemma for Theorem $a.b$. We also use scholia in the same manner as in the *Principia*: generally useful ancillary results that are not of the central importance of a theorem. The scholia and equations are numbered like the theorems. The sections of each chapter are numbered serially. A compendium of the main notations we use is given in Appendix A. References are to the Bibliography following the text. For basic functional analytic results we refer primarily to Segal and Kunze (1978), which is cited throughout as SK.

Introduction to Algebraic and Constructive Quantum Field Theory

1

The Free Boson Field

1.1. Introduction

Much of the quantum field theory is of a very general character independent of the nature of space-time. Indeed, a universal formalism applies whether or not there exists an underlying "space" in the usual geometrical sense. In its primary form, this universal part of quantum field theory depends only on a given underlying (complex) Hilbert space, say **H**. Colloquially, **H** is often called the *single-particle space*.

Thus, for a nonrelativistic particle in three-dimensional euclidean space $\mathbf{R}^3$, **H** is the space $L_2(\mathbf{R}^3)$ consisting of all square-integrable complex-valued functions on $\mathbf{R}^3$, in the usual formalism of elementary quantum mechanics. For a relativistic field or particle as usually treated, **H** is the space of "normalizable" wave functions. Here the norm derives from a Lorentz-invariant inner product in the solution manifold of the corresponding wave equation. For possible more exotic types of fields, the situation is much the same.

This chapter presents the mathematical theory of one of the most fundamental quantum field constructs from a given complex Hilbert space **H**, without at all concerning itself with the origin of **H**. This theory has close relations to integration and Fourier analysis in Hilbert space, and can in part be interpreted as the extension of analysis in euclidean n-space to the case in which n is allowed to become infinite. We call this universal construct the *free boson* (more properly, Bose-Einstein) *field over* **H**.

The next section of the chapter begins the rigorous mathematical development. From time to time, items logically outside the mathematical development, labeled Lexicon, will interrupt in order to correlate the somewhat abstract treatment with physical usage and intuition. Mathematical Examples, in the nature of special cases, will also be provided. Readers interested primarily in the mathematics may largely ignore the Lexicon items. Those who would like to appreciate at this point how the conventional treatment of relativistic fields can be subsumed under the universal Hilbert space formulation of this chapter will find an explicit treatment in Appendix B.

1.2. Weyl and Heisenberg systems

In any Hilbert space, the imaginary part of the inner product provides a real antisymmetric (real-) bilinear form. A more general type of space, in which only such a form is given, also plays an important part in boson theory.

DEFINITION. *A symplectic vector space* is a pair $(\mathbf{L}, A)$ consisting of a real topological vector space $\mathbf{L}$, together with a continuous antisymmetric, "nondegenerate" bilinear form A on $\mathbf{L}$. To be more explicit, "nondegenerate" means that if $A(x, y) = 0$ for all $x \in \mathbf{L}$, then $y = 0$.

When $\mathbf{L}$ is a real finite-dimensional vector space, it is easily seen that it can be given the structure of real topological vector space in one and only one way. A real infinite-dimensional vector space is a topological vector space relative to the topology in which a set is open if and only if its intersection with each and every finite-dimensional subspace is open relative to this subspace. A space with this topology will be said to be "topologized algebraically." The continuity condition on A is then easily seen to be vacuous.

EXAMPLE 1.1. Let $\mathbf{M}$ be a finite-dimensional real vector space, and $\mathbf{M}^*$ its dual, i.e., the space of all linear functionals on $\mathbf{M}$. Let $\mathbf{L}$ denote the direct sum $\mathbf{M}\oplus\mathbf{M}^*$, and let A denote the form

$$A(x\oplus f, x'\oplus f') = f(x') - f'(x),$$

for arbitrary $x\oplus f$ and $x'\oplus f'$ in $\mathbf{L}$. Then it is easily verified that $(\mathbf{L}, A)$ is a symplectic vector space; it will be called the *symplectic vector space built from* $\mathbf{M}$. More generally, suppose $\mathbf{M}$ and $\mathbf{N}$ are arbitrary given real topological vector spaces, and $B(x, f)$ is a given continuous nondegenerate bilinear form on $\mathbf{M} \times \mathbf{N}$. The *symplectic vector space built from* $(\mathbf{M}, \mathbf{N}, B)$ is defined as the space $(\mathbf{L}, A)$ where $\mathbf{L} = \mathbf{M}\oplus\mathbf{N}$ and $A(x\oplus f, x'\oplus f') = B(x', f) - B(x, f')$.

DEFINITION. Let $(\mathbf{L}, A)$ be a given symplectic vector space. A *Weyl system* over $(\mathbf{L}, A)$ is a pair $(\mathbf{K}, W)$ consisting of a complex Hilbert space $\mathbf{K}$ and a continuous map W from $\mathbf{L}$ to the unitary operators on $\mathbf{K}$ (taken as always, unless otherwise specified, in their strong operator topology) such that for all z and z' in $\mathbf{L}$,

$$W(z)W(z') = e^{\frac{1}{2}iA(z,z')}W(z + z'). \tag{1.1}$$

Equations 1.1 are known as the *Weyl relations*.

If **H** is a given complex pre-Hilbert space ("pre" signifying that completeness is not assumed), and A denotes the form $A(z, z') = \mathrm{Im}\langle z, z'\rangle$, then the pair $(\mathbf{H}^{\#}, A)$, where $\mathbf{H}^{\#}$ denotes **H** as a real vector space with the same topology, is a symplectic vector space, a Weyl system over which is called simply a *Weyl system over* **H**. Here and below we shall follow mathematical convention and take the complex inner product $\langle\cdot,\cdot\rangle$ to be complex-linear in the *first* argument.

EXAMPLE 1.2. Let **L** denote the space **C** of all complex numbers as a real two-dimensional space, and let $A(z, z') = \mathrm{Im}(z\bar{z}')$. Let **K** denote the space $L_2(\mathbf{R})$ of all complex-valued square-integrable functions on the real line **R**; here and later when the measure in euclidean space is unspecified, it is understood to be Lebesgue measure. For arbitrary z in **C** of the form $z = x + iy$, where x and y are real, let $W(z)$ denote the operator on **K**,

$$W(z): f(u) \mapsto e^{-iyu - ixy/2} f(u + x).$$

It is easy to check that $(\mathbf{K}, W)$ is Weyl system over $(\mathbf{L}, A)$; it is known as the "Schrödinger system" or the "Schrödinger representation of the Weyl relations."

More generally, let **H** denote a finite-dimensional complex Hilbert space. Let $e_1, e_2, \ldots, e_n$ denote an arbitrary orthonormal basis for **H**, and let $\mathbf{H}'$ denote the *real* span of the e_j, that is, the real subspace consisting of all real linear combinations of the e_j. Relative to the restriction of the given inner product $\langle\cdot,\cdot\rangle$ to $\mathbf{H}'$, $\mathbf{H}'$ forms a euclidean space. Let **K** denote $L_2(\mathbf{H}')$, and for arbitrary z in **H**, of the form $z = x + iy$, where x and y are in $\mathbf{H}'$, let $W(z)$ denote the operator on **K**:

$$W(z): f(u) \mapsto e^{-i\langle y,u\rangle - i\langle x,y\rangle/2} f(u + x).$$

Just as in the one-dimensional case, $(\mathbf{K}, W)$ is easily seen to form a Weyl system over **H**.

The Weyl relations are a regularized form of the Heisenberg relations, which are essentially the infinitesimal form of the Weyl relations. Such an infinitesimal form, like the infinitesimal representation associated with a group representation (of which the Weyl systems are in fact a special case; cf. Problems following this section) is more effective in algebraic contexts than the global form, although the latter is more cogent for rigorous analytical purposes.

More specifically, if $(\mathbf{K}, W)$ is a Weyl system over a given symplectic vector space $(\mathbf{L}, A)$, then the map $t \mapsto W(tz)$ is for any fixed $z \in \mathbf{L}$ a continuous one-parameter unitary group whose selfadjoint generator (whose existence is asserted by Stone's theorem) is denoted as $\phi(z)$. The map $z \mapsto \phi(z)$ from $\mathbf{L}$ into the selfadjoint operators on $\mathbf{K}$ will be called a *Heisenberg system*.

THEOREM 1.1. *Let ϕ denote the Heisenberg system for the Weyl system $(\mathbf{K}, W)$ over the symplectic vector space $(\mathbf{L}, A)$. Then for arbitrary vectors x and y in $\mathbf{L}$, and nonzero $t \in \mathbf{R}$, the following conclusions can be made:*

i) $\phi(tx) = t\phi(x)$;
ii) $\phi(x) + \phi(y)$ *has closure* $\phi(x + y)$;
iii) for arbitrary u in the dense domain $\mathbf{D}(\phi(x)\phi(y)) = \mathbf{D}(\phi(y)\phi(x))$, $[\phi(x),\phi(y)]u = -iA(x,y)u$; *and*
iv) $\phi(x) + i\phi(y)$ *is closed.*

PROOF. The definition of ϕ makes i) clear. For the rest, we use two lemmas. Here and later the notation $\mathbf{D}(T)$ for an operator T denotes the domain of T.

LEMMA 1.1.1. *Let x and y be arbitrary in $\mathbf{L}$, and suppose that $u \in \mathbf{D}(\phi(x))$. Then $W(y)u \in \mathbf{D}(\phi(x))$ and*

$$\phi(x)W(y)u = W(y)[\phi(x) + A(x, y)]u.$$

PROOF. It follows from the Weyl relations that for arbitrary nonzero real t,

$$-it^{-1}[W(tx) - I]W(y)u = -it^{-1}[W(tx)W(y) - W(y)]u$$
$$= -it^{-1}W(y)[W(tx)e^{itA(tx,y)} - I]u.$$

Letting $t \rightarrow 0$, the lemma follows. □

LEMMA 1.1.2. *Let $\mathbf{D}'$ denote the set of all finite linear combinations of vectors of the form of the weak integral*

$$\int W(sx + ty)v\, F(s, t)\, dsdt,$$

where v is arbitrary in $\mathbf{K}$ and F is arbitrary in $C_0^\infty(\mathbf{R}^2)$. Then $\mathbf{D}'$ is dense in $\mathbf{K}$, $\mathbf{D}' \subset \mathbf{D}(\phi(x)) \cap \mathbf{D}(\phi(y))$, and $\mathbf{D}' \subset \mathbf{D}(\phi(x)\phi(y))$.

PROOF. Here and elsewhere, integrals are extended over all values of the variables of integration, unless otherwise indicated, and the notation $C_0^\infty(S)$ for an arbitrary manifold S denotes the set of all C^∞ complex-valued functions of compact support on S.

The density of $\mathbf{D}'$ in $\mathbf{K}$ follows from the choice of a sequence $\{F_n\}$ suitably approximating the Dirac measure. To show that $\mathbf{D}' \subset \mathbf{D}(\phi(x)\phi(y))$ it suffices to show that $\mathbf{D}' \subset \mathbf{D}(\phi(y))$ and that $\phi(y)\mathbf{D}' \subset \mathbf{D}'$. To show that $\mathbf{D}' \subset \mathbf{D}(\phi(y))$ is to show that if

$$u = \int W(sx + ty)v\, F(s, t)\, dsdt,$$

then $\lim_{\varepsilon\to 0} \varepsilon^{-1}[W(\varepsilon y) - \mathrm{I}]u$ exists. Now

$$\begin{aligned}&\varepsilon^{-1}[W(\varepsilon y) - \mathrm{I}]u \\ &= \varepsilon^{-1} \int [W(\varepsilon y + sx + ty)e^{\frac{1}{2}i\varepsilon sA(y,x)} - W(sx + ty)]v\, F(s, t)\, dsdt,\end{aligned}$$

using the Weyl relations. In the integrand,

$$\begin{aligned}W(\varepsilon y + sx + ty)e^{\frac{1}{2}i\varepsilon sA(y,x)} - W(sx + ty) &= e^{\frac{1}{2}i\varepsilon sA(y,x)}(W(\varepsilon y + sx + ty) \\ &\quad - W(sx + ty)) + (e^{\frac{1}{2}i\varepsilon sA(y,x)} - 1)W(sx + ty),\end{aligned}$$

and making a translation in t,

$$\begin{aligned}\varepsilon^{-1}[W(\varepsilon y) - I]u &= \int e^{\frac{1}{2}i\varepsilon sA(y,x)}W(sx + ty)v\, (F(s, t - \varepsilon) - F(s, t))/\varepsilon\, dsdt \\ &\quad + \int (e^{\frac{1}{2}i\varepsilon sA(y,x)} - \mathrm{I})/\varepsilon\, W(sx + ty)v\, F(s, t)\, dsdt;\end{aligned}$$

letting $\varepsilon \to 0$ on the right side shows that the limit of the left side exists and equals

$$-\int W(sx + ty)v\, \partial_t F(s, t)\, dsdt + \tfrac{1}{2}isA(y, x) \int W(sx + ty)v\, F(s, t)\, dsdt.$$

This shows that $\mathbf{D}' \subset \mathbf{D}(\phi(y))$ and that $\phi(y)\mathbf{D}' \subset \mathbf{D}(\phi(x)\phi(y))$. The proof that $\mathbf{D}' \subset \mathbf{D}(\phi(x))$ is analogous to that of $\mathbf{D}' \subset \mathbf{D}(\phi(y))$. □

To see that $\phi(x) + \phi(y) \subset \phi(x + y)$, note that

$$W(tx)W(ty)u = e^{(it^2/2)A(x,y)}W(t(x + y))u$$

so that

$$-it^{-1}[W(t(x + y)) - \mathrm{I}]u = -it^{-1}[e^{-(it^2/2)A(x,y)}W(tx)W(ty) - \mathrm{I}\,]u.$$

Now taking u in $\mathbf{D}(\phi(x))\cap\mathbf{D}(\phi(y))$, the usual argument for treating the derivative of a product shows that the right hand side of the above equation converges as $t \to 0$ to $(\phi(x) + \phi(y))u$. The inclusion in question follows now from Stone's theorem.

To show that $\phi(x + y)$ is the closure of $\phi(x) + \phi(y)$ is thus equivalent to showing that $\phi(x + y)$ is the closure of its restriction to $\mathbf{D}(\phi(x))\cap\mathbf{D}(\phi(y))$. To prove this, recall the general criterion: a selfadjoint operator A is the closure of its restriction to a domain $\mathbf{D}$ if: a) $\mathbf{D}$ is dense; and b) $e^{itA}\mathbf{D} \subset \mathbf{D}$ for all $t \in \mathbf{R}$. Taking $A = \phi(x + y)$ and $\mathbf{D} = \mathbf{D}(\phi(x))\cap\mathbf{D}(\phi(y))$, then both (a) and (b) follow from Lemma 1.1.2.

Note that by Lemma 1.1.1 for arbitrary $w \in \mathbf{D}(\phi(x)\phi(y))$,

$$[\phi(y)W(sx) - W(sx)\phi(y)]w = A(y, sx)W(sx)w,$$

or

$$\phi(y)W(sx)w = W(sx)\phi(y)w + A(y, sx)W(sx)w.$$

Differentiating with respect to s on the right side and setting $s = 0$ yields $i\phi(x)\phi(y)w + A(y, x)w$. It follows that the left side is differentiable at $s = 0$, and since $\phi(y)$ is closed, the limit is $\phi(y)\phi(x)w$. Thus

$$i[\phi(y), \phi(x)]w = A(y, x)w.$$

Conclusion iii) now follows from Lemma 1.1.2.

It also follows that

$$\|(\phi(x) + i\phi(y))w\|^2 = \|\phi(x)w\|^2 + \|\phi(y)w\|^2 + A(x, y)\|w\|^2.$$

To show that $\phi(x) + i\phi(y)$ is closed, let $\{u_n\}$ be a sequence of vectors in $\mathbf{D}(\phi(x)) \cap \mathbf{D}(\phi(y))$ such that $u_n \to u$ and $(\phi(x) + i\phi(y))u_n \to v$. Then

$$\|\phi(x)(u_m - u_n)\| \to 0 \text{ and } \|\phi(y)(u_m - u_n)\| \to 0$$

as $m, n \to \infty$. It follows in turn that $u \in \mathbf{D}(\phi(x)) \cap \mathbf{D}(\phi(y))$, so that $\phi(x) + i\phi(y)$ is closed, proving (iv). □

EXAMPLE 1.3. A map ϕ from $\mathbf{L}$ to operators on a Hilbert space $\mathbf{K}$ that satisfies conditions i)–iii) is not in general a Heisenberg system. In other terms, the operators $W(z)$ defined as $e^{i\phi(z)}$ do not necessarily satisfy the Weyl relations. Sufficient conditions for this to be the case follow from the theory of analytic vectors for unitary group representations due to Nelson (1959); related conditions were derived earlier in the special case of the Weyl relations by Rellich and later by Dixmier (1958).

Nelson's criterion implies that for a given mapping ϕ_0 from $\mathbf{L}$ to operators on a Hilbert space $\mathbf{K}$ to be essentially a Heisenberg system, in the sense that $\phi_0(z)$ has a closure $\phi(z)$, for all $z \in \mathbf{L}$, such that ϕ is a Heisenberg system, the following is sufficient:

(i) All $\phi_0(z)$ have the same domain $\mathbf{D}$, which is dense in $\mathbf{K}$, and left invariant by all the $\phi_0(z)$, which are hermitian.

(ii) Every finite-dimensional subspace of $\mathbf{L}$ has a basis $z_1, z_2, \ldots, z_n$ such that $\phi_0(z_1)^2 + \phi_0(z_2)^2 + \cdots + \phi_0(z_n)^2$ is essentially selfadjoint.

This applies in particular to the simplest case of the "particle representation." In this, $\mathbf{K}$ is the space ℓ_2 of all sequences $(a_0, a_1, \ldots)$ for which the sum $|a_0|^2 + |a_1|^2 + \cdots$ is convergent, with the inner product $\langle A, B\rangle = a_0\bar{b}_0 + a_1\bar{b}_1 + \cdots$. The domain $\mathbf{D}$ is that of all finite sequences, and ϕ_0 is determined as a map on $\mathbf{L} = \mathbf{C}$ by the specification of $\phi_0(1)$ and $\phi_0(i)$, which may be denoted as P and Q. Setting e_n for the vector in ℓ_2 whose only nonvanishing component is a 1 in the nth position,

$$Pe_n = 2^{-1/2}(n^{1/2}e_{n-1} + (n+1)^{1/2}e_{n+1}); \qquad (1.2)$$
$$Qe_n = i2^{-1/2}(n^{1/2}e_{n-1} - (n+1)^{1/2}e_{n+1}).$$

On observing that $(P^2 + Q^2)e_n = (2n+1)e_n$, the requisite essential selfadjointness follows.

This particle representation, in its global form, is unitarily equivalent to the Schrödinger representation, and the same is true of its analog for **L** of arbitrary finite dimension. A generalization to the case when **L** is a Hilbert space of arbitrary dimension is developed later in this chapter.

The preceding example illustrates the formulation of a Weyl system in terms of a "canonical pair," consisting of suitable hermitian operators P and Q satisfying the relation $[P, Q] \subset -iI$. The essential equivalence of this type of formulation, which is more familiar and elementary, but typically less invariant, with the explicitly symplectic formulations, will now be developed.

DEFINITION. A linear *dual couple* is a system $(\mathbf{M}, \mathbf{N}, \langle\cdot,\cdot\rangle)$ consisting of real topological vector spaces **M** and **N**, together with a real bilinear continuous function $\langle x, \lambda\rangle$ on $\mathbf{M} \times \mathbf{N}$ that is nondegenerate. A *Weyl pair* over a given dual couple is a system $(\mathbf{K}, U, V)$ consisting of a complex Hilbert space **K** together with continuous unitary representations on **K**, U, and V of the additive groups of **M** and **N** respectively, satisfying the relations

$$U(x)V(\lambda) = e^{i\langle x,\lambda\rangle}V(\lambda)U(x), \qquad x \in \mathbf{M}, \qquad \lambda \in \mathbf{N}. \qquad (1.3)$$

Equations 1.3 are called the *restricted* Weyl relations.

EXAMPLE 1.4. A variant of the Schrödinger representation is as follows. Let **M** denote a finite-dimensional real vector space, $\mathbf{M}^*$ its dual space, and define $\langle x, \lambda\rangle = \lambda(x)$ for $x \in \mathbf{M}$ and $\lambda \in \mathbf{M}^*$. Then $(\mathbf{M}, \mathbf{M}^*, \langle\cdot,\cdot\rangle)$ is a dual couple and is said to be *built* from **M**. Let m denote an arbitrary regular measure on **M** that is *quasi-invariant*, meaning that its null sets are invariant under vector translations in **M**; or equivalently that m and its translate m_x through x, defined by the equation $m_x(E) = m(E + x)$ for any Borel set E, are mutually absolutely continuous for all $x \in \mathbf{M}$. For arbitrary $x \in \mathbf{M}$, let $U(x)$ denote the operation on $\mathbf{K} = L_2(\mathbf{M}, m)$

$$U(x): f(u) \mapsto f(u+x)[dm_x/dm]^{1/2} \qquad (f \in \mathbf{K}). \qquad (1.4a)$$

For arbitrary $\lambda \in \mathbf{M}^*$ let $V(\lambda)$ denote the operation

$$V(\lambda): f(u) \mapsto e^{i\lambda(u)}f(u) \qquad (f \in \mathbf{K}). \qquad (1.4b)$$

It is not difficult to verify that $(\mathbf{K}, U, V)$ is a Weyl pair over $(\mathbf{M}, \mathbf{M}^*, \langle\cdot,\cdot\rangle)$.

Any quasi-invariant measure m on $\mathbf{M}$ is mutually absolutely continuous with Lebesgue measure on $\mathbf{M}$, by a result of Mackey (1952). From this it follows that the Weyl pair just constructed is unitarily equivalent to that obtained when m is Lebesguc measure. It will be seen later that there is an analogous construction in the case when $\mathbf{M}$ is infinite-dimensional, in which case the conclusion of Mackey's theorem is not at all valid. Even in the finite-dimensional case the construction will be convenient on occasion, especially with m taken as a Gaussian measure.

The relation between Weyl systems and pairs is noted in

THEOREM 1.2. *Let* $(\mathbf{M}, \mathbf{N}, \langle\cdot,\cdot\rangle)$ *be a dual couple and* $\mathbf{K}$ *be a complex Hilbert space. Let* U *and* V *denote given continuous mappings from* $\mathbf{M}$ *and* $\mathbf{N}$ *respectively to the unitary operators on* $\mathbf{K}$. *Then* $(\mathbf{K}, U, V)$ *is a Weyl pair over* $(\mathbf{M}\oplus\mathbf{N}, \langle\cdot,\cdot\rangle)$ *if and only if the mapping* W *from* $\mathbf{M}\oplus\mathbf{N}$ *to the unitary operators on* $\mathbf{K}$:

$$W(z) = U(x)\, V(-\lambda)\, e^{\frac{1}{2}i\langle x,\lambda\rangle}, \qquad z = x\oplus\lambda,$$

is in conjunction with $\mathbf{K}$ *a Weyl system over the symplectic space built from* $(\mathbf{M}, \mathbf{N}, \langle\cdot,\cdot\rangle)$.

PROOF. The proof is left as an exercise. □

As in the case of a Weyl system, for any Weyl pair $(\mathbf{K}, U, V)$ over a given dual couple $(\mathbf{M}, \mathbf{N}, \langle\cdot,\cdot\rangle)$, there exist unique mappings P and Q from $\mathbf{M}$ and $\mathbf{N}$ to the selfadjoint operators in $\mathbf{K}$ that are given by the equations:

$$U(tx) = e^{itP(x)}, \qquad V(t\lambda) = e^{itQ(\lambda)}, \qquad t \in \mathbf{R}.$$

DEFINITION. A *Heisenberg pair* $(\mathbf{K}, P, Q)$ *over a dual couple* $(\mathbf{M}, \mathbf{N}, \langle\cdot,\cdot\rangle)$ consists of a complex Hilbert space $\mathbf{K}$ and mappings P and Q from $\mathbf{M}$ and $\mathbf{N}$, respectively, to the selfadjoint operators in $\mathbf{K}$, having the property that if $U(x) = e^{iP(x)}$ and $V(\lambda) = e^{iQ(\lambda)}$, then $(\mathbf{K}, U, V)$ is a Weyl pair over $(\mathbf{M}, \mathbf{N}, \langle\cdot,\cdot\rangle)$.

DEFINITION. A collection of selfadjoint operators on a Hilbert space is said to be *strictly commutative* if the spectral projections of the operators are mutually commutative.

COROLLARY 1.2.1. *If* $(\mathbf{K}, P, Q)$ *is a Heisenberg pair over the dual couple* $(\mathbf{M}, \mathbf{N}, \langle\cdot,\cdot\rangle)$, *then for arbitrary* $x, y \in \mathbf{M}$ *and* $\lambda, \mu \in N$, *and nonzero* $t \in \mathbf{R}$,

i) $P(tx) = tP(x)$; $\quad Q(t\lambda) = tQ(\lambda)$;

ii) $P(x) + P(y)$ *has closure* $P(x + y)$; $Q(\lambda) + Q(\mu)$ *has closure* $Q(\lambda + \mu)$; *and*

iii) $P(x)$ *and* $P(y)$ *commute strictly, and* $Q(\lambda)$ *and* $Q(\mu)$ *commute strictly;* $[P(x), Q(\lambda)]$ *has closure* $-i\langle x, \lambda\rangle$.

PROOF. The corollary follows from Theorems 1.1 and 1.2 by specialization of z and z'. □

DEFINITION. A collection of bounded linear operators on a Hilbert space is *irreducible* ("topologically," as will be understood unless otherwise indicated) if the only closed linear subspaces invariant under all of them are the trivial ones (i.e., the entire space, or that consisting only of 0).

THEOREM 1.3. *The Schrödinger system over a finite-dimensional space is irreducible.*

PROOF. Let $\mathbf{K}'$ denote an arbitrary closed invariant subspace. Being invariant under multiplications by the $e^{i\lambda(u)}$, where the notation of Example 1.4 is used, it is invariant under multiplications by arbitrary finite linear combinations of these functions. But an arbitrary bounded measurable function on $\mathbf{R}$ is a w^*-limit of such linear combinations (in L_∞ as the dual of L_1). It follows that $\mathbf{K}'$ is invariant under multiplications by arbitrary bounded measurable functions.

If $\mathbf{K}'$ is not all of $\mathbf{K}$, there exists a vector $f \in \mathbf{K}$ that is orthogonal to every vector in $\mathbf{K}'$: $\int \overline{f}(x)g(x)dx = 0$ for all $g \in \mathbf{K}'$. By replacing g by $sgn(f\overline{g})g$ (where $sgn(x) = x/|x|$ if $x \neq 0$ and $sgn(0) = 0$), and noting the invariance of $\mathbf{K}'$ under translations, it follows that $\int |f(x)|\,|g(x + a)|dx = 0$ for all real a. Now $\mathbf{K}'$ is invariant under convolution by arbitrary integrable functions, by virtue of its assumed invariance features, so every element of $\mathbf{K}'$ may be approximated arbitrarily closely by a continuous function in $\mathbf{K}'$. It follows that if $\mathbf{K}'$ does not consist only of zero, then $f(x)$ must vanish on a nonempty open set. But the class of open sets on which f vanishes is translation-invariant, so f vanishes identically, a contradiction that completes the proof. □

LEXICON. In the physics literature, hermitian operators P and Q satisfying the relation $[P, Q] = -iI$, first used by Heisenberg, are said to be "canonically conjugate" or to form a canonical pair. More generally, operators P_1, $P_2, \ldots$ and $Q_1, Q_2, \ldots$ satisfying the relations $[P_j, Q_k] = -i\delta_{jk}$, $[P_j, P_k] = 0 = [Q_j, Q_k]$, are said to satisfy the "canonical commutation relations" or "CCRs." In field theory it is common for the index j to be continuous, so that in more precise mathematical terms one is dealing with operator-valued distributions. Operator-fields $P(x)$ and $Q(x)$ defined over a manifold M are then said to be canonically conjugate or to satisfy canonical commutation relations if formally $[P(x), Q(y)] = -i\delta(x - y)$, $[P(x), P(y)] = 0 = [Q(x), Q(y)]$. But the precise meaning of such relations requires a formulation in terms of Weyl systems.

The Heisenberg relations descend in part from relations between classical quantities rather than operator-valued quantities. The word "classical" is used often to mean "numerically-valued" or, in Dirac's term, a "c-number." In contrast, a "quantized" quantity, meaning effectively one represented by an operator, is called a "q-number." Lie's theory of contact transformations made extensive use of Poisson bracket-relations between classical canonically conjugate variables P_j and Q_j; these take the form $\{P_j, Q_k\} = \delta_{jk}$, $\{P_j, P_k\} = 0 = \{Q_j, Q_k\}$ $(j, k = 1, 2, \ldots)$. Although the close parallel with the Heisenberg relations is no accident and has been particularly strongly developed by Dirac, it appears to be impossible to make precise or effective in a general way, notwithstanding repeated attempts to do so. A key problem is the importance in quantum mechanics that the energy be represented by an operator that is bounded below (for the treatment of stable systems), but this is never the case for the corresponding classical motion induced in the space of square integrable functions over phase space, which space moreover is roughly twice the size of the quantum mechanical Hilbert space. The phase space is often just the cotangent bundle of the *configuration space* **M** that figures earlier in this section, and so has twice the dimension. As first noted by Koopman (1931), a classical motion induces a one-parameter group resembling the one-parameter unitary group representing the quantum mechanical motion, but is quite distinct from it; at most it may be possible to deduce the quantum mechanical group by restriction of Koopman's group to a suitable invariant subspace. As yet, however, this has been shown only in extremely limited cases, such as that of the harmonic oscillator.

It is only in nonrelativistic theory that there is an invariant distinction between the P's and Q's—the "momenta" on the one hand and the "coordinates" on the other. In relativistic theory a change of Lorentz frame mixes the P's and Q's, and the full Weyl relations display symplectic invariance that is lacking in the formulation in terms of canonical conjugate variables (or the restricted Weyl relations).

The Weyl relations serve to suppress irrelevant pathology that is permitted by the original loose formulation of the Heisenberg relations (cf. Problems) and have been shown by experience to be generally appropriate. The unitarity of the Weyl operators, like the hermitian character of the canonical P's and Q's, has been a universal assumption deriving from standard quantum phenomenology and physically based on conservation of probability and stability conceptions. However, unstable systems are the rule rather than exception, and nonunitary Weyl systems should perhaps not be excluded from consideration.

Problems

1. Show that a finite-dimensional symplectic vector space $(\mathbf{L}, A)$ is isomorphic to a symplectic space built over a given vector space $\mathbf{M}$. (Two symplectic vector spaces $(\mathbf{L}, A)$ and $(\mathbf{L}', A')$ are isomorphic if there exists a linear isomorphism T of $\mathbf{L}$ onto $\mathbf{L}'$ as topological vector spaces such that $A'(Tx, Ty) = A(x, y)$ for all x and y in $\mathbf{L}$.) An equivalent statement of the result is that there exist coordinates $x_1, \ldots, x_n, y_1, \ldots, y_n$ in $\mathbf{L}$ such that

$$A(z, z') = \sum_{j=1}^{n} (x_j' y_j - x_j y_j').$$

2. Let $(\mathbf{L}, A)$ be a finite-dimensional symplectic space, and let W be a mapping from $\mathbf{L}$ to the unitary operators on a pre-Hilbert space $\mathbf{K}$, satisfying the Weyl relations. Show that $(\mathbf{K}, W)$ is a Weyl system if and only if $W(tz)$ is a continuous function of $t \in \mathbf{R}$ for every fixed z in $\mathbf{L}$. (Hint: Show by induction that if W satisfies the Weyl relations over $(\mathbf{L}, A)$, then for arbitrary $z_1, \ldots, z_n$ in $\mathbf{L}$,

$$W(z_1)\cdots W(z_n) = \exp[\tfrac{1}{2} i \sum_{j<k} A(z_j, z_k)]\, W(z_1 + \cdots + z_n).)$$

3. Let $\mathbf{M} = \mathbf{R}$, let $\mathbf{K} = L_2(\mathbf{R})$; $Q(0) = 0$ and if $a \neq 0$, let $Q(a)$ be the operator in $\mathbf{K}$ given by: $f(x) \mapsto axf(x)$ on the domain of all $f \in \mathbf{K}$ such that $xf(x) \in \mathbf{K}$; $P(0) = 0$ and if $a \neq 0$, $P(a)$ is given by: $f(x) \mapsto -iaf'(x)$, on the domain of all absolutely continuous functions in $\mathbf{K}$ whose derivative is again in $\mathbf{K}$. Show that $(\mathbf{K}, P, Q)$ is a Heisenberg pair over the symplectic vector space built over $\mathbf{M}$.

4. Let $\mathbf{L}$ denote the space of real C^∞ functions on the circle S^1, and define

$$A(f, g) = \int f(\theta) dg(\theta);$$

topologize $\mathbf{L}$ algebraically. Show that $(\mathbf{L}, A)$ is not a symplectic vector space, but becomes such when $\mathbf{L}$ is replaced by its quotient modulo the subspace of constant functions, and A is replaced by the result of applying the given A to the representative functions in the residue classes. Show also that this symplectic structure is invariant under the induced action of arbitrary diffeomorphisms on S^1, where such a diffeomorphism T acts by sending $f(p)$ into $f(\mathrm{T}^{-1}(p))$, $p \in S^1$.

5. Let **L** denote the space of real C^∞ solutions of the wave equation $\varphi_{tt} - \varphi_{xx} = 0$ in $\mathbf{R}^2$ that have compact support at any fixed time. Let A denote the form $A(\varphi, \psi) = \int [\varphi(t, x)\partial_t\psi(t, x) - \psi(t, x)\partial_t\varphi(t, x)]\, dx$.

a) Show that this form is independent of t and defines a linear symplectic structure on **L** in its algebraic topology.

b) A Lorentz transformation T is defined as one preserving the form $dt^2 - dx^2$. It acts on solutions φ of the wave equation by sending $\varphi(p)$ into $\varphi(\mathrm{T}^{-1}(p))$. Show that this action does indeed carry a solution of the wave equation into another solution, and that A is invariant under this action.

c) Show that the foregoing A is the unique one (within proportionality via a real constant) that is Lorentz invariant and continuous in the topology of convergence at any one time (and, by implication, at all times) in $C_0^\infty(\mathbf{R})$. Here and elsewhere, for any manifold M, $C_0^\infty(M)$ is topologized by the convergence of derivatives of every order $(0, 1, 2, \ldots)$ on some compact set inclusive of the support of the limit. (Hint: cf. Poulsen, 1972.)

6. Show that there exists a map $z \mapsto W(z)$ from **C** to unitary operators on a Hilbert space **K** that satisfies the Weyl relations, but is not a Weyl system. (Hint: let G denote the compact dual group to the discrete group $\mathbf{R}_d$, consisting of the reals under addition with the discrete topology. Let $\mathbf{K} = L_2(G, m)$, where m is Haar measure on G. Then $\mathbf{R}_d$ is canonically embedded in G, by mapping x in $\mathbf{R}_d$ into the character $y \to e^{ixy}$ of $\mathbf{R}_d$. Let $U(x)$ be the operator $f(u) \to f(u + x)$, and let $V(x)$ denote the opeartor $f(u) \to e^{ixu}f(u), f \in \mathbf{K}$. Show that the restricted Weyl relations are satisfied by the pair (U, V), but that the continuity required for a Weyl system is lacking.)

7. Let $(L_2(\mathbf{R}), U, V)$ denote the Weyl pair for the one-dimensional Schrödinger representation. Show that the linear subspace of all C_0^∞ functions in $L_2(0, \infty)$ is invariant under P and Q, but that its closure is not invariant under the associated Weyl system.

8. Let $(\mathbf{L}, A)$ be a symplectic vector space, and let G denote the set of all pairs (z, a) with $z \in \mathbf{L}$, and $a \in \mathbf{R}$, with the multiplication law $(z, a)(z', a') = (z + z', a + a' + A(z, z'))$. Show that G is a topological group (called the *Heisenberg group*) and that any continuous unitary representation U such that $U(0, a) = e^{ia}\,\mathrm{I}$ defines a Weyl system by the equation $W(z) = U(z, 0)$. Show conversely that every Weyl system arises in this way.

9. Let M be a C^∞ manifold and m a C^∞ nonvanishing form of maximal degree; let $\mathbf{K} = L_2(M, m)$. Let **P** denote the set of all C^∞ vector fields that are generators of global continuous one-parameter groups of diffeomorphisms of **M**, and **Q** the set of all C^∞ functions on M. Define $P(X)$ as the selfadjoint generator of the unitarized action of e^{tX} in **K**, where $X \in \mathbf{P}$, and $Q(f)$ as the operation of multiplication by $f \in \mathbf{Q}$. (If G is a group of measurable transformations on a measure space (M, m), its *unitarized action* on $L_2(M, m)$ is defined as the unitary representation U of G given by the equation: $(U(g)f)(x) =$

$D(g^{-1})^{1/2}f(g^{-1}(x))$, where $D(g^{-1})$ is the Radon-Nikodym derivative, or Jacobian, of the transformed measure $m_g(E) = m(g^{-1}(E))$ with respect to m.) Show that if $X \in \mathbf{P}, f \in \mathbf{Q}$, then $P(X)$ and $Q(f)$ leave invariant the domain $\mathbf{D} = C_0^\infty(M)$, and that their restrictions to $\mathbf{D}$ satisfy the relations $[P(X), P(Y)] = iP[X, Y]$ (where on the right side $[X, Y]$ denotes the commutator, or Lie bracket, of X and Y), $[P(X), Q(f)] = -iQ(Xf)$, $[Q(f), Q(f')] = 0$. Apply these results to euclidean space and obtain commutation relations for the cases in which X or Y is an infinitesimal rotation or (as earlier) translation.

10. Give an example of a pair of selfadjoint operators p and q in Hilbert space that leave invariant a common dense domain $\mathbf{D}$ and satisfy the relations $[p, q]u = -iu$ for $u \in \mathbf{D}$, but which do not generate groups satisfying the restricted Weyl relations. (Hint: consider the Schrödinger p and q, restricted to the submanifold of functions satisfying periodic boundary conditions [or vanishing outside an interval, which is slightly more complex]. In $L_2(S^1)$, e.g., take p as $-i\partial_\theta$, and q as multiplication by θ on the domain of C^∞ functions vanishing in some neighborhood of $\theta = 0$, and compute the commutator $e^{isp}e^{itq}e^{-isp}e^{-itq}$.)

11. Let $\mathbf{H}$ denote an n-dimensional complex Hilbert space, $n < \infty$, and let $\mathbf{K}$ denote $L_2(\mathbf{H}^\#_r\, dg)$, where $\mathbf{H}^\#$ denotes $\mathbf{H}$ as a real euclidean space of dimension $2n$ and $dg = e^{-|x|^2/2}\,dx$. For arbitrary $z \in \mathbf{H}$ let $W(z)$ denote the operator on $\mathbf{K}$:

$$f(u) \mapsto f(u - z)\, e^{\langle z,u\rangle/2 - \langle z,z\rangle/4}.$$

Show that $(\mathbf{K}, W)$ is a Weyl system over $\mathbf{H}$, and that it is not irreducible. (Hint: show that the subspace $\mathbf{S}$ of antiholomorphic functions on $\mathbf{H}$ that are in $\mathbf{K}$ is a closed invariant subspace.)

12. In problem 11, suppose $n = 1$, and represent $\mathbf{H}$ as $\mathbf{C}$. Let the selfadjoint generators of the groups $W(t)$ and $W(-it)$, $t \in \mathbf{R}$, be denoted as P and Q. Show that $2^{-1/2}(P \pm iQ)$ act on the space of antiholomorphic functions as multiplication by $-i2^{-1/2}\bar{z}$ and $i2^{1/2}\,\partial/\partial\bar{z}$, respectively.

13. Show that the restriction to $\mathbf{S}$ of the Weyl system of Problem 11 is irreducible.

1.3. Functional integration

The existence of a Weyl system over a finite-dimensional Hilbert space follows from either the Schrödinger representation or, in a somewhat more complex way, from the particle representation on a Hilbert space of sequences rather than functions, by exponentiation of a Heisenberg system. Neither of these constructions applies to the case of an infinite-dimensional Hilbert space, as formulated, but it will be seen that appropriate analogs can be developed. This section begins the treatment of appropriate substitutes for Lebesgue measure in the infinite-dimensional context.

There simply is no nontrivial measure on a Hilbert space (or any Banach space) of infinite dimension that is translation-invariant. There is however a substitute that is unitarily invariant: in a Hilbert space of large finite dimension, the unitary group is much larger than the translation group, so this substitute is relatively well suited to the expression of physical symmetries. For short, we call it the *isonormal (probability) distribution* in the Hilbert space, which, while quite effective, involves a serious technical problem, which impeded its recognition and development. Specifically, the corresponding measure in Hilbert space is not countably additive, which appears at first glance to render conventional abstract Lebesgue integration theory inapplicable. But there is an abstract measure space in the background which is roughly a direct limit of the corresponding probability spaces over n-dimensional Hilbert space as $n \to \infty$ (a notion that can in fact be made precise).

The most direct and succinct way to treat the generalized notion of integration needed is by an algebraic formulation that can be interpreted physically as more operational than abstract Lebesgue integration theory, but consistent with it. The formulation of modern probability theory by Kolmogorov (1933) in terms of measure theory, the chief tenet being the axiom that a random variable is a measurable function on a probability space, is effective technically but is less intuitive than the original approach of the Bernoullis and their successors. This can now be precisely formulated and legitimized by the line of representation theory due to Gelfand and Stone. In these terms, the random variables are assumed simply to form an associative algebra on which a (partially defined) expectation value function is given, satisfying the natural properties of positivity and normalization. In the presence of boundedness or reasonable growth conditions, it follows from representation theory that this concept of random variable is equivalent (in respects interpretable as observable or operationally meaningful) to the Kolmogorov concept. But there is little specificity to the measure space involved in this definition, and the actions of physical symmetries on the points of the space are relatively implicit, while the algebraic and group transformation properties of the random variables are rather explicit, and particularly effective technically in the case of the isonormal distribution on a Hilbert space.

The simplest functions on a Hilbert space are the coordinates $x_1, x_2, \ldots$, or functions $f(x_1, \ldots, x_n)$ of a finite number of such (e.g., *polynomials* on Hilbert space, defined as such functions with f a polynomial in n numerical variables). In the algebraic approach it is natural to begin with the assignment of an expectation value functional to the algebra of all such functions. The coordinates are of course linear functionals over the Hilbert space, and this algebraic approach boils down to the self-consistent assignment of a random variable to each such linear functional. This is the essential idea of the treatment that follows. Its invariance and operational features compensate quite adequately for its abstraction relative to the classical Kolmogorov approach, and it is also readily extendable to noncommuting generalized random variables such as those that occur in connection with fermion fields.

DEFINITION. Given an arbitrary measure space **P**, a *measurable* on **P** is an equivalence class of complex-valued measurable functions on **P** modulo null functions. Let **L** be a given locally convex topological vector space. A *predistribution* in **L** is a linear mapping d from the dual $\mathbf{L}^*$ of **L** to the measurables on a probability measure space **P**. Two such predistributions d and d' are *equivalent* in case the joint distribution of $d(f_1), \ldots, d(f_n)$ is the same as that of $d'(f_1), \ldots, d'(f_n)$, for every finite ordered set $f_1, \ldots, f_n$ of vectors in $\mathbf{L}^*$. An equivalence class of predistributions is a *distribution*. To avoid circumlocution, however, the term distribution will normally be used for both predistributions and equivalence classes of such, since it will usually be clear from the context, or on occasion immaterial, which concept is appropriate.

If T is a continuous linear transformation on **L**, and $\bar{d}$ is a distribution on **L**, the transform of $\bar{d}$ by T, denoted $\bar{d}^T$, is defined by the predistribution $d(T^*x)$, where d is a predistribution that represents $\bar{d}$. Note that this is independent of the choice of the representing predistribution, and that $\bar{d}^{ST} = (\bar{d}^T)^S$.

EXAMPLE 1.5. If **L** is finite-dimensional, then a distribution is tantamount to a probability measure on **L**. On the one hand, given a regular such measure m, any vector f in $\mathbf{L}^*$ represents a measurable function $f(x)$, $x \in \mathbf{L}$, so that $d(f) = f(\cdot)$ defines a predistribution. Conversely, given a predistribution, the corresponding probability measure m is definable on an arbitrary Borel subset B of **L** by the equation $m(B) = Pr[(d(f_1), d(f_2), \ldots, d(f_n)) \in B']$, where B' is the subset of $\mathbf{R}^n$ into which B is carried by the coordinate functions $(f_1, \ldots, f_n)$ corresponding to some basis of **L**.

When **L** is infinite-dimensional, a regular probability measure m on **L** determines a distribution on **L**, just as in the finite-dimensional case. In general, a distribution on **L** will not be of this form; in the special case that it is, it will be called *strict*.

If **L** is a real Hilbert space, for example, it will be seen that given any bounded positive selfadjoint operator C on **L**, there exists a unique distribution d such that: i) $d(x_1), \ldots, d(x_n)$ are jointly normally distributed for arbitrary $x_1, \ldots, x_n$ (where **L** is being identified with its dual in the usual way); and ii) for arbitrary x and y in **L**, $E(d(x)d(y)) = \langle Cx, y\rangle$ and $E(d(x)) = 0$, where here and henceforth E denotes *expectation* (i.e., integral with respect to the probability measure). This distribution, called the centered *normal distribution of* *covariance* C, is strict if and only if the operator C is of trace class. If for example $C = cI$, the distribution is called the *isonormal distribution of variance parameter* c on $\mathbf{L}$. This distribution will be denoted by g_c, where the subscript c will be dropped when immaterial or clear from the context. The isonormal distribution is invariant under arbitrary orthogonal transformations U on $\mathbf{L}$, in the sense that $g^U = g$. However, it is by no means strict unless $\mathbf{L}$ is finite-dimensional, in which case it corresponds to the Gaussian measure $dg_c = (2\pi c)^{-n/2} e^{-\langle x,x\rangle/2c}\, dx$, where n is the dimension.

The existence of a distribution d having specified joint distributions for $d(f_1),\ldots,d(f_n)$ for arbitrary finite sets of vectors $f_1,\ldots,f_n$ in $\mathbf{L}^*$ follows by a counterpart to a theorem of Kolmogorov on the existence of stochastic processes with preassigned joint distributions at arbitrary, finite sets of times. This counterpart is as follows in the simple and central case in which $\mathbf{M}$ is a real Hilbert space.

THEOREM 1.4. *Suppose that for each finite-dimensional subspace* $\mathbf{F}$ *of a real Hilbert space* $\mathbf{H}$ *there is given a predistribution* $d_{\mathbf{F}}$, *with the property that if* $\mathbf{G}$ *is a linear subspace of* $\mathbf{F}$, *then* $d_{\mathbf{F}}|\mathbf{G}$ *is equivalent to* $d_{\mathbf{G}}$. *Then there exists a unique distribution* d *on* $\mathbf{H}$ *such that for every finite dimensional subspace* $\mathbf{F}$, $d|\mathbf{F}$ *is equivalent to* $d_{\mathbf{F}}$.

PROOF. In order to reduce this theorem to general representation theory, it is convenient to make the

DEFINITION. Let $\mathbf{L}$ be a real topological vector space. A *tame* function on $\mathbf{L}$ is one of the form $F(x) = f(\lambda_1(x),\ldots,\lambda_n(x))$, where f is a complex-valued Borel function on $\mathbf{R}^n$, and $\lambda_1,\ldots,\lambda_n$ are vectors in $\mathbf{L}^*$ ($n < \infty$). If Λ is any linear subspace of $\mathbf{L}^*$ that contains $\lambda_1,\ldots,\lambda_n$, F is said to be *based on* Λ.

Referring now to the theorem, let $\mathbf{A}$ denote the collection of all bounded tame functions on the given Hilbert space $\mathbf{H}$. Noting that a tame function based on a subspace Λ is also based on any finite-dimensional subspace that contains Λ, it follows that the product and sum of tame functions is again tame. It follows in turn that the bounded tame functions on $\mathbf{H}$ form a *-algebra with F^* defined as the complex conjugate of F.

For any bounded tame function F based on the finite-dimensional subspace $\mathbf{N}$ of $\mathbf{H}$, the expectation $E_{\mathbf{N}}$ is well defined (as the integral of F over $\mathbf{R}^n$ with respect to the probability measure on $\mathbf{R}^n$ corresponding to $d|\mathbf{N}$). If $\mathbf{N}'$ is any other finite-dimensional subspace on which F is based and which contains $\mathbf{N}$, then $E_{\mathbf{N}'}(F) = E_{\mathbf{N}}(F)$, since $d_{\mathbf{N}'}|\mathbf{N}$ is equivalent to $d_{\mathbf{N}}$. But it follows from this that even if $\mathbf{N}'$ does not necessarily contain $\mathbf{N}$, then still $E_{\mathbf{N}'}(F) = E_{\mathbf{N}}(F)$, for $\mathbf{N}$ and $\mathbf{N}'$ are both contained in their finite-dimensional linear span $\mathbf{N}''$, so

$E_{N''}(F) = E_N(F)$ and $E_{N''}(F) = E_{N'}(F)$. Setting $E(F)$ equal to $E_N(F)$ for any finite-dimensional subspace **N** on which F is based, it follows that E is a well-defined linear functional on **A**.

Observe next that the system (**A**, E, *) consisting of the algebra **A**, the linear functional E, and the adjunction operator *, satisfies conditions characterizing algebras of bounded random variables, apart from the essentially trivial possibility that $E(|F|^2)$ may vanish for some class of functions F in **A** (which in the usual measure-theoretic description of random variables are represented by null functions).

Specifically, the system (**A**, E, *) forms a *commutative integration algebra*, meaning that: i) $E(F^*) = \overline{E(F)}$; ii) $E(F^*F) \geq 0$; and iii) $|E(G^*FG)| \leq c(F)$ $E(G^*G)$, for all F and G in **A**, where $c(F)$ is a positive F-dependent constant. Conditions i) and ii) are obvious, and iii) is satisfied with $c(F)$ taken as the supremum of F.

It follows (SK, sec. 8.4) that there exists a probability measure space **P** and a *-preserving homomorphism ψ of **A** into bounded measurable functions on **P** (where the * on functions is defined throughout as the complex conjugate) such that: i) ψ is an isomorphism modulo the ideal of $F \in$ **A** for which $E(F^*F) = 0$, for which $\psi(F)$ are null functions; and ii) $E(F) = \mathbf{I}(\psi(F))$, where **I** denotes the integral over **P**. The requisite mapping d can now be defined by setting $d(\lambda)$ equal to the unique measurable such that $\psi(F) = F(d(\lambda_1),\ldots, d(\lambda_n))$ for the bounded tame functions F on an arbitrary finite-dimensional subspace **F** in **H**, on which $\lambda_1,\ldots,\lambda_n$ are linear functionals. The existence of this mapping d, which effectively serves to extend ψ from the bounded tame functions to arbitrary tame functions, which are in any event Borel functions of a finite number of the $\lambda_j(x)$, follows by elementary functional analysis and is left as an exercise. □

Example 1.6. Let **H** be a real Hilbert space, and C a bounded positive selfadjoint operator on **H**; for arbitrary $x, y \in$ **H**, let $S(x, y) = \langle Cx, y\rangle$. On any finite-dimensional subspace **F** of **H**, there is a unique normal distribution of vanishing mean and covariance form equal to $S|\mathbf{F}$, say $d_{\mathbf{F}}$. This unicity implies that if **G** is a finite-dimensional subspace of **F**, then $d_{\mathbf{F}}|\mathbf{G}$ is equivalent to $d_{\mathbf{G}}$. It follows that there exists a unique distribution n_C on **H** whose restriction to any finite-dimensional subspace is normal with the indicated mean and covariance operator.

Theorem 1.4 is formulated for Hilbert space for simplicity; the argument applies equally well to general cases. The argument also shows that the concept of distribution on a given topological vector space **L** is equivalent to that of a normalized positive linear functional E on the algebra **A**(**L**) of all bounded tame functions over **L**, having the property that $|E(G^*FG)| \leq c(F)\ E(G^*G)$

for all F and G in $\mathbf{A}(\mathbf{L})$, where $c(F)$ is a finite and depends only on F. The general case is more easily stated in terms of the latter concept of an algebra equipped with an expectation-value form.

COROLLARY 1.4.1. *Let E be a linear functional on the algebra $\mathbf{A}$ of all bounded tame functions on the given real topological vector space $\mathbf{L}$, having the properties:*

$$E(1) = 1, \qquad E(F^*F) \geq 0, \qquad |E(G^*FG)| \leq c(F)\, E(G^*G)$$

for arbitrary F and G in $\mathbf{A}$, where $c(F)$ is finite. There then exists a probability measure space $\mathbf{P}$ and a predistribution d mapping $\mathbf{L}^*$ to the measurables on $\mathbf{P}$ with the following properties:

*1) d extends to a *-homomorphism (also denoted d) from $\mathbf{A}$ into the algebra of all measurables on $\mathbf{P}$, such that if $F \in \mathbf{A}$ has the form $F(x) = f(\lambda_1(x), \ldots, \lambda_n(x))$, where f is a Borel function on $\mathbf{R}^n$, then*

$$d(F) = f(d(\lambda_1), \ldots, d(\lambda_n)).$$

2) $E(F) = \mathbf{I}(d(F))$, where $\mathbf{I}$ denotes the integral over $\mathbf{P}$. Moreover, $\mathbf{P}$ can be chosen so that $d(\mathbf{A})$ is dense in $L_p(\mathbf{P})$ for every $p \in [1, \infty)$, and the pair $(d, \mathbf{P})$ is then unique within isomorphism. More specifically, given two such pairs $(d, \mathbf{P})$ and $(d', \mathbf{P}')$, there is an isomorphism α from the measurables on $\mathbf{P}$ to the measurables on $\mathbf{P}'$ such that $d'(F) = \alpha(d(F))$ for every $F \in A$.

PROOF. This proof is similar to that for Theorem 1.4 and is left as an exercise, with the remark that density of a *-algebra of bounded functions in L_p ($p < \infty$) of a probability measure space is equivalent to the property that the minimal σ-algebra of sets relative to which every member of the algebra is measurable includes all measurable sets, apart from possible null sets. This property guarantees uniqueness up to isomorphism of $(d, \mathbf{P})$ and serves to avoid redundancy in $\mathbf{P}$. (Cf. Glossary, Stone-Weierstrass Theorem for probability space.) □

A probabilistic concept familiar in the finite-dimensional case is extended in the

DEFINITION. If d is a distribution on the real topological space $\mathbf{L}$, its *characteristic function* μ is the function on $\mathbf{L}^*$: $\mu(\lambda) = E(e^{id(\lambda)})$.

Its characterization is similar to that in the finite-dimensional case:

COROLLARY 1.4.2. *A distribution is uniquely determined by its character-*

istic function. A given function $\mu(\lambda)$ *on* $\mathbf{L}^*$ *is the characteristic function of a distribution if and only if:*

1) μ *is positive definite and* $\mu(0) = 1$;

2) the restriction of μ *to an arbitrary finite-dimensional subspace of* $\mathbf{L}^*$ *is continuous.*

PROOF. We recall that for μ to be positive definite means that $\Sigma_{1 \le j,k \le n} \mu(\lambda_j - \lambda_k)\alpha_j\bar{\alpha}_k \ge 0$ for arbitrary $\lambda_1, \ldots, \lambda_n$ in $\mathbf{L}^*$ and complex $\alpha_1, \ldots, \alpha_n$. Both 1) and 2) obviously hold if μ is a characteristic function. On the other hand, if μ satisfies 1) and is continuous on the finite-dimensional subspace $\mathbf{N}$ of $\mathbf{L}^*$, then by Bochner's theorem there exists a unique regular probability measure $\Pi_\mathbf{N}$ on $\mathbf{N}^*$ of which $\mu|\mathbf{N}$ is the characteristic function. This in turn determines an expectation form $E_\mathbf{N}$ on the algebra of all bounded Borel functions of an arbitrary basis $\lambda_1, \ldots, \lambda_n$ for $\mathbf{N}$, regarded as measurable functions with respect to $\Pi_\mathbf{N}$. Now allowing $\mathbf{N}$ to vary, and forming the union of the algebras of bounded tame functions for each $\mathbf{N}$, yields the algebra $\mathbf{A}$ of all bounded tame functions over $\mathbf{L}$. It provides also a unique expectation value functional E extending each $E_\mathbf{N}$, and satisfying the criteria for a distribution in Corollary 1.4.1, with $c(F)$ taken as the supremum of $|F|$. □

CAVEAT. Some care is needed in the infinite-dimensional case in distinguishing between functions on the underlying space $\mathbf{L}$, residue classes of tame functions modulo the ideal $\mathbf{J}$ of all functions F for which $E(F^*F) = 0$, and elements of function spaces over the probability space $\mathbf{P}$. This is a result of the weakness of the distribution that may be in question, i.e., the lack of full countable additivity. As a consequence of this, a vector in $L_2(\mathbf{P})$ need not be representable by a function on $\mathbf{L}$; the Riesz-Fischer theorem carries no such implication. Conversely, simple operations on the functions on $\mathbf{L}$ do not necessarily carry over to the residue classes mod $\mathbf{J}$. For example, the induced action on $\mathbf{A}$ of translations $x \mapsto x + a$ on $\mathbf{L}$ is well defined: $F(x) \mapsto F(x + a)$, but is naturally extendable to the quotient $\mathbf{A}/\mathbf{J}$ only if $\mathbf{J}$ is invariant under such translations, which is in general not the case. In practice, an elaborate notation is not required to eliminate confusion in these matters and will be avoided, but on occasion it will be useful to use the notation θ for the canonical homomorphism from $\mathbf{A}$ to $\mathbf{A}/\mathbf{J}$. It is also useful to make the following definitions:

DEFINITION. Let d be a predistribution in the vector space $\mathbf{L}$, extended to a *-homomorphism from the algebra $\mathbf{A}$ of bounded tame functions on $\mathbf{L}$ to measurables on $\mathbf{P}$. Then the space of *random variables* in $(\mathbf{L}, d)$, denoted $\mathbf{M}(\mathbf{L}, d)$, is defined to be the smallest subspace of the space of measurables on $\mathbf{P}$ containing $d(\mathbf{A})$ and closed under the operation of taking pointwise a.e. limits of sequences. The subspace of random variables f for which $E(|f|^p) < \infty$ is denoted as $L_p(\mathbf{L}, d)$.

We leave as an exercise to check that if d and d' are equivalent predistributions on $\mathbf{L}$, there is a unique sequentially continuous isomorphism α from $\mathbf{M}(\mathbf{L}, d)$ to $\mathbf{M}(\mathbf{L}, d')$ such that $\alpha(d(F)) = d'(F)$ for all $F \in \mathbf{A}$. Thus given a distribution e on $\mathbf{L}$, we may define the algebra of *random variables* in $(\mathbf{L}, e)$ to be $\mathbf{M}(\mathbf{L}, d)$ for any predistribution d representing e, and the dependence on d is inessential. Similarly, we define $L_p(\mathbf{L}, e)$ to be $L_p(\mathbf{L}, d)$ for any predistribution d representing e.

The following Scholium constructs (the easy) half of the Schrödinger or real wave representation for distributions on an infinite-dimensional space.

SCHOLIUM 1.1. *For any distribution d on the real topological vector space* $\mathbf{L}$, *there exists a unique unitary representation V of the additive group of* $\mathbf{L}^*$ *on* $\mathbf{K} = L_2(\mathbf{L}, d)$ *such that* $V(\lambda)\theta(F) = e^{id(\lambda)}\theta(F)$ *for all bounded tame functions F, and which is continuous in the algebraic topology on* $\mathbf{L}$. *Moreover* 1 *is a cyclic vector in* $\mathbf{K}$ *for V.*

PROOF. It is straightforward to check that the mapping, say $V_0(\lambda)$, that carries $\theta(F)$ into $\theta(e^{id(\lambda)}F)$, where $F \in \mathbf{A}$, is independent of the choice of representative for $\theta(F)$. In addition, $V_0(\lambda)$ preserves the inner product in $\mathbf{K} = L_2(\mathbf{L}, d)$, and $V_0(\lambda)V_0(\lambda') = V_0(\lambda + \lambda')$, by easy arguments. Thus $V_0(\lambda)$ has the inverse $V_0(-\lambda)$ and can be extended uniquely to a unitary operator $V(\lambda)$ on the closure $L_2(\mathbf{L}, d)$ of $\theta(\mathbf{A})$. It follows that V is a representation of the additive group of $\mathbf{L}^*$ on $\mathbf{K}$. To demonstrate the continuity, it suffices to show that $V(\lambda)w$ is a continuous function of λ for a dense set of vectors w in $\mathbf{K}$, for example, the $\theta(\mathbf{A})$, in which case the requisite continuity follows by dominated convergence.

To show that 1 is a cyclic vector for the $V(\lambda)$, it suffices to show that the $\theta(e^{id(\lambda)})$ span $\theta(\mathbf{A})$ in $\mathbf{K}$. To this end it is adequate to treat the case in which λ ranges over a finite-dimensional subspace $\mathbf{M}$ of $\mathbf{L}^*$ and $\mathbf{A}(\mathbf{M})$ is the corresponding algebra of bounded tame functions, since the totality of the $\mathbf{A}(\mathbf{M})$ is dense in $\mathbf{K}$. Since this is a finite-dimensional question, θ may be omitted. The span of the $V(\lambda)1$ in the finite-dimensional case includes all functions on $\mathbf{L}$ of the form $\int e^{i\lambda(x)}f(\lambda)\, d\lambda$, where f is integrable, by a simple argument, and hence all $\hat{f}(\lambda_1(x), \ldots, \lambda_n(x))$, where $\hat{f}$ denotes the Fourier transform of f. Such $\hat{f}$ are dense in the algebra of all continuous functions of x that vanish at infinity on $\mathbf{R}^n$, and hence in the algebra of continuous functions of compact support, which in turn is dense in $L_2(\mathbf{R}^n)$. □

Notions of the mean, variance, and continuity of distributions extend from strict to general distributions, as in the

Definition. If d is a distribution in **L** such that $d(\lambda)$ is integrable for all $\lambda \in \mathbf{L}^*$, and if there exists a vector $a \in \mathbf{L}$ such that $E[d(\lambda)] = \lambda(a)$ for all $\lambda \in \mathbf{L}^*$, a is called the *mean* of the distribution d. If $d(\lambda)$ is square-integrable for all λ and if $E(d(\lambda)d(\lambda')) = C(\lambda, \lambda')$, the form $C(\cdot,\cdot)$ is called the *covariance form* of the distribution. When **L** is a real Hilbert space, the covariance *operator* of the distribution is an operator C such that $C(\lambda, \lambda') = \langle C\lambda, \lambda' \rangle$, with the usual identification between **L** and $\mathbf{L}^*$ in the Hilbert space case. A *bounded distribution* on a Banach space **B** is a distribution d such that $\|d(\lambda)\|_2 \leq c\|\lambda\|$, where c is a constant, and the notation $\|\cdot\|_p$ applied to a random variable indicates its L_p norm. A *continuous distribution* is a distribution d on **L** that is continuous from $\mathbf{L}^*$ to random variables, relative to a given topology on the space of the latter. For example, let **H** be a real Hilbert space and A a densely defined operator in **H** with domain **D**. If A is positive and selfadjoint, an argument given previously shows that there exists a normal distribution on **D** with mean 0 and covariance operator A. If A and its inverse are bounded, this distribution is continuous in the L_2-norm on random variables.

Problems

1. Evaluate the characteristic function of the normal distribution of mean 0 and covariance operator C, where C is bounded, positive, and selfadjoint.

2. Show that every vector a in the real Hilbert space **H** is the mean of a unique normal distribution d whose restriction to each finite-dimensional subspace has centered covariance operator I, where the centered covariance operator, say B, is defined by the equation

$$\langle Bx, y \rangle = E[(d(x) - d(b))(d(y) - d(b))],$$

where b is the mean of the restriction. Evaluate the characteristic function of this distribution.

3. Evaluate the characteristic function of the normal distribution on the given real Hilbert space **H** of given mean in **H** and positive selfadjoint bounded centered covariance.

4. Let $\mathbf{H} = L_2(\mathbf{R})$. Let c_t denote the characteristic function of the subset $[0, t]$ of **R**, let g denote the isonormal distribution of variance 1 on **H**, and set $x(t) = g(c_t)$.

a) Show that for arbitrary positive $t_1 < t_2 < \cdots < t_n$, the random variables $x(t_{j+1}) - x(t_j)$ are stochastically independent ($j = 0,1,\ldots,n$; $t_0 = 0$). Show also that for arbitrary positive t and t', $x(t) - x(t')$ is normally distributed with mean 0 and variance $|t - t'|$. (The stochastic process $x(t)$ is thus equivalent to Wiener Brownian motion.)

b) Show that $x(t)$ is a continuous function of t, in the L_2-topology on random variables. (A similar celebrated theorem of Wiener asserts that each $x(t)$ can be represented by a measurable function $x(t, \cdot)$ on a measure space **W**, in such a way that for almost all ω, $x(t, \omega)$ is a continuous function of t [cf., e.g., Doob, 1965].)

5. Show that a bounded distribution on a Hilbert space always has a mean and a bounded covariance operator.

6. Let $e_1,e_2,\ldots$ denote a complete orthonormal set in the Hilbert space **H**, and let $u_1,u_2,\ldots$ denote a sequence of independent measurables on a probability measure space **P** with uniformly bounded second moments. Show that there exists a bounded distribution f on **H** such that $f(e_j) = u_j$ $(j = 1,2,\ldots)$.

7. A group of measure-preserving transformations on a probability measure space is said to be *ergodic* if the only measurables that are left fixed by the group are the constants. Show that the orthogonal group **O(H)** of a real infinite-dimensional Hilbert space **H** acts ergodically on the space of random variables over $(\mathbf{H}, g)$, where g is the isonormal distribution. (This is in contrast to the action of the orthogonal group on the isonormal distribution in the finite-dimensional case, which is not at all ergodic. [Hint: **O(H)** leaves invariant the joint distributions of the $n(x_1), \ldots, n(x_m)$ for arbitrary $x_1, \ldots, x_m$.] It follows that to each $T \in \mathbf{O(H)}$ there corresponds a *-algebraic automorphism $a(T)$ of the algebra $\mathbf{M}(\mathbf{H}, g)$ of all random variables, which moreover leaves invariant the expectation. [This is the operational form of a measure-preserving transformation. Representation by a point transformation is superfluous: it can always be attained by a suitable choice of **P**, cf. SK, but this is superfluous here.] If there is a nontrivial invariant element of $\mathbf{M}(\mathbf{H}, g)$, then there is an invariant bounded such random variable, which can be approximated arbitrarily closely by a vector in $L_2(\mathbf{H}, g)$, which in turn can be approximated arbitrarily closely in this space by a tame function. Now observe that **O(H)** cannot leave any such element of $L_2(\mathbf{H}, g)$ invariant, for it can carry the function into one based on an orthogonal subspace.)

8. A tame subset S of a real topological vector space **L** is defined as one whose characteristic function c_S is tame. Show that for any distribution the functional $m(S) = E(c_S)$ is additive on the ring of all tame subsets.

9. Show that in the case of the isonormal distribution in an infinite-dimensional real Hilbert space **H**, the associated finitely additive measure m of Problem 8 is not countably additive. (Hint: let $x_1,x_2,\ldots$ be coordinates in **H** relative to an orthonormal basis, let c_n be a sequence $(n = 1, 2, \ldots)$ which is monotone increasing to ∞, and consider the tame sets $S_n = \{x: x_1^2 + x_2^2 + \cdots + x_n^2 > c_n\}$, whose intersection is empty. Show that for suitable choice of c_n, $m(S_n)$ is bounded away from 0, using, for example, the central limit theorem.)

10. Let n denote the integral $\int_0^\infty g_c \, d\mu(c)$, where g_c is the isonormal distribution of variance c on **H** and $d\mu(c)$ is a probability measure on $(0, \infty)$. Show that n is invariant under **O(H)**. Are all orthogonally invariant distributions of

this form for some μ? Does $\mathbf{O}(\mathbf{H})$ act ergodically on the space of random variables over $(\mathbf{H}, n)$ if $\mathbf{H}$ is infinite-dimensional?

1.4. Quasi-invariant distributions

Although there is no nontrivial measure on an infinite-dimensional Banach space that is invariant under all translations by vectors in the space, there exist distributions that are "quasi-invariant" in the sense that, roughly speaking, the effect on them of such translation leaves their class of null sets intact. In a finite-dimensional space this has a precise meaning, since distributions are then strict, and implies that they are equivalent in the sense of absolute continuity to Lebesgue measure (Mackey, 1952). Equivalently, quasi-invariant distributions are the indefinite integrals of functions vanishing only on a Lebesgue null set.

Quasi-invariant distributions play a central role in the extension of the Schrödinger representation to the infinite-dimensional case. Their structure is much more complex than in the finite-dimensional case, and they are not at all necessarily mutually absolutely continuous in any sense approximating the usual one. This section begins the extension of the absolute continuity concept to distributions and its application to the treatment of quasi-invariance.

DEFINITION. Let d and e be given distributions in the real topological vector space $\mathbf{L}$. For any distribution d in $\mathbf{L}$, let θ_d denote the canonical homomorphism of the algebra $\mathbf{A}$ of all bounded tame functions on $\mathbf{L}$ modulo the ideal $\mathbf{J}$ of null functions, for the distribution d (or simply θ when d is clear from the context). Then e is *absolutely continuous with respect to* d (symbolically, $e \ll d$) if there exists $D \in L_1(\mathbf{L}, d)$ such that for all $F \in \mathbf{A}$,

$$E_e[\theta_e(F)] = E_d[\theta_d(F)D].$$

D is called the *derivative* of e with respect to d and is denoted $D(e, d)$.

SCHOLIUM 1.2. *Every element B of $L_\infty(\mathbf{L}, d)$ is the limit almost everywhere of a sequence of the form $\theta(F_n)$, where $F_n \in \mathbf{A}$ and the range of F_n lies in the essential range of B.*

PROOF. For an arbitrary element $u \in L_2(\mathbf{L}, d)$, there exists a sequence $\{F_n\}$ of tame functions such that $\theta_d(F_n) \rightarrow u$ in $L_2(\mathbf{L}, d)$. Taking a subsequence if necessary, it may be assumed that $\theta_d(F_n) \rightarrow u$ almost everywhere. If f is the identity function on the essential range of u and takes the value c elsewhere on $\mathbf{C}$, where c lies in the essential range of u, then f is a Borel function and $f \circ$

$\theta_d(F_n) \to f \circ u$ almost everywhere; and $f \circ \theta_d(F_n) = \theta_d(f \circ F_n)$ while $f \circ u = u$. □

SCHOLIUM 1.3. *The derivative $D(e, d)$ is unique and nonnegative.*

PROOF. If $E_d(\theta_d(F)D) = E_d(\theta_d(F)D')$ for some D', then, choosing an appropriate sequence and using dominated convergence, it follows that $E_d(BD) = E_d(BD')$ for all $B \in L_\infty$, implying that $D = D'$. If B is an arbitrary nonnegative element of L_∞, the same argument shows that $E_d(BD) \geq 0$, showing that $D \geq 0$.

EXAMPLE 1.7. Let R be a set, $\mathbf{R}$ a sigma-ring of subsets including R, and r and s two probability measures on $\mathbf{R}$, with $s \ll r$. Let d be a linear mapping from the given topological vector space $\mathbf{L}^*$ to the measurables on the probability measure space $(R, \mathbf{R}, r)$, and for any λ in $\mathbf{L}^*$, let $e(\lambda)$ denote $d(\lambda)$ as a measurable on the probability measure space $(R, \mathbf{R}, s)$. Then e is absolutely continuous with respect to d in the foregoing sense, by the Radon-Nikodym theorem.

The definition just made is further justified by

SCHOLIUM 1.4. *With the above notation, $e \ll d$ if and only if e and d can be simultaneously represented, within equivalence, in the form given in Example 1.7.*

PROOF. The "if" part has already been disposed of, so assume that e and d are given predistributions over $\mathbf{L}$ such that $e \ll d$. It is no essential loss of generality to suppose that d is represented in the form given in Example 1.7. Now let e' be the predistribution which carries an arbitrary element λ of $\mathbf{L}^*$ into $d(\lambda)$ as a measurable on $(R, \mathbf{R}, s)$ where s is the probability measure on $(R, \mathbf{R})$ given by $ds = Ddr$. To complete the proof it suffices to show that e' is equivalent to e; this means that if b is any bounded Borel function of n real variables, then

$$\int b(d(\lambda_1), \ldots, d(\lambda_n))\, ds = E_e[\theta_e(b(\lambda_1(\cdot), \ldots, \lambda_n(\cdot)))].$$

By hypothesis, the expression on the right equals

$$E_d[\theta_d(b(\lambda_1(\cdot), \ldots, \lambda_n(\cdot)))D] = \int b(d(\lambda_1), \ldots, d(\lambda_n))\, Ddr,$$

which is the same as the expression on the left, since $ds = Ddr$. □

DEFINITION. Two distributions d and e in a given real topological vector space are *equivalent in the sense of absolute continuity*, or *algebraically equivalent*, if $e \ll d$ and $d \ll e$.

A convenient criterion for weak equivalence in Hilbert space terms is given by

SCHOLIUM 1.5. *In order that two given distributions d and e in a given real topological vector space* **L** *be algebraically equivalent, it is necessary and sufficient that there exist a (fixed) unitary transformation U from* $L_2(d)$ *onto* $L_2(e)$ *which transforms the operation of multiplication by* $d(\lambda)$ *into that of multiplication by* $e(\lambda)$, *for all* $\lambda \in \mathbf{L}^*$:

$$U^{-1}M_{e(\lambda)}U = M_{d(\lambda)},$$

where $M_{d(\lambda)}$ *denotes the selfadjoint operator in* $L_2(d)$ *consisting of multiplication by* $d(\lambda)$, *and* $M_{e(\lambda)}$ *is the similar operator on* $L_2(e)$. *Moreover, U is unique if it is additionally required to be positivity-preserving.*

PROOF. Suppose first that d and e are algebraically equivalent. By virtue of the absolute continuity of e with respect to d, they may be jointly represented as in Example 1.7. Since d is absolutely continuous with respect to e, there exists also an element D' of $L_1(e)$ such that

$$E_d[\theta_d(F)] = E_e[\theta_e(F)D']$$

for all $F \in \mathbf{A}$. Setting r' for the probability measure on $(R, \mathbf{R})$ such that $dr' = D'ds$, then by the argument in the proof of Scholium 1.4, the predistribution d' mapping λ to $d(\lambda)$ as a measurable on $(R, \mathbf{R}, r')$ is equivalent to d. This means that if f is any measurable function on $(R, \mathbf{R})$ of the form $b(d(\lambda_1), \ldots, d(\lambda_n))$, where b is a bounded Borel function on $\mathbf{R}^n$ and $\lambda_1, \ldots, \lambda_n$ are arbitrary in $\mathbf{L}^*$, then

$$\int f dr' = \int f dr.$$

Since these f form a dense subset of both $L_1(r)$ and $L_1(r')$, it follows that $r = r'$. Hence $DD' = 1$ a.e.*

In particular, D is invertible as a measurable, and the transformation $U: f \mapsto D^{-1/2}f$ is unitary from $L_2(d)$ into $L_2(e)$. It is immediate that U has the required transformation property on the dense domain of functions f supported by a set on which D and D^{-1} are bounded. By a conventional approximation argument this follows for all relevant f in $L_2(d)$.

If, on the other hand, there exists a unitary operator U having the indicated property, then by the unitary invariance of the operational calculus for commuting selfadjoint operators,

$$U^{-1}M_{b(e(\lambda_1), \quad, e(\lambda_n))}U = M_{b(d(\lambda_1), \quad, d(\lambda_n))},$$

b being an arbitrary bounded Borel function and $\lambda_1, \ldots, \lambda_n$ any finite subset of

* The abbreviation "a.e." means "almost everywhere"—i.e., except on a set of measure zero.

$\mathbf{L}^*$, where M_k denotes the operation of multiplication by the measurable k. It follows that if $\alpha(T)$ denotes $U^{-1}TU$ for any operator T in the maximal abelian W^*-algebra $\mathbf{M}(e)$ generated by the $M_{b(e(\lambda_1), \ldots, e(\lambda_n))}$, then α is a $*$-isomorphism of $\mathbf{M}(e)$ onto the corresponding algebra $\mathbf{M}(d)$ for d. Evidently, α is continuous in the weak operator topology. It follows (SK, Schol. 6.5) that $\langle \alpha(T)1, 1\rangle_{L_2(d)}$ has the form $\langle Tx_1, y_1\rangle + \cdots + \langle Tx_n, y_n\rangle$ for some finite set of vectors $x_1, y_1, \ldots, x_n, y_n$ in $L_2(e)$. (Alternatively, the Radon-Nikodym theorem may be applied in conjunction with the $*$-isomorphisms of $\mathbf{M}(e)$ and $\mathbf{M}(d)$ with the L_∞ algebras of probability measure spaces.) This means in particular that $E_d(\theta(F)) = E_e(\theta(F)D)$ for all $F \in \mathbf{A}$, where $D = \Sigma x_i y_i$ is an element of $L_1(e)$. By symmetry, the same is true when d and e are interchanged, so they are algebraically equivalent.

It remains only to show the unicity of U when it is required to be positivity-preserving; the existence of such a U is shown by the construction above. If U_1 and U_2 both have the required transformation property and are positivity-preserving, then $U_1{}^{-1}U_2$ commutes with all the $M_{b(d(\lambda_1), \ldots, d(\lambda_n))}$; but these generate a maximal abelian algebra of operators on $L_2(d)$, and hence $U_1^{-1}U_2$ is itself a multiplication operator. Since it is positivity-preserving as well as unitary, it must be the identity. □

DEFINITION. The distribution d is called *quasi-invariant* in case for all a in $\mathbf{L}$, d is algebraically equivalent to the distribution d_a, defined by the equation

$$d_a(\lambda) = d(\lambda) + \langle a, \lambda\rangle$$

for all $\lambda \in \mathbf{L}^*$. For such a process, the measurable $D(d_a, d)^{1/2}$ is called the *unitarizer* for the transformation $x \mapsto x + a$.

EXAMPLE 1.8. If $\mathbf{L}$ is finite-dimensional, d is quasi-invariant if and only if the corresponding probability measure m on $\mathbf{L}$ is quasi-invariant in the classical sense—that any translate m_a by a vector a in $\mathbf{L}$ has the same null sets as m, in which case m is equivalent to Lebesgue measure in the sense of absolute continuity.

The relevance of quasi-invariant processes to Weyl systems is indicated by

THEOREM 1.5. *Let d be a given quasi-invariant distribution in the real topological vector space* $\mathbf{L}$, *algebraically topologized. Then there exists a unique Weyl pair $(U, V, \mathbf{K})$ over the dual couple* $(\mathbf{L}, \mathbf{L}^*)$, *with V and* $\mathbf{K}$ *as in Scholium 1.1, and with U given as follows: If F is any tame function, and a is arbitrary in* $\mathbf{L}$, *then $U(a)\theta(F) = \theta(F_a)G(a)$, where $F_a(x) = F(x + a)$ for all $x \in$* $\mathbf{L}$, *and $G(a)$ is the unitarizer for the transformation $x \mapsto x + a$ on* $\mathbf{L}$.

PROOF. Observe first that $\theta(F_a)$ is indeed determined by $\theta(F)$, precisely by virtue of the quasi-invariance of d. To show this, it suffices by linearity to show that if $\theta(F) = 0$, then $\theta(F_a) = 0$. Now

$$E_d[|\theta_d(F_a)|^2] = E_{d_a}[|\theta_{d_a}(F)|^2] = E_d[|\theta_d(F)|^2\, D(d_a,d)] = 0.$$

It is easily verified that the transformation $U_0(a)$ defined on the tame measurables in the indicated fashion is isometric, and so extends to a unique isometric mapping $U(a)$ from all of $L_2(d)$ into itself. Since $U_0(a)U_0(b) = U_0(a + b)$ by the chain rule for differentiation, $U(a)U(a) = U(a + b)$, from which it follows that the $U(a)$ are unitary. The restricted Weyl relation follows in a similar fashion, by first checking it on the dense domain of bounded tame measurables, and it remains only to establish the continuity of $U(\cdot)$.

By virtue of the unitarity of the $U(a)$, it suffices to show that $\langle U(a)x, x'\rangle$ is a continuous function for arbitrary fixed bounded tame measurables x and x', say $x = \theta(F)$ and $x' = \theta(F')$. Let $\mathbf{G}$ denote a finite dimensional subspace of $\mathbf{L}^*$ on which F and F' are based. Let d_0 denote the restriction of d to $\mathbf{G}$. Then d_0 is again a quasi-invariant distribution in $\mathbf{L}/\mathbf{G}^\perp$, and by Example 1.8 corresponds to an absolutely continuous probability measure of nonvanishing density on $\mathbf{L}/\mathbf{G}^\perp$.

Now note that if $e \ll d$, so that $E_e[\theta_e(H)] = E_d[\theta_d(H)D]$ with $D \in L_1(d)$ for all bounded tame functions H, then for H restricted to be based on the linear subspace $\mathbf{G}$ of $\mathbf{L}^*$,

$$E_e[\theta_e(H)] = E_d[\theta_d(H)D_{\mathbf{G}}],$$

where $D_{\mathbf{G}} = D(e|\mathbf{G}, d|\mathbf{G})$. It follows that $\langle U(a)x, x'\rangle = \langle U_{\mathbf{G}}(a)x, x'\rangle_{\mathbf{G}}$, where the subscript $\mathbf{G}$ indicates that the Hilbert space $L_2(d|\mathbf{G})$ is involved. Thus it suffices to establish the continuity in the case in which $\mathbf{L}$ is finite-dimensional; and in this case it is an easy variant of the classical result of Lebesgue, to the effect that $\int_{\mathbf{R}^n}|f(x + a) - f(x)|^2dx \to 0$ as $a \to 0$ for any $f \in L_2(\mathbf{R}^n)$. □

COROLLARY 1.5.1. *With the notation of Theorem 1.5, there exists a representation U of the additive group of $\mathbf{L}$ by unitary operators on $\mathbf{K}$ that is continuous relative to each finite-dimensional subspace of $\mathbf{L}$, and satisfies the restricted Weyl relation with $V(\cdot)$, if and only if d is quasi-invariant; and in this case, U may be further restricted, and is then unique, by the requirement that it be positivity-preserving.*

PROOF. That d is necessarily quasi-invariant if U is of the indicated type follows from Theorem 1.5. That the $U(x)$ may all be chosen to be positivity-preserving follows from the construction in the proof of Theorem 1.4. If $U'(\cdot)$ is another mapping from $\mathbf{L}$ to the unitaries on $\mathbf{K}$ having the same properties, then for every x, $U(x)^{-1}U'(x)$ is a unitary operator which is both positivity-preserving and commutes with all $V(\lambda)$; by the argument given earlier, it must therefore be the identity. □

Problems

1. Show that there exists no translation-invariant distribution on a Banach space. (Hint: reduce the problem to the finite-dimensional case.)
2. Calculate the unitarizer of a Gaussian distribution with given covariance.
3. Show that a σ-finite Borel measure on $\mathbf{R}^n$ that is absolutely continuous relative to its translates is mutually absolutely continuous with Lebesgue measure. (Hint: use the Fubini theorem.)

1.5 Absolute continuity

It has now been shown that for any given quasi-invariant distribution there is a corresponding Weyl pair. This serves to show the existence of Weyl pairs in the case when $\mathbf{L}$ is infinite-dimensional once quasi-invariant distributions are obtained. On the other hand, the latter are not so simple to come by as one might naively anticipate. Indeed, in an infinite-dimensional Banach space there are no quasi-invariant distributions that are strict. This means that one has to use weak distributions, for which the criteria for absolute continuity, and hence quasi-invariance, have to be developed.

Earlier on, Cameron and Martin (1944) proved that Wiener measure on the space $\mathbf{W}$ of continuous real functions on $[0, 1]$ vanishing at 0 is quasi-invariant relative to displacements by sufficiently smooth elements of $\mathbf{W}$. Using this, a corresponding Weyl system acting on $L_2(\mathbf{W})$ exists by Theorem 1.5. However, this Weyl system lacks natural transformation properties relative to typical physical symmetry groups. In this section we develop criteria for absolute continuity that are adapted to quite general classes of Weyl systems. As a by-product, the Cameron-Martin result follows in a maximally sharp form, exemplifying the Hilbert space character of the underlying theory. We first recall the

DEFINITION. If $\mathbf{P} = (M, \mathbf{M}, m)$ is a probability measure space with σ-ring $\mathbf{M}$ of measurable sets, and $\mathbf{N}$ is any sub-σ-ring of $\mathbf{M}$ (always assumed inclusive of $\varnothing$ and M unless otherwise specified), the *conditional expectation* of $f \in L_1(\mathbf{P})$ with respect to $\mathbf{N}$, denoted on occasion as $E[f|\mathbf{N}]$, is the unique measurable f' with respect to $\mathbf{P}' = (M, \mathbf{N}, m|\mathbf{N})$ such that for all bounded measurables g on $\mathbf{P}'$, $E[fg] = E[f'g]$.

THEOREM 1.6. *Let* $\mathbf{P} = (M, \mathbf{M}, m)$ *be a probability measure space, and let* $\{M_\lambda: \lambda \in \Lambda\}$ *be a directed system of sub-sigma-rings of* $\mathbf{M}$, *with* $M_\lambda \subset M_{\lambda'}$, *for* $\lambda \leq \lambda'$, *and such that* $\mathbf{M}$ *is generated by the union of the* M_λ. *Then a given* *probability measure n on $\mathbf{M}$ is absolutely continuous with respect to m if and only if*

i) $n|M_\lambda$ is absolutely continuous with respect to $m|M_\lambda$; and

ii) denoting as D_λ the derivative of $n|M_\lambda$ with respect to $m|M_\lambda$, then $\{D_\lambda\}$ is convergent in $L_1(\mathbf{P})$.

When i) and ii) hold, dn/dm is the limit of D_λ in $L_1(\mathbf{P})$.

PROOF. Note first that if f is any integrable with respect to $\mathbf{P}$, and if f_λ denotes the conditional expectation of f with respect to $\mathbf{M}_\lambda$, then the net $\{f_\lambda\}$ is L_1-convergent to f. To show this it suffices, since conditional expectation is a contraction, to establish the conclusion for a dense set of integrables f. But all integrables measurable with respect to some $\mathbf{M}_\lambda$ constitute such a dense set, and the stated L_1-convergence is trivially valid for any such f. This establishes the "only if" part of Theorem 1.6.

To prove the "if" part, let D denote the limit in L_1 of the net $\{D_\lambda\}$. Now if f is bounded and measurable with respect to some $\mathbf{M}_\lambda$, $n(f) = m(fD_{\lambda'})$ for $\lambda' \geq \lambda$, from which it follows on letting $\lambda' \uparrow$, that $n(f) = m(fD)$. Since any bounded measurable f is the limit a.e. of a uniformly bounded sequence of measurables with respect to the $\mathbf{M}_\lambda$, it follows in turn that $n(f) = m(fD)$ for all bounded measurables f. Thus n is absolutely continuous with respect to m. □

COROLLARY 1.6.1. *Condition ii) of Theorem 1.6 is equivaltent to the condition*

ii') $\{D_\lambda^{1/2}\}$ is convergent in $L_2(\mathbf{P})$.

PROOF. The inequality $|a^{1/2} - b^{1/2}|^2 \leq |a - b|$ for arbitrary nonnegative real numbers a and b shows that ii) implies ii'). On the other hand, from the relation $|a - b| = |a^{1/2} - b^{1/2}||a^{1/2} + b^{1/2}|$, it follows via the Schwarz inequality that $\|f - g\|_1 \leq \|f^{1/2} - g^{1/2}\|_2 \|f^{1/2} + g^{1/2}\|_2$ for arbitrary probability densities f and g with respect to m. But in this case $\|f^{1/2} + g^{1/2}\|_2 \leq 2$ and the stated equivalence follows. □

DEFINITION. An indexed collection $\{\mathbf{M}_\sigma; \sigma \in \Sigma\}$ of sub-σ-rings of the measure ring of a probability measure space is *stochastically independent* in case for every finite subset $\sigma_1, \ldots, \sigma_n$ of Σ and bounded measurables $f_1, \ldots, f_n$ with respect to $\mathbf{M}_{\sigma_1}, \ldots, \mathbf{M}_{\sigma_n}$ (respectively),

$$E(f_1 \cdots f_n) = E(f_1) \cdots E(f_n).$$

Terminology regarding infinite products will follow that of Titchmarsh (1952), chapter 1, as extended to the case in which the index set is not necessarily countable, in a parallel fashion to the case of infinite sums treated in SK. In particular, a product can converge only if its limit is nonvanishing.

COROLLARY 1.6.2. *Let* $\mathbf{P} = (M, \mathbf{M}, m)$ *be a probability measure space, and let* $\{\mathbf{N}_\sigma; \sigma \in \Sigma\}$ *be an indexed collection of sub-sigma-rings of* $\mathbf{M}$, *which generate* $\mathbf{M}$, *and are stochastically independent both with respect to m and another given probability measure n on* $\mathbf{M}$. *Then* $n \ll m$ *if and only if* $n|\mathbf{N}_\sigma \ll m|\mathbf{N}_\sigma$, *and the following product is convergent:*

$$\prod_{\sigma\in\Sigma} m(D_\sigma^{1/2}),$$

where $D_\sigma = D(n|N_\sigma, m|N_\sigma)$.

PROOF. Let Λ denote the directed system consisting of the finite subsets of Σ, ordered by inclusion, and $\mathbf{M}_\lambda$ the σ-ring generated by the elements of the N_σ with $\sigma \in \lambda$; the hypotheses of Theorem 1.6 are then satisfied. Now, if λ is any finite subset of Σ, $D(n|\mathbf{M}_\lambda, m|\mathbf{M}_\lambda) = \Pi_{\sigma\in\lambda} D(n|\mathbf{N}_\sigma, m|\mathbf{N}_\sigma)$ by virtue of the stochastic independence assumption. Thus it is necessary and sufficient for the absolute continuity of n with respect to m, setting $D(n|\mathbf{N}_\sigma, m|\mathbf{N}_\sigma) = D_\sigma$, that

$$\| \prod_{\sigma\in\lambda} D_\sigma^{1/2} - \prod_{\sigma\in\lambda'} D_\sigma^{1/2} \|_2 \to 0 \text{ in } L_2(m) \text{ as } \lambda,\lambda' \uparrow.$$

The squared inner product in question here is easily evaluated as

$$2 - 2\{ \prod_{\sigma\in\lambda\Delta\lambda'} m(D_\sigma^{1/2})\}$$

(where Δ denotes the symmetric difference). Thus it is necessary and sufficient that

$$\prod_{\sigma\in\lambda\Delta\lambda'} m(D_\sigma^{1/2}) \to 1.$$

Noting that $m(D_\sigma^{1/2}) \le 1$ by Schwarz's inequality, this is the same as requiring that $\Pi_\sigma m(D_\sigma^{1/2})$ be convergent. □

COROLLARY 1.6.3. *With the notation of Corollary 1.6.2, suppose that* $n|N_\sigma \approx m|N_\sigma$ *for all* σ. *Then* $n \ll m$ *if and only if* $m \ll n$. *(Thus absolute continuity implies equivalence.)*

PROOF. The infinite product is unchanged when n and m are interchanged, since $m(D_\sigma^{1/2}) = n(D_\sigma^{-1/2})$. □

Note that the condition that $\Pi_\sigma\, m(D_\sigma^{1/2})$ be convergent is easily seen to be equivalent to the condition that $\Sigma_\sigma\, (1 - m(D_\sigma^{1/2}))$ be convergent (cf. Titchmarsh).

SCHOLIUM 1.6. *The centered normal distribution in a real Hilbert space* **H** *with given covariance operator B that is bounded and has bounded inverse is quasi-invariant.*

PROOF. Note first that it suffices to consider the case of the isonormal process with covariance parameter 1. For, suppose the conclusion has been attained in this case. If d is the given centered process of covariance operator B, set

$$e(x) = d(B^{-1/2}x);$$

then e is isonormal, so for any vector $a \in \mathbf{H}$, there exists a unitary operator U on $L_2(d)$ such that

$$U M_{e(x)} U^{-1} = M_{e(x)} + \langle x, a\rangle I,$$

$$U M_{d(x)} U^{-1} = M_{d(x)} + \langle x, B^{1/2}a\rangle I,$$

showing that d is quasi-invariant.

Now suppose that g is isonormal with variance parameter $c = 1$. Let $\{e_\sigma\}$ be an orthonormal basis in **H**; then the $g(e_\sigma)$ are mutually stochastically independent. If a is a fixed vector, it has the form $a = \Sigma_\sigma a_\sigma e_\sigma$, and by Corollary 1.6.2 there is quasi-invariance as stated if and only if the infinite product

$$\prod_\sigma \int [\{(2\pi)^{-1/2} \exp(-\tfrac{1}{2}(x + a_\sigma)^2)\} \{(2\pi)^{-1/2} \exp(-\tfrac{1}{2}x^2)\}]^{1/2}\, dx$$

is convergent. The σ-th factor in this product is $\exp\left(-\frac{1}{8}a_\sigma^2\right)$, so the product is convergent. □

EXAMPLE 1.9. If e and d are weakly equivalent quasi-invariant distributions on a real topological vector space **L**, it is not difficult to show that the associated Weyl systems (given by Theorem 1.5) are unitarily equivalent, via the operation of multiplication by the square root of the derivative of the one distribution with respect to the other. (The details are left as an exercise.) Conversely, if these Weyl systems are unitarily equivalent, then e and d must be weakly equivalent by Scholium 1.5.

The preceding scholium shows the existence of Weyl systems over any real Hilbert space algebraically retopologized (but this change in the topology will soon be eliminated). It also shows the nonunicity of these Weyl systems, within unitary equivalence, in contrast with the unicity in the finite-dimensional case established by the Stone–von Neumann Theorem. This is implied by

EXAMPLE 1.10. Let g denote the isonormal distribution of unit variance on the real infinite-dimensional Hilbert space **H**, and let α and α' denote any positive constants. Then the distributions αg and $\alpha' g$ are weakly equivalent only if $\alpha = \alpha'$. For, by Corollary 1.6.2, there is absolute continuity only if the infinite product

$$\prod_{\sigma \in \Sigma} \int [\{(2\pi\alpha^2)^{-1/2} \exp(-\tfrac{1}{2}x^2/\alpha'^2)\} \{(2\pi\alpha^2)^{-1/2} \exp(-\tfrac{1}{2}x^2/\alpha^2)\}]^{1/2}\, dx$$

is convergent. The factors in this product are independent of σ, so there is convergence only if all the factors are 1, which a simple computation shows is not the case unless $\alpha = \alpha'$. (A byproduct of this observation is the result of Cameron and Martin [1945] to the effect that the transformation $f \mapsto \alpha f$ ($\alpha > 0$) on Wiener space is absolutely continuous only if $\alpha = 1$.)

A useful extension of the earlier notion of characteristic function of a weak process is given by the

DEFINITION. Let E be a state of a C^*-algebra containing the operators $W(z)$ of a Weyl system W over **L**. The *generating function* of E relative to W is defined as the function μ such that:

$$\mu(z) = E[W(z)], \qquad z \in \mathbf{L}.$$

EXAMPLE 1.11. If v is any unit vector in the representation space of W, the generating function of the state it determines is $\mu(z) = \langle W(z)v, v\rangle$. If W is associated with a quasi-invariant process as in Theorem 1.5 and $v = 1$, the designation of the state will ordinarily be omitted.

In the case in which **L** is built from a real Hilbert space **H**, the notation $x \oplus y$ will be used for an element of the direct sum $\mathbf{H} \oplus \mathbf{H}$.

THEOREM 1.7. *The generating function of the Weyl system associated via Theorem 1.5 with the centered normal process of covariance operator B on the real Hilbert space* **H** *has the form*

$$\mu(x \oplus y) = \exp(-\langle B^{-1}x, x\rangle/8 - \langle By, y\rangle/2) \qquad (x, y \in \mathbf{H}).$$

PROOF. By the method of proof of Scholium 1.6, the proof reduces to that for the case $B = I$. Now $W(x \oplus y) = U(x)V(-y)e^{i\langle x,y\rangle/2}$, so

$$\mu(x \oplus y) = e^{i\langle x,y\rangle/2}\langle V(-y)1, U(-x)1\rangle.$$

Here $V(-y)1$ is $\theta(e^{-i\langle u,y\rangle})$, u being a dummy variable; $U(-x)$ is multiplication by the unitarizer for the displacement $u \mapsto u - x$. The inner product in question here is thus the expectation of the tame function $\theta(e^{-i\langle u,y\rangle}$

$e^{\frac{1}{2}\langle u,x\rangle - \frac{1}{4}\langle x,x\rangle})$, which is based on the at most 2-dimensional subspace spanned by x and y. This is an elementary Gaussian integral which is readily evaluated from the formula

$$(2\pi)^{-\frac{1}{2}} \int \exp(vx - \tfrac{1}{2}v^2)dx = \exp(\tfrac{1}{2}v^2) \tag{1.5}$$

as $\exp(\frac{1}{2}(-iy_1 + \frac{1}{2}x_1)^2 + \frac{1}{2}(-iy_2 + \frac{1}{2}x_2)^2 - \frac{1}{4}(x_1^2 + x_2^2))$ in terms of the components of x and y in this space. The given expression for $\mu(x \oplus y)$ follows. □

The case $B = \frac{1}{2}I$, corresponding to unit variance for the complex random variable $x + iy$, is the most symmetric one.

DEFINITION. If **H** is a given real Hilbert space, the *normal Weyl pair* over **H** is the Weyl pair $(\mathbf{K}, U, V)$ given by Theorem 1.5 for the isonormal distribution in **H** with variance parameter $c = \frac{1}{2}$. The *normal Weyl system* over **H** is the corresponding Weyl system $(\mathbf{K}, W)$.

COROLLARY 1.7.1. *The generating function of the normal Weyl system (relative to the state vector 1) is*

$$\mu(x \oplus y) = \exp(-\tfrac{1}{4}(\|x\|^2 + \|y\|^2)).$$

PROOF. This follows from Theorem 1.7. □

DEFINITION. If **H** is a real Hilbert space, its *complexification* is the complex space $\mathbf{H}^C$ of all pairs (x, y) with $x, y \in \mathbf{H}$, with the complex structure and inner product

$$i(x, y) = (-y, x)$$
$$\langle (x, y), (x', y') \rangle = \langle x, x' \rangle + \langle y, y' \rangle + i(\langle y, x' \rangle - \langle x, y' \rangle).$$

The *normal Weyl system* over $\mathbf{H}^C$ is defined as that over **H**. It will be seen that for a given complex Hilbert space **H**, the various choices of real subspaces $\mathbf{H}'$ such that $\mathbf{H} = \mathbf{H}'^C$ give rise to unitarily equivalent Weyl systems.

COROLLARY 1.7.2. *The normal Weyl system over the complexification* $\mathbf{H}^C$ *of a real Hilbert space has the generating function*

$$E(W(z)) = \exp(-\|z\|^2/4);$$

and the system is continuous not only in the algebraic topology on $\mathbf{H}^C$, *but in its Hilbert space topology.*

PROOF. To show that W is continuous in the Hilbert topology, it suffices to show that $\langle W(z)u, u'\rangle$ is a continuous function of $z \in \mathbf{H}^c$ for u and u' in a dense subset of the representation space $\mathbf{K}$. Let $\mathbf{D}$ denote the algebraic span of the $W(z)1$; since v is cyclic for W, $\mathbf{D}$ is dense in $\mathbf{K}$. It suffices therefore to show that

$$\langle W(z)W(z_1)1, W(z_2)1\rangle$$

is a continuous function of z, z_1 and z_2 being fixed. Employing the Weyl relations and the form of the generating function, this follows. □

COROLLARY 1.7.3. *A sufficient condition that there exist a Weyl system over a given symplectic vector space* $(\mathbf{L}, A)$ *is that there exists a real positive definite symmetric bilinear form* S *on* $\mathbf{L}$ *such that*

i) convergence in $\mathbf{L}$ *implies convergence relative to the S-norm,* $\|x\| = S(x, x)^{1/2}$; *and*

ii) $|A(x, y)| \le c\,\|x\|\,\|y\|$.

PROOF. Let $\bar{\mathbf{L}}$ denote the completion of $\mathbf{L}$ with respect to S, and let $\mathbf{K} = L_2(g)$ where g is the isonormal distribution on $\bar{\mathbf{L}}$. For arbitrary $z \in \mathbf{L}$ and bounded tame function f on $\bar{\mathbf{L}}$, let $W_0(z)\theta(f)$ be defined as

$$\theta(e^{iA(\cdot,z)/2})\, U(z)\theta(f),$$

where U is as given in Theorem 1.5. Then $W_0(z)$ is isometric on a dense linear subspace of $L_2(g)$, and so extends uniquely to an isometry $W(z)$ on all of $L_2(g)$.

It is readily verified that for arbitrary $z, z' \in \mathbf{L}$,

$$W_0(z)\, W_0(z') = e^{iA(z,z')/2}\, W_0(z + z'),$$

from which it follows that W satisfies the Weyl relations. In particular the $W(z)$ are unitary. To show that the map $z \mapsto W(z)$ is continuous from $\mathbf{L}$ to the unitaries on $\mathbf{K}$, it suffices to show continuity from $\bar{\mathbf{L}}$ in the $\|\cdot\|$ topology; this follows from Corollary 1.6.1 and the continuity of A on $\bar{\mathbf{L}}$. □

Problems

1. Prove the existence of the conditional expectation $f' = [f|\mathbf{S}]$ of an integrable f of a probability measure space $P = (R, \mathbf{R}, r)$ with respect to a sub-σ-ring $\mathbf{S}$ of $\mathbf{R}$. Show that

a) $\|f'\|_p \le \|f\|_p$ for $p = 1, 2$, or ∞;

b) $(f')' = f'$;

c) if g is a bounded measurable with respect to $\mathbf{S}$, then $(fg)' = f'g$; and

d) if f is square-integrable, then f' is the projection of f onto the subspace of $L_2(P)$ consisting of all the elements of $L_2(P)$ that are measurable with respect to $\mathbf{S}$.

2. Let $(R_j, \mathbf{R}_j)$ $(j = 1,2,\ldots)$ be a sequence of sets R_j and σ-rings of subsets $\mathbf{R}_j$ of R_j; let m_j and n_j be countably-additive probability measures on $\mathbf{R}_j$ that are mutually absolutely continuous for all j. Show that the infinite product measures $\Pi_j m_j$ and $\Pi_j n_j$ are mutually absolutely continuous if and only if $\Pi \int (dm_j\, dn_j)^{1/2}$ is convergent, where for absolutely continuous measures m and n, $\int (dm\, dn)^{1/2}$ means $\int (dm/dn)^{1/2}\, dn$. (This result is due to Kakutani, 1948.)

3. Show that there are no quasi-invariant strict distributions in an infinite-dimensional Banach space. (Cf. Feldman, 1966.)

1.6. Irreducibility and ergodicity

The construction used in the proof of Corollary 1.7.3 does not result in an irreducible Weyl system when $\mathbf{L}$ is finite-dimensional. The Schrödinger representation is however irreducible. This section treats related aspects of irreducibility and ergodicity.

DEFINITION. Let $(M, \mathbf{M}, m)$ be a given measure space; a *nonsingular transformation* T on this space is a mapping of M into M that carries null sets into null sets, and whose induced action on the measure ring (of all measurable sets modulo the null sets) is an automorphism. A set of such transformations is called *ergodic* if their induced actions leave invariant no elements of the measure ring other than the images of $\varnothing$ and M.

SCHOLIUM 1.7. *Let G be a given group of nonsingular transformations on the finite measure space $\mathbf{P} = (M, \mathbf{M}, m)$. Let U be the representation of G on $\mathbf{H} = L_2(\mathbf{P})$ given by the equation*

$$U(a): f(x) \to f(a^{-1}(x))\, [dm_a/dm]^{1/2},$$

where $m_a(E) = m(a^{-1}E)$ for $E \in \mathbf{M}$. Then G acts ergodically on $\mathbf{P}$ if and only if the totality of the $U(a)$, together with the totality of all multiplication operators $M_k: f \to kf$ $(k \in L_\infty(\mathbf{P}))$, is irreducible on $\mathbf{H}$.

PROOF. It suffices to show the more general result that in any event, any bounded linear operator on $\mathbf{H}$ that commutes with all $U(a)$ and M_k is of the form M_h, where h is G-invariant; i.e., $h(a^{-1}x) = h(x)$ a.e., for each a in G. If h is invariant, then for any Borel subset B of $\mathbf{C}$, $h^{-1}(B)$ is also invariant, so if G acts ergodically it follows that h must be constant a.e.

Now a bounded linear operator A that commutes with all M_k, $k \in L_\infty$, is itself of the form M_h for some h (cf. SK). Noting that

$$U(a)\, M_k\, U(a)^{-1} = M_{k_a};\ k_a(x) = k(a^{-1}x),$$

for arbitrary k, it follows that M_h commutes with all M_k and $U(a)$ if and only if it is G-invariant. □

SCHOLIUM 1.8. *Let* $\mathbf{P} = (M, \mathbf{M}, m)$ *be a given finite measure space; let* $\mathbf{M}_\lambda$ *be a generating net of sub-σ-rings of* $\mathbf{M}$, *such that* $\mathbf{M}_\lambda \subset \mathbf{M}_{\lambda'}$ *if* $\lambda \le \lambda'$; *and let G be a given group of nonsingular transformations on* $\mathbf{P}$. *In order that G be ergodic on* $\mathbf{P}$, *it is sufficient that for every* λ, *the subgroup* G_λ *of all transformations T in G which leave* $\mathbf{M}_\lambda$ *invariant, modulo null sets, and are such that* dm_T/dm *is measurable with respect to* $\mathbf{M}_\lambda$, *acts ergodically on* $(M, \mathbf{M}_\lambda, m)$ *(or its measure ring).*

PROOF. Suppose that h is a G-invariant element of $L_\infty(\mathbf{P})$, and let h_λ denote its conditional expectation with respect to $(M, \mathbf{M}_\lambda, m)$. This means that

$$\textstyle\int_{\mathbf{P}} hf = \int_{\mathbf{P}} h_\lambda f$$

for all f in $L_1(M, \mathbf{M}_\lambda, m)$. Applying $T \in G_\lambda$,

$$\textstyle\int_{\mathbf{P}} h_T f_T\, dm_T/dm = \int_{\mathbf{P}} (h_\lambda \circ T^{-1}) f_T\, dm_T/dm,$$

so that

$$\textstyle\int_{\mathbf{P}} h f_T\, dm_T/dm = \int_{\mathbf{P}} (h_\lambda \circ T^{-1}) f_T\, dm_T/dm.$$

Since dm_T/dm is measurable with respect to $\mathbf{M}_\lambda$, and since T leaves $\mathbf{M}_\lambda$ invariant, the product $f_T\, dm_T/dm$ is measurable with respect to $\mathbf{M}_\lambda$, and ranges over $L_1(M, \mathbf{M}_\lambda, m)$ as f ranges over this same space. It follows that $h_\lambda \circ T^{-1}$ satisfies the defining equation for h_λ, that is, h_λ is an invariant element of $L_\infty(M, \mathbf{M}_\lambda, m)$ under T, and hence under all of G_λ. By the ergodicity of G_λ, h_λ must be a constant; but $h_\lambda \to h$ in $L_1(P)$ as $\lambda\uparrow$, so that h is a limit of constants and hence itself constant. □

DEFINITION. A quasi-invariant distribution d is *ergodically quasi-invariant* (or simply *ergodic* when the relevant transformation group is clearly that induced from the vector translations in $\mathbf{L}$) if there exists no nonconstant measurable K in $L_\infty(d)$ such that

$$U(x) M_K U(x)^{-1} = M_K$$

for all $x \in \mathbf{L}$, $U(x)$ being as in Theorem 1.5.

This is the same as requiring that a model exists for d in which the $d(x)$ are measurables on a probability measure space $\mathbf{P}$, and in which the automorphisms of the measure ring of $\mathbf{P}$ induced from translations of $\mathbf{L}$ act ergodically on the measure ring.

SCHOLIUM 1.9. *A normal distribution over a real Hilbert space with bounded invertible covariance operator is ergodic.*

PROOF. By the argument given at the beginning of the proof of Scholium 1.6, it suffices to treat the case of the unit-variance isonormal process g on the given real Hilbert space **H**. If **N** is an arbitrary finite-dimensional subspace of **H**, any vector displacement by an element a of **N** leaves **N** invariant. The derivative $D(g_a, g)$ is $\exp(-\ \tfrac{1}{2}\langle x, a\rangle - \tfrac{1}{4}\langle a, a\rangle)$ and so is a tame function based on **N**. Consequently Scholium 1.8 implies that ergodic quasi-invariance in question follows from the same result for the finite-dimensional restrictions $g|\mathbf{N}$ of g. This finite-dimensional ergodicity is equivalent to the characterization of Lebesgue measure as the unique translation-invariant regular measure on $\mathbf{R}^n$. □

Thus irreducible Weyl systems exist over infinite-dimensional spaces, but there is no unicity within unitary equivalence, unlike the finite-dimensional situation. There is, however, a considerable degree of unicity in the theory, which comes from essentially three different sources: (i) group invariance; (ii) a positive spectrum condition; and (iii) C^*-algebraic phenomenology. Aspects of the first of these sources will be considered here, and the others later.

DEFINITION. A *symplectic group representation* is a system $(\mathbf{L}, A, G, S)$ where $(\mathbf{L}, A)$ is a symplectic vector space as earlier defined; G is a topological group; and S is a continuous representation of G by invertible linear transformations on **L** leaving the form A invariant. It is a *semirepresentation* if, for each element of G, A is either invariant or transformed into its negative.

EXAMPLE 1.12. Let **H** be a complex Hilbert space, let $A(x, y) = \operatorname{Im}\langle x, y\rangle$; let G denote the group of all unitaries on **H**; and for $U \in G$, let $S(U) = U$. Then $(\mathbf{H}^{\#}, A, G, S)$ is a symplectic group representation. If V is an antiunitary operator on **H**, it transforms A into its negative; replacing G by the larger group of all unitary or antiunitary operators on **H**, a semirepresentation is obtained.

DEFINITION. The *symplectic group* on a symplectic vector space $(\mathbf{L}, A)$, denoted $Sp(\mathbf{L},A)$, is the group of all real-linear invertible continuous transformations on $(\mathbf{L}, A)$ that leave A invariant; the *extended symplectic group* includes, in addition, those transforming A into its negative.

Given a symplectic group representation or semirepresentation $(\mathbf{L}, A, G, S)$, a *(G, S)-covariant Weyl system* over $(\mathbf{L}, A)$ is a triple $(\mathbf{K}, W, \Gamma)$, where $(\mathbf{K}, W)$ is a Weyl system over $(\mathbf{L}, A)$ as earlier, and Γ is a continuous representation of G by unitary or antiunitary operators on **K** having the property that

$$\Gamma(g)W(z)\Gamma(g)^{-1} = W(S(g)z)$$

for all $g \in G$ and $z \in \mathbf{L}$. A vector $v \in \mathbf{K}$ such that $\|v\| = 1$ and $\Gamma(g)v = v$ for all g is called an *invariant* or *equilibrium state vector*; and $(\mathbf{K}, W, \Gamma, v)$ is called a *G-covariant boson field* over $(\mathbf{L}, A, G, S)$.

THEOREM 1.8. *Let* $\mathbf{H}$ *be a given real Hilbert space,* $\mathbf{H}^C$ *its complexification,* G *the group of all unitary and antiunitary operators on* $\mathbf{H}^C$*, and* $S(U) = U$ *for* $U \in G$*. There is a unique* (G, S)*-covariant Weyl system* $(\mathbf{K}, W, \Gamma)$ *over* $\mathbf{H}^C$*, such that* $(\mathbf{K}, W)$ *is the normal Weyl system over* $\mathbf{H}^C$*, and for all* $U \in G$*,* $\Gamma(U)1 = 1$.

PROOF. Note first that if any such covariant system exists at all, it is necessarily unique by the irreducibility of the normal Weyl system. For if Γ_1 and Γ_2 both satisfy the conditions on Γ, then for $z \in \mathbf{H}^C$, $\Gamma_2(U)W(z)\Gamma_2(U)^{-1} = \Gamma_1(U)W(z)\Gamma_1(U)^{-1}$. It follows that $\Gamma_2(U)\ \Gamma_1(U)^{-1}$ commutes with all the $W(z)$ and so must be a scalar λI; but since $\lambda v = v$, $\lambda = 1$.

Now let U be an arbitrary unitary in G and define $\Gamma_0(U)$ as the transformation in the space $\mathbf{K}_0$, consisting of the algebraic span of the $W(z)1$, given by the equation

$$\Gamma_0(U)\colon \Sigma\, \alpha_i W(z_i)1 \mapsto \Sigma\, \alpha_i\, W(Uz_i)1;$$

that this transformation is well defined, and in fact isometric, follows from the computation

$$\begin{aligned}\langle \Sigma\, \alpha_i W(z_i)1, \Sigma\, \beta_i W(z_i)1\rangle &= \sum_{i,j} \alpha_i\bar{\beta}_j\langle W(-z_j)W(z_i)1, 1\rangle \\ &= \sum_{i,j} \alpha_i\bar{\beta}_j \exp(i\, \mathrm{Im}\langle z_i, z_j\rangle/2)\exp(-\|z_i - z_j\|^2/4);\end{aligned}$$

the replacement of the z_i by the Uz_i does not affect the value of the last expression. Therefore $\Gamma_0(U)$ has a unique isometric extension from $\mathbf{K}_0$ to all of $\mathbf{K}$ denoted as $\Gamma(U)$. From the easily verified result that $\Gamma_0(U_1)\Gamma_0(U_2) = \Gamma_0(U_1U_2)$, it follows that the same is true of Γ, so that Γ is a representation. To show that it is continuous, it suffices to show that $\langle\Gamma(U)w, w'\rangle$ is a continuous function of U, when w and w' range over a dense subset of $\mathbf{K}$; e.g., the algebraic span of the $W(z)1$, and a slight modification of the computation just made, shows this.

If U is antiunitary on G, let $\Gamma_0(U)$ be defined as the representation

$$\Sigma\, \alpha_i W(z_i)1 \to \Sigma\, \bar{\alpha}_i W(Uz_i)1;$$

then $\Gamma_0(U)$ is an antilinear isometric transformation, and an argument similar to the foregoing shows that there is a unique antiunitary transformation $\Gamma(U)$ on $\mathbf{K}$ extending $\Gamma_0(U)$. Now setting Γ for the extension of the mappings Γ defined on the unitaries and antiunitaries separately to the full group, in which the unitaries form a subgroup of index 2, the conclusion of the theorem follows. □

DEFINITION. Given a complex Hilbert space $\mathbf{H}$, a *conjugation* on $\mathbf{H}$ is a conjugate-linear norm-preserving map $\varkappa$ from $\mathbf{H}$ to itself such that $\varkappa^2 = I$. The *real* subspace $\mathbf{H}_\varkappa$ is defined as the space of all vectors $z \in \mathbf{H}$ such that $\varkappa z = z$; such a subspace is called a *real part* of $\mathbf{H}$.

For any conjugation $\varkappa$ there is a natural isomorphism between $\mathbf{H}$ and the complexification of $\mathbf{H}_\varkappa$, mapping the pair (x, y) in the complexification of $\mathbf{H}_\varkappa$ to the vector $x + iy \in \mathbf{H}$. Conversely, any orthonormal basis $\{e_j\}$ in $\mathbf{H}$ determines a conjugation $\varkappa$ such that $\varkappa e_j = e_j$ and $\varkappa(ie_j) = -ie_j$. $\mathbf{H}_\varkappa$ is then the real closed subspace spanned by the e_j.

COROLLARY 1.8.1. *Let* $\mathbf{H}$ *be a given complex Hilbert space, and let* $\mathbf{H}_1$ *and* $\mathbf{H}_2$ *be any two real parts of* $\mathbf{H}$. *Then the normal Weyl system over* $\mathbf{H}$ *as the complexification of* $\mathbf{H}_1$ *is the same, within unitary equivalence, as that over* $\mathbf{H}$ *as the complexification of* $\mathbf{H}_2$.

PROOF. If $(W_i, \mathbf{K}_i)$ $(i = 1, 2)$ are the two Weyl systems in question, the mapping

$$W_1(z)v_1 \mapsto W_2(z)v_2,$$

where v_1 and v_2 denote the measurable 1 in $\mathbf{K}_1$ and $\mathbf{K}_2$ respectively, extends uniquely to the required unitary equivalence, by the argument in the proof of Theorem 1.8. □

DEFINITION. *The normal Weyl system* over a given complex Hilbert space $\mathbf{H}$ is the system $(\mathbf{K}, W)$, unique within unitary equivalence, that is the normal one over any real part of $\mathbf{H}$. The *free boson field over* $\mathbf{H}$ is the system $(\mathbf{K}, W, \Gamma, v)$, likewise unique within unitary equivalence, where $(\mathbf{K}, W)$ is the normal system just defined, say over the real subspace $\mathbf{H}_\varkappa$; v is the measurable 1 on $\mathbf{H}_\varkappa$; and Γ is the representation given by Theorem 1.8. The representation Γ will be called the *free boson representation* of the extended unitary group on $\mathbf{H}$, and the vector v will be called the *vacuum vector*.

1.7. The Fourier-Wiener transform

Having obtained the representation Γ of the extended unitary group, it is natural to consider the questions of the more explicit appearance of this representation and of its decomposition into irreducible constituents. To begin with, the simplest nontrivial case will be considered; that in which the real Hilbert space $\mathbf{H}$ is one-dimensional. $\mathbf{H}^C$ may then be identified with $\mathbf{C}$, relative to the inner product $\langle\alpha, \beta\rangle = \alpha\bar{\beta}$. In this case, the decomposition of $\mathbf{K}$ under the action of $\Gamma(U(1))$, where $U(1)$ is the unitary group in 1 dimension, is equivalent to the expansion into Hermite functions in $L_2(\mathbf{R})$. However, to develop the infinite-dimensional theory, it is helpful to proceed along more invariant lines.

Let g denote the probability measure on $\mathbf{R}$:

$$dg = \pi^{-1/2} \exp(-x^2)\, dx.$$

The isonormal distribution of variance 1/2 on the real subspace $\mathbf{R}$ of $\mathbf{C}$ is strict and corresponds to the measure g. Consequently, for the normal Weyl system over $\mathbf{R}$, $\mathbf{K} = L_2(\mathbf{R}, g)$, while W is the system derived from the Weyl pair

$$U(x): F(u) \mapsto F(u + x)\, e^{-\langle u,x\rangle - \frac{1}{2}\langle x,x\rangle},$$

$$V(y): F(u) \mapsto F(u)\, e^{i\langle u,y\rangle}.$$

The vacuum v is represented by the function identically 1 on $\mathbf{R}$. The general unitary operator U on $\mathbf{H}$ corresponds to the operator $z \to e^{i\theta} z$ on $\mathbf{C}$, θ being a fixed real number. The transformation $\Gamma(e^{i\theta})$ can be given by a singular kernel, i.e., in the form

$$f(x) \mapsto \lim_{\varepsilon\to 0} \int K_\varepsilon(x, y) f(y)\, dy,$$

where $K_\varepsilon(x, y)$ is explicitly computable; in a different form it was given, on a formal basis, by Mehler about 100 years ago. More exactly, by Stone's theorem, $\Gamma(e^{i\theta})$ has the form $\Gamma(e^{i\theta}) = e^{i\theta N}$ for some selfadjoint operator N (the "harmonic oscillator Hamiltonian"); and $e^{-\theta N}$ is for $\theta > 0$ the integral operator with kernel

$$(1 - a^2)^{1/2} \exp[-(1 - a^2)^{-1} (a^2(x^2 + y^2) - 2axy)],$$

where $a = e^{-\theta}$. On analytic continuation in θ, this gives the singular kernel indicated. In the case $\theta = \pi/4$, however, the transformation is very familiar in another form: $\Gamma(i)$ is similar to the Fourier transform.

Specifically, the unitary transformation of multiplication by $\pi^{-1/4} \exp(-\frac{1}{2}x^2)$, from $L_2(\mathbf{R}, g)$ onto $L_2(\mathbf{R})$, transforms $\Gamma(i)$ into the Fourier transform and transforms the pair (U, V) into the pair (U_0, V_0) given by the equations

$$U_0(x): f(u) \mapsto f(u + x),$$

$$V_0(y): f(u) \mapsto e^{iuy} f(u).$$

The verification of this is left as an exercise.

The conventional Fourier transform on $L_2(\mathbf{R}^n)$,

$$f(x) \mapsto (2\pi)^{-n/2} \int e^{i\langle x,y\rangle} f(y)dy,$$

cannot be rigorously extended to the case of a Hilbert space ($n \sim \infty$), irrespective of how the constants involved may be adjusted. This may be ascribed, intuitively speaking, to the absence of an invariant measure in Hilbert space relative to translations. One might conclude from this that there is no natural extension of the Plancherel theory to the additive group of Hilbert space, as is perhaps to be expected in view of its not being locally compact, but this would be too hasty. There is a quite natural theory, invariantly associated with any given Hilbert space, with the properties: (a) it is essentially equivalent to the Plancherel theory when the Hilbert space is finite-dimensional; and (b) it is in a certain sense even more invariant than the Plancherel theory. This transforms simply under the euclidean group and fairly simply under the general linear group. But the theory attached to a real Hilbert space is invariant not only under the orthogonal group, but transforms simply under the unitary group on the complexification. At the same time it extends and provides an invariant reinterpretation of the transform in Wiener space previously developed by Cameron and Martin.

In order to motivate the "Fourier-Wiener" (or "Wiener") transform, as the rigorous counterpart to the Fourier-Plancherel transform in infinitely many dimensions will be called, consider how the Fourier transform appears when euclidean measure is replaced by the Gaussian measure $dg = (2\pi c)^{-n/2} e^{-\langle x,x\rangle/2c}\,dx$ in $\mathbf{R}^n$. If $\mathbf{T}$ denotes the Fourier transform on $L_2(\mathbf{R}^n)$, scaled as follows: $f \mapsto (4\pi)^{-n/2} \int e^{i\langle x,y\rangle/2} f(y)\,dy$; and if Z denotes the unitary transformation

$$f(x) \to f(x)\,[(2\pi c)^{-n/2} e^{-\langle x,x\rangle/2c}]^{1/2}$$

from $L_2(\mathbf{R}^n, g)$ onto $L_2(\mathbf{R}^n)$, then $\mathbf{F} = Z^{-1}\mathbf{T}Z$ is the operator on $L_2(\mathbf{R}^n; g)$ which corresponds to the Fourier transform on $L_2(\mathbf{R}^n)$. It is not difficult to compute the action of the operator $\mathbf{F}$ on sufficiently simple functions, e.g., polynomials. The result is that for any polynomial p on $\mathbf{R}^n$, $\mathbf{F}p$ is the function

$$P(y) = \int p(2^{1/2}x + iy)\,dg(x)$$

(the proof is left as an exercise). Observing that the form of this transformation is independent of the dimension n (and also of the "variance" parameter c), it is natural to consider the possible extension of the transform to the case of an infinite-dimensional Hilbert space.

To facilitate the statement of a formal result to this effect, note that if f is a complex polynomial on a real Banach space $\mathbf{B}$, that is, a conventional polynomial in a finite number of linear functionals on the space, then it has a unique extension to the complexification of the space that is holomorphic as a function on a complex space; thus $f(x + iy)$ makes good sense for $x, y \in \mathbf{B}$ when f is a polynomial, although for a general tame function it is not defined.

THEOREM 1.9. *Let* $\mathbf{H}$ *be a given real Hilbert space, and let* $\mathbf{P}$ *denote the algebra of all (complex-valued) polynomials over* **H**. *Let* $\mathbf{F}_0$ *denote the mapping from* **P** *to* **P** *given by the equation*

$$\mathbf{F}_0 p = P; \qquad P(y) = \int_{\mathbf{H}} p(2^{1/2}x + iy)\, dg(x),$$

where g is the isonormal process over **H** *with variance parameter* $c > 0$. *Then i)* $\mathbf{F}_0$ *extends uniquely to a unitary transformation* **F** *on all of* $L_2(g)$, *the* Wiener Transform; *and ii)* $\mathbf{F}_0$ *is onto* **P** *and has the inverse*

$$\mathbf{F}_0^{-1} P = p; \qquad p(y) = \int_{\mathbf{H}} P(2^{1/2}x - iy)\, dg(x).$$

PROOF. It should be noted first of all that any polynomial on a Hilbert space actually lies in $L_2(g)$ as a tame measurable. The algebra $\theta(\mathbf{P})$ is thus contained in $L_2(g)$. Moreover, since **P** contains all linear functionals, $\theta(\mathbf{P})$ is measure-theoretically separating (i.e., the minimal σ-ring with respect to which all elements of $\theta(\mathbf{P})$ are measurable includes all measurable sets, modulo null sets). If the elements of $\theta(\mathbf{P})$ were bounded, this would imply the density of $\theta(\mathbf{P})$ in $L_2(g)$, by a result described earlier; however they are unbounded, and in general it is false (even in the simple case of a polynomial algebra on **R**) that a measure-theoretically separating subalgebra of L_2 of a finite measure space is dense in L_2. It is nevertheless true that $\theta(\mathbf{P})$ is dense in $L_2(g)$; this is an infinite-dimensional generalization of the density of the Hermite functions in $L_2(\mathbf{R})$. This follows from the following result (which is best possible of its type, as shown by examples in the theory of moments).

LEMMA 1.9.1. *Let* **A** *be an algebra of measurables on a finite measure space* **M**, *which is measure-theoretically separating, contains the identity function* 1, *and has a set of (algebraic) generators* **G** *all of which are such that* $e^{|f|} \in L_p(M)$ *for all* $p < \infty$. *Then* **A** *is dense in* $L_2(M)$.

PROOF. Suppose K is orthogonal to every element of **A**. If $f_1, \ldots, f_r$ are among the given set **G** of generators of **A**, which it is no essential loss of generality to assume to be real, then

$$\int f_1^{n_1} \cdots f_r^{n_r} \overline{K} = 0$$

for all nonnegative integers $n_1, \ldots, n_r$ (and f^0 is defined as 1). Multiplying by $(ia_1)^{n_1} \cdots (ia_r)^{n_r}/[(n_1)! \cdots (n_r)!]$, summing, and using dominated convergence—noting that $\exp[|a_1||f_1| + \cdots + |a_r||f_r|] \in L_1$ by Hölder's inequality and the assumption that each $\exp(|f_i|)$ is in all L_p for $p < \infty$,—it follows that

$$\int \exp[i(a_1 f_1 + \cdots + a_r f_r)]\, \overline{K} = 0.$$

Setting $F(b_1, \ldots, b_r)$ for the integral of $\overline{K}$ over the set $A = \{x \in M: f_1(x) <$

$b_1, \ldots, f_r(x) < b_r\}$, it follows from the general transformation properties of the integrals that

$$\int \exp[i(a_1 b_1 + \cdots + a_r b_r)]\, dF(b_1,\ldots,b_r) = 0.$$

By the unicity theorem for Fourier-Stieltjes transforms, F vanishes identically. Thus $\int_A \overline{K} = 0$ for all sets A of the indicated type. By the measure-theoretical separation hypothesis, these generate the full ring of all measurable sets, modulo null sets; it follows that $\int_A \overline{K} = 0$ for all measurable sets A, so that $K = 0$. □

PROOF OF THEOREM 1.9. There is no difficulty in verifying that $\mathbf{F}_0$ is a linear map from $\mathbf{P}$ to $\mathbf{P}$. To show that $\mathbf{F}_0$ is an isometry, it suffices to show that $E[p\bar{q}] = E[P\overline{Q}]$, where $P = \mathbf{F}_0 p$ and $Q = \mathbf{F}_0 q$, for p and q ranging over a set $\mathbf{D}$ that spans $\mathbf{P}$ (linearly); in particular, for the set of all functions p of the form $p(x) = \langle x, e_1\rangle^{n_1} \cdots \langle x, e_r\rangle^{n_r}$, $e_1,\ldots,e_r$ being an arbitrary finite orthogonal subset of $\mathbf{H}$, and the n_i being nonnegative integers. By linearity, it is no essential loss of generality to take the orthonormal sets relative to which p and q have monomial expressions of the indicated type to be the same, so that $p(x) = h_1(x)^{m_1} \cdots h_r(x)^{m_r}$, $q(x) = h_1(x)^{n_1} \cdots h_r(x)^{n_r}$, where $h_j(x) = \langle x, e_j\rangle$. Now denoting the expectation of a polynomial in two variables x and y ranging over $\mathbf{H}$ with respect to the variable x as E_x, it follows from the stochastic independence of the h_j that

$$P(y) = E_x[h_1(2^{1/2}x + iy)^{m_1}] \cdots E_x[h_r(2^{1/2}x + iy)^{m_r}],$$

and that $Q(y)$ is given by a similar expression. Using stochastic independence again, it results that

$$E[P\overline{Q}] = E_y[E_x(h_1(2^{1/2}x + iy)^{m_1})\, E_x(h_1(2^{1/2}x - iy)^{n_1})] \cdots$$
$$E_y[E_x(h_r(2^{1/2}x + iy)^{m_r})\, E_x(h_r(2^{1/2}x - iy)^{n_r})].$$

Noting that

$$E[p\bar{q}] = E_x[h_1(x)^{m_1+n_1}] \cdots E_x[h_r(x)^{m_r+n_r}],$$

it follows that it suffices to show that

$$E_y[E_x(h_j(2^{1/2}x + iy)^{m_j}) E_x(h_j(2^{1/2}x - iy)^{n_j})] = E_x[h_j(x)^{m_j+n_j}].$$

Since all the functions involved in the last equation are tame functions based on the one-dimensional subspace spanned by e_j, the validity of the equation reduces to the equality of two integrals over a one-dimensional space; more specifically to the question of the validity of the equations involving two independently (identically) distributed normal random variables x and y with zero mean and variance parameter c,

$$E_x(x^{m+n}) = E_y[E_x((2^{1/2}x + iy)^m)\,E_x((2^{1/2}x - iy)^n)]$$
$$(m, n = 0, 1, 2, \ldots) \tag{1.6}$$

where E denotes expectation. To establish equations 1.6, it suffices to show that the products of either side with $s^m t^n/m!n!$, summed over m and $n = 0, 1, \ldots$, are finite and equal for all real s and t. Using dominated convergence, there is obtained in this way on the left side the expression

$$(2\pi c)^{-1/2} \int \exp[(s + t)x - x^2/2c]\,dx;$$

and on the right side the expression

$$(2\pi c)^{-3/2} \iiint \exp[s(2^{1/2}x + iy) + t(2^{1/2}u - iy) - (x^2 + y^2 + u^2)/2c]dxdydu.$$

These Gaussian integrals are easily evaluated using the formula 1.5, and are equal, concluding the proof that $\mathbf{F}_0$ is isometric.

That $\mathbf{F}_0$ is onto* follows once the expression for $\mathbf{F}_0^{-1}$ is established. By an argument similar to that used in the preceding paragraph, it suffices to obtain the expression for the functionals $p(x) = x^n$ on a one-dimensional space. Thus it suffices to show that if $P(y) = E_x((2^{1/2}x + iy)^n)$, then $y^n = E_x(P(2^{1/2}x - iy))$. Now

$$E_x(P(2^{1/2}x - iy)) = E_x[E_u((2^{1/2}u + y + i2^{1/2}x)^n)],$$

so that the required equation can be expressed as

$$y^n = (2\pi c)^{-1} \iint (2^{1/2}u + y + i2^{1/2}x)^n \exp(-(u^2 + x^2)/2c)dxdu \; (n = 0, 1, \ldots).$$

To verify these, it suffices to show that after multiplication of either side by $s^n/n!$ and summation over $n = 0, 1, \ldots$, the same function of $s \in \mathbf{R}$ is obtained. This follows by dominated convergence and the evaluation of simple Gaussian integrals. Noting that Lemma 1.9.1 shows that $\mathbf{P}$ is dense in $L_2(g)$, the proof of Theorem 1.9 is complete. □

Problems

1. Stieltjes shows that there exist distinct probability distributions on $\mathbf{R}$ having the same moments. Use this to show that there exists a measure-theoretically separating subalgebra of L_2 of a finite measure space that is not dense in L_2.

2. Prove Mehler's formula by the generating function technique of the preceding section.

* A mapping is *onto* (also known as surjective or exhaustive) if every member of its a priori range is the image of some point in its domain.

3. Let f be an arbitrary real function in $\mathbf{H}' = L_2(\mathbf{R})$, let n be the isonormal distribution on $\mathbf{H}'$, and set $y(t) = n(f_t)$, where $f_t(x) = f(x + t)$. Show that $y(t)$ is a stationary Gaussian stochastic process, determine its autocovariance function $\varphi(t) = E(y(s)y(s + t))$, and show that φ is the Fourier transform of an L_1 function.

4. Show conversely that any stationary Gaussian stochastic process whose autocovariance function is the Fourier transform of an L_1 function is of the form given in problem 3.

1.8. The structure of Γ and wave-particle duality

In formulating a simple structure theorem giving the reduction into irreducible components of the normal boson field representation Γ of the unitary group, some aspects of tensor products of Hilbert spaces are needed. The basic concept will be assumed known. In the case of identical Hilbert spaces, there is, as is straightforward to verify, a canonical unitary representation R of the group S_n of permutations of $\{1, 2, \ldots, n\}$, on the n-fold tensor product $\mathbf{H}^n = \mathbf{H}\otimes\cdots\otimes\mathbf{H}$ of the complex Hilbert space $\mathbf{H}$ with itself, uniquely determined by the condition

$$R(a): x_1\otimes\cdots\otimes x_n \mapsto x_{a^{-1}(1)}\otimes\cdots\otimes x_{a^{-1}(n)}.$$

The elements of $\mathbf{H}^n$ are called *covariant n-tensors* over $\mathbf{H}$. The elements u of $\mathbf{H}^n$ such that $R(a)u = u$ (resp. $R(a)u = \mathrm{sgn}(a)u$) for all $a \in S_n$ are called *symmetric* (resp. *antisymmetric*). The set of all symmetric elements of $\mathbf{H}^n$ is a closed linear subspace $\mathbf{K}_n$; it is spanned, as a Hilbert space, by the symmetric tensors of the form $x\otimes\cdots\otimes x$, having identical factors.

If T is a given operator on $\mathbf{H}$, the tensor product $T\otimes\cdots\otimes T$ (the uniquely determined bounded linear transformation on $\mathbf{H}^n$ that carries $x_1\otimes\cdots\otimes x_n$ into $Tx_1\otimes\cdots\otimes Tx_n$) will be denoted as $\Omega_n(T)$. The map $T \mapsto \Omega_n(T)$ is a representation of the general linear group on $\mathbf{H}$ into the general linear group on $\mathbf{H}^n$. It is easily verified that $R(a)$ and $\Omega_n(T)$ commute for all $a \in S_n$ and $T \in GL(\mathbf{H})$, showing that $\Omega_n(\cdot)$ leaves invariant the subspace of all symmetric and all antisymmetric n-tensors. The transformation $\Omega_n(S)|\mathbf{K}_n$ will be denoted as $\Gamma_n(S)$; thus $\Gamma_n(\cdot)$ is a representation of $GL(\mathbf{H})$ on the space of all symmetric n-tensors over $\mathbf{H}$.

When $\mathbf{H}$ is finite-dimensional, it was shown by Schur that the only bounded linear operators on $\mathbf{H}^n$ commuting with all $\Omega_n(T)$ are the linear combinations of the $R(a)$; from this it follows that the irreducibly invariant subspaces of $\mathbf{H}^n$ under Ω_n are those of the form $P\mathbf{H}^n$, where P is a minimal projection in the algebra of linear combinations of the $R(a)$. This algebra is a homomorphic image of the "group algebra" of S_n, and the irreducible constituents of $\mathbf{H}^n$

under Ω_n can be read off from Young's determination of the irreducible representations of S_n. These results can be extended to the case when **H** is infinite-dimensional, but we treat here only the symmetric (and in Chapter 2 the antisymmetric) subspace involved in physical quantum field theory.

It will be convenient to define the space $\mathbf{H}^0$ (resp. $\mathbf{K}_0$) of all 0-tensors (resp. all symmetric 0-tensors) as **C**, relative to the inner product: $\langle\alpha, \beta\rangle = \alpha\bar{\beta}$, and to define $\Gamma_0(U)$ to be the identity map on **C** if U is unitary, and complex conjugation if U is antiunitary. The Hilbert space direct sum $\mathbf{K}' = \bigoplus_0^\infty \mathbf{K}_n$ of the spaces of all symmetric covariant tensors over the given complex Hilbert space **H**—of all ranks $n = 0, 1, \ldots$—will be called the (Hilbert) space of all (covariant) *symmetric tensors* over **H**, and may be denoted as $e^{\mathbf{H}}$. If U is unitary (or antiunitary) on **H**, the direct sum $\Gamma'(U)$ of the $\Gamma_n(U)$ $(n = 0, 1, \ldots)$ is unitary (or antiunitary) on $\mathbf{K}'$. Evidently Γ' is a representation of the extended unitary group on **H**. Each $\Gamma_n(\cdot)$ is continuous both in the uniform and strong operator topologies; $\Gamma'(\cdot)$, however, is continuous only in the strong operator topology.

For any $x \in \mathbf{H}$, *(left) tensor multiplication by* x, as an operator on $\mathbf{H}^n$, is the unique bounded linear transformation from $\mathbf{H}^n$ to $\mathbf{H}^{n+1}$ that carries $x_1 \otimes \cdots \otimes x_n$ into $x \otimes x_1 \otimes \cdots \otimes x_n$ for arbitrary $x_1, \ldots, x_n$ in **H**. For any n, the *symmetrization* (resp. *antisymmetrization*) operator S (resp. A) on $\mathbf{H}^n$ is defined to be the projection onto the space of symmetric (resp. antisymmetric) tensors. These operators are determined by

$$S(x_1 \otimes \cdots \otimes x_n) = 1/n! \sum_{\lambda \in S_n} x_{\lambda(1)} \otimes \cdots \otimes x_{\lambda(n)};$$

$$A(x_1 \otimes \cdots \otimes x_n) = 1/n! \sum_{\lambda \in S_n} sgn(\lambda)\, x_{\lambda(1)} \otimes \cdots \otimes x_{\lambda(n)}.$$

Symmetrized tensor multiplication by x is the result of following tensor multiplication by symmetrization. Note that symmetrized multiplication by x in **H** gives the same result whether left or right multiplication by x is used.

The complex-linearity of the underlying Hilbert space makes possible the introduction of creation and annihilation operators, which are algebraically quite convenient and much used in physical practice. If W is a Weyl system over the complex inner product space **H**, and ϕ is the associated Heisenberg system, the *creation operator* $C(z)$ for an element z of **H** is defined as

$$2^{-1/2}(\phi(z) - i\phi(iz));$$

the *annihilation operator* $C^*(z)$ as

$$2^{-1/2}(\phi(z) + i\phi(iz)).$$

SCHOLIUM 1.10. *For any $z \in \mathbf{H}$, $C(z)$ and $C^*(z)$ exist, are closed and densely defined, and satisfy the relations*

$$C(\alpha z) = \alpha\, C(z), \qquad C^*(\alpha z) = \bar{\alpha}\, C^*(z)$$

for arbitrary nonzero $\alpha \in \mathbf{C}$. For any $z \in H$, $C(z)$ and $C^*(z)$ are mutually adjoint on $\mathbf{D}(\phi(z)) \cap \mathbf{D}(\phi(iz))$. For arbitrary z and z' in $\mathbf{H}$,

$$[C(z), C^*(z')]u = -\langle z, z' \rangle u$$

for all u for which the left side is defined.

PROOF. This follows from Theorem 1.1; details are left as an exercise. □

In the statement of Theorem 1.10, primes are introduced to differentiate between the two Hilbert spaces $\mathbf{K}$ and the two representations Γ introduced in connection with integration in Hilbert space on the one hand, and tensor spaces on the other. As a result of the theorem, however, the primed and unprimed structures are seen to be unitarily equivalent, with the effect that the primes may be dropped and the coincident notations justified.

THEOREM 1.10. *Let $\mathbf{H}$ be a given complex Hilbert space, and let $(\mathbf{K}, W, \Gamma, v)$ denote the free boson field over $\mathbf{H}$. Let $\mathbf{K}'$ denote the Hilbert space $e^{\mathbf{H}} = \bigoplus_0^\infty \mathbf{K}_n$. For any unitary or antiunitary operator U on $\mathbf{H}$, let $\Gamma'(U)$ denote the direct sum of the operators $\Gamma_n(U)$ on $\mathbf{K}_n$, where $\Gamma_n(U)$ is the restriction to $\mathbf{K}_n$ of the n-fold product $U \otimes \cdots \otimes U$ $(n = 0, 1, \ldots)$. Let v' denote the 0-tensor 1. Then there exists a unique unitary operator T from $\mathbf{K}'$ onto $\mathbf{K}$ with the properties*

1) $T^{-1}\Gamma(U)T = \Gamma'(U)$;

2) $Tv' = v$;

3) if $C(z)$ is the creation operator for z corresponding to the normal Weyl system, then $T^{-1}C(z)T$ is the closed direct sum of the operators $(n + 1)^{1/2}$ times symmetrized tensor multiplication by z, acting on $\mathbf{K}_n$.

PROOF. The unicity of T will follow from the irreducibility of W. The existence of T will be shown by an explicit construction. To this end and for later purposes some further aspects of tensor algebra over Hilbert space are noted.

Let $\mathbf{H}$ be a given complex Hilbert space. If $u = \bigoplus_0^\infty u_n$ with u_n in $\mathbf{H}^n$, and if u_n vanishes for all sufficiently large n, then u is said to be of *finite rank k*, where k is the maximal index n such that $u_n \neq 0$ (0 is of rank $-\infty$); if all $u_n = 0$ except when $n = k$, u is called a *pure tensor* of rank k. If u and u' are tensors of finite rank their product $u \otimes u'$ is defined as the tensor $\bigoplus_0^\infty w_n$, where $w_n = \Sigma_{j+k=n} u_j \otimes u_k'$.

SCHOLIUM 1.12. *The tensors of finite rank over a given Hilbert space form an associative algebra with unit over* **C**, *and multiplication is continuous, relative to any subset of bounded rank.*

PROOF. The proof is left as an exercise. □

DEFINITION. The *symmetric* (resp. *antisymmetric*) *product* of two tensors of finite rank is defined as their ordinary product, followed by symmetrization (resp. antisymmetrization). This result is the same as the projection of the product on the subspace of all symmetric (resp. antisymmetric) tensors, where the tensor u is said to be *symmetric* (resp. *antisymmetric*) if each of its components is such. The symmetric product will be denoted by the symbol $\vee$ and the antisymmetric product by the symbol $\wedge$.

SCHOLIUM 1.13. *The symmetric (resp. antisymmetric) tensors of finite rank over a given Hilbert space* **H** *form an associative algebra with unit over* **C**; *relative to symmetric (resp. antisymmetric) multiplication.*

The symmetric algebra is commutative. In the antisymmetric algebra $x \wedge y = -y \wedge x$ *for arbitrary* $x, y \in \mathbf{H}$.

PROOF. Left as an exercise. □

The multiplication of tensors can be extended to be continuous when only one of the tensors is of finite rank, or to put it another way:

SCHOLIUM 1.14. *The set of all tensors over* **H** *(resp. symmetric or antisymmetric tensors) form an associative left and right module over the algebra of tensors of finite rank (resp. symmetric or antisymmetric tensors) in a unique fashion extending the earlier defined multiplication so that multiplication is jointly continuous in the two factors, when the factor of finite rank is restricted to have bounded rank.*

PROOF. Straightforward and again left as an exercise. □

An *algebraic tensor* over **H** is defined as a tensor u of finite rank such that $u_n \in \mathbf{H}^{n,\mathrm{alg}}$, where "alg" indicates the algebraic tensor product. It is evident that the set of all algebraic tensors is a subalgebra in the algebra of all tensors of finite rank, and is dense in the full tensor algebra (and the same is true with symmetric or antisymmetric restrictions). In the symmetric case, the algebra of all algebraic tensors is identifiable as follows, where the notation x^n indicates the n-fold product $x \vee x \vee \cdots \vee x$.

SCHOLIUM 1.15. *Let* $\mathbf{H}$ *denote the complexification of the real Hilbert space* $\mathbf{H}_\kappa$, *and for any algebraic symmetric tensor* u *over* $\mathbf{H}$, *let* f_u *denote the functional on* $\mathbf{H}_\kappa$: $f_u(x) = \Sigma_n \langle u_n, x^n \rangle$. *Then the mapping* $M: u \mapsto f_u$ *is an algebraic isomorphism of the algebra of all algebraic symmetric tensors over* $\mathbf{H}$ *onto the algebra of all polynomials over* $\mathbf{H}_\kappa$, *and is the unique such isomorphism that carries any element* $x \in \mathbf{H}_\kappa$ *into the corresponding linear function on* $\mathbf{H}_\kappa$.

PROOF. This is for the most part a well-known algebraic fact. The proof is again left as an exercise. □

The gist of the next lemma is the existence of the operator T on a dense domain.

LEMMA 1.10.1. *Let* $\mathbf{P}'$ *denote the set of all algebraic symmetric tensors. For arbitrary* $c > 0$ *let* Z_c *denote the linear operator on* $\mathbf{P}'$ *that divides a tensor of rank* n *by* $(2c)^n(n!)^{1/2}$. *Then the map* $T_0 = \mathbf{F}_c M Z_c$, *where* $\mathbf{F}_c$ *denotes the Wiener transform for the isonormal distribution of variance parameter* c, *is isometric into* $L_2(\mathbf{H}_\kappa, g_{2c})$, *where* g_{2c} *denotes the isonormal distribution of variance parameter* $2c$, *and extends uniquely to a unitary transformation from* $\mathbf{K}'$ *onto* $L_2(\mathbf{H}_\kappa, g_{2c})$.

PROOF. Note that the Wiener transform is involved here with respect to the isonormal process whose variance is half that of the process with respect to which the cited L_2-space is formed; it would be impossible to have the same variance in both cases, for the Wiener transform is then unitary, while the factor MZ_c is not. Note also that although the Wiener transform $\mathbf{F}_c$ has no meaning in the entire space $L_2(\mathbf{H}_\kappa, g_{2c})$, it is well defined on polynomials, and maps into polynomials, which are in all the spaces $L_2(\mathbf{H}_\kappa, g_k)$ for arbitrary k.

To establish the lemma it suffices to show that the operator $T_0 = \mathbf{F}_c M Z_c$ carries an orthonormal basis for $\mathbf{P}'$ into an orthonormal basis for $L_2(\mathbf{H}_\kappa, g_{2c})$, which may be denoted here as $\mathbf{K}$. To this end, let $\{e_\lambda\}$ be an indexed orthonormal basis for $\mathbf{H}_\kappa$, where λ varies over the index set Λ; let $n(\cdot)$ denote an arbitrary function from Λ to the nonnegative integers, having the property that $n(\lambda) = 0$ except for at most finitely many λ; and let $\mathbf{N}$ denote the set of all such functions $n(\cdot)$. For each $n(\cdot) \in \mathbf{N}$, let $x_{n(\cdot)}$ denote the polynomial in the symmetric algebra $(n!/\Pi_\lambda n(\lambda)!)^{1/2}\, \Pi_\lambda e_\lambda^{n(\lambda)}$; basis vectors e_λ having zero exponents are defined as 1, and so may be deleted; and $n = \Sigma_\lambda n(\lambda)$. The elements $x_{n(\cdot)}$ span $\mathbf{P}'$ algebraically, by an essentially finite-dimensional argument. They are also orthonormal. To see this, note that for arbitrary $y_1, \ldots, y_k$ in $\mathbf{H}$,

$$\begin{aligned}\langle x_1\vee\cdots\vee x_k, y_1\vee\cdots\vee y_k\rangle &= \langle S(x_1\otimes\cdots\otimes x_k), S(y_1\otimes\cdots\otimes y_k)\rangle\\ &= \sum_{\lambda,\lambda'\in S_k}\langle x_{\lambda(1)}\otimes\cdots\otimes x_{\lambda(k)}, y_{\lambda'(1)}\otimes\cdots\otimes y_{\lambda'(k)}\rangle/(k!)^2\\ &= 1/k!\sum_{\lambda\in S_k}\langle x_{\lambda(1)}\otimes\cdots\otimes x_{\lambda(k)}, y_1\otimes\cdots\otimes y_k\rangle\\ &= 1/k!\sum_{\lambda\in S_k}\langle x_{\lambda(1)}, y_1\rangle\cdots\langle x_{\lambda(k)}, y_k\rangle.\end{aligned}$$

Applying this computation to the case in which the x_i and y_i are elements of the orthonormal basis $\{e_\lambda\}$ it follows that $x_{n(\cdot)}$ is orthogonal to $x_{n'(\cdot)}$ unless $n(\cdot) = n'(\cdot)$, and that $\langle x_{n(\cdot)}, x_{n(\cdot)}\rangle = (n!/\Pi_\lambda n(\lambda)!)\,(n!)^{-1}\Pi_\lambda n(\lambda)! = 1$.

Thus, to establish the lemma, it suffices to show that the $T_0 x_{n(\cdot)}$ are orthogonal and span **K** (topologically). The latter follows from (i) the fact that if $p(x)$ is any polynomial on $\mathbf{H}_\times$ in the inner products $\langle x, e_\lambda\rangle$, then $p = T_0 u$, where u is the algebraic element $Z_c^{-1}M^{-1}\mathbf{F}_c^{-1}p$ of $\mathbf{K}'$; and (ii) the observation that polynomials over $\mathbf{H}_\times$ in the inner products with the elements of a fixed orthonormal set are dense in $L_2(\mathbf{H}_\times, g)$, by the same argument as in the case of all polynomials. It is only necessary to observe that the subalgebra in question is measure-theoretically separating. The linear functionals $\langle x, a\rangle$, a being arbitrary in $\mathbf{H}_\times$, are separating. If $a_n \to a$, $\theta(\langle x, a_n\rangle) \to \theta(\langle x, a\rangle)$ in L_2, so the $\langle x, a\rangle$ are separating when a ranges over a dense set, and hence also if a ranges over any orthonormal set, such as the e_λ.

To show that the $T_0 x_{n(\cdot)}$ are orthonormal, note that $T_0 x_{n(\cdot)}$ is the polynomial $\mathbf{F}_c(\Pi_\lambda p_{\lambda,n(\lambda)})$, where $p_{\lambda,j}$ is the polynomial on $\mathbf{H}_\times$: $p_{\lambda,j}(x) = \langle x, e_\lambda\rangle^j r_j(c)$, where $r_j(c) = [j!\,(2c)^j]^{1/2}$ with the convention that $p_{\lambda,0}(x)$ is defined as being identically 1 for all λ. The Wiener transform (of any variance parameter) carries a product of polynomials based on orthogonal subspaces into the product of their Wiener transforms by inspection of the definition and the stochastic independence of polynomials based on orthogonal subspaces. Thus $\mathbf{F}_c(\Pi_\lambda p_{\lambda,n(\lambda)}) = \Pi_\lambda \mathbf{F}_c(p_{\lambda,n(\lambda)})$, showing that it suffices to show that the $\mathbf{F}_c(p_{\lambda,n(\lambda)})$ form an orthonormal set in **K** as λ and $n(\cdot)$ vary, or simply that the $\mathbf{F}_c(p_{\lambda,j})$ are such for any fixed λ as j varies. The latter polynomials are based on the one-dimensional subspace spanned by e_λ, so the problem reduces to the treatment of the case in which $\mathbf{H}_\times$ is one-dimensional.

It is sufficient therefore to show that if $\mathbf{H}_\times$ is $\mathbf{R}$ with the inner product $\langle a, b\rangle = ab$, and if $p_j(x) = r_j(c)x^j$, then the $\mathbf{F}_c p_j$ are orthonormal in $L_2(\mathbf{R}, g_{2c})$, where

$$dg_c(x) = (2\pi c)^{-1/2}\exp(-x^2/2c)\,dx.$$

To this end, the expression

$$\sum_{j,k=0}^{\infty}\langle \mathbf{F}_c p_j, \mathbf{F}_c p_k\rangle\, s^j t^k/[j!\,k!\,r_j(2c)\,r_k(2c)]$$

can be evaluated by dominated convergence as $\langle \mathbf{F}_c e^{sx}, \mathbf{F}_c e^{tx} \rangle$ (all inner products being in $L_2(\mathbf{R}, g_{2c})$, and x being a dummy variable). This is readily evaluated in turn as $e^{2cst} = \sum_{j=0}^{\infty} (2cst)^j/j!$, implying the stated orthonormality.

Definition. For arbitrary $z \in \mathbf{H}$, $C_0'(z)$ denotes the linear operator in $\mathbf{K}'$ with domain $\mathbf{P}'$ consisting of the algebraic symmetric tensors, such that if $u \in \mathbf{P}'$ and if u is pure of rank n, then

$$C_0'(z) = (n + 1)^{1/2} z \vee u.$$

$\phi_0'(z)$ is defined by the equation

$$\phi_0'(z) = 2^{-1/2}(C_0'(z) + C_0'(z)^*);$$

it follows that $C_0'(z) = 2^{-1/2}(\phi_0'(z) - i\phi_0'(iz))$. (In this connection, note that $C_0'(z)$ is bounded from the algebraic n-tensors to the algebraic $(n + 1)$-tensors. $C_0'(z)^*$ is consequently bounded from the algebraic $(n + 1)$-tensors to the algebraic n-tensors, and in particular its domain contains $\mathbf{P}'$; thus $\phi_0'(z)$ has domain $\mathbf{P}'$.)

For $x \in \mathbf{H}_\varkappa$, $\phi_0'(x)$ is denoted as $P_0'(x)$ and $-\phi_0'(ix)$ as $Q_0'(x)$; thus $2^{1/2}C_0'(x) = P_0'(x) + iQ_0'(x)$.

If W is an arbitrary Weyl system over the complex inner product space $\mathbf{H}$, with associated Heisenberg system ϕ, then relative to a given real subspace $\mathbf{H}_\varkappa$ corresponding to a conjugation $\varkappa$ on $\mathbf{H}$, $P(\cdot)$ and $Q(\cdot)$ are defined as follows: for arbitrary $x \in \mathbf{H}_\varkappa$, $P(x) = \phi(x)$ and $Q(x) = -\phi(ix)$. In the case of the free boson field, the restrictions of $\phi(z)$, $P(x)$, and $Q(x)$ to the polynomial algebra $\mathbf{P}$ will be denoted as $\phi_0(z)$, $P_0(x)$, and $Q_0(x)$ (resp.).

Lemma 1.10.2. *Let h be a real measurable function on a measure space M, and let $\mathbf{E}$ denote a dense subset of $L_2(M)$ such that $hf \in \mathbf{E}$ for all $f \in \mathbf{E}$. If $e^{\varepsilon|h|}f \in L_2(M)$ for all sufficiently small $\varepsilon > 0$ and $f \in \mathbf{E}$, then the operator $f \mapsto hf$ in $L_2(M)$ is essentially selfadjoint in $\mathbf{E}$.*

Proof. Otherwise there would exist a nonzero element g in $L_2(M)$ such that either $\langle hf + if, g\rangle = 0$ for all $f \in \mathbf{E}$, or $\langle hf - if, g\rangle = 0$ for all $f \in \mathbf{E}$. The following argument applies in either case (for specificity, suppose $\langle hf - if, g\rangle = 0$ for all $f \in \mathbf{E}$): defining h^0 as 1, $\langle h^k f, g\rangle = i^k\langle f, g\rangle$ for $k = 0, 1, \ldots$, so that

$$\sum_{k=0}^{\infty} \langle h^k f, g\rangle (i\varepsilon)^k/k! = \sum_{k=0}^{\infty} (-\varepsilon)^k/k! \, \langle f, g\rangle.$$

The assumption that $e^{\varepsilon|h|}f \in L_2(M)$ for all $f \in \mathbf{E}$ validates the interchange of summation and integration, and it results that

$$\int e^{ieh} f \bar{g} = e^{-\varepsilon} \int f \bar{g}.$$

Since $\mathbf{E}$ is dense, this equation must hold for all $f \in L_2(M)$. It follows that $e^{ieh}\bar{g} = e^{-\varepsilon}\bar{g}$ almost everywhere, for all sufficiently small positive ε. It follows in turn that $ih + 1$ vanishes a.e. on the set where g is nonzero; as h is real valued, this set must be of measure zero, i.e., $g = 0$. □

LEMMA 1.10.3. *Let x be arbitrary in $\mathbf{H}_\kappa$, and let $\{e_\lambda\}$ be any maximal orthonormal set in $\mathbf{H}_\kappa$; the restrictions of $\phi(x)$ and of $\phi(ix)$ to the algebra of all polynomials $p(y)$ on $\mathbf{H}_\kappa$ in the inner products $\langle y, x\rangle$ and $\langle y, e_\lambda\rangle$ are then essentially selfadjoint.*

PROOF. The algebra in question is dense in $\mathbf{K}$, and invariant under $\phi_0(ix)$, which is multiplication by $\langle y, x\rangle$ within a constant factor. Furthermore, $e^{\varepsilon|\langle y,x\rangle|}$ is in $L_p(\mathbf{H}_\kappa, g)$ for all $p < \infty$, where g denotes the isonormal distribution on $\mathbf{H}_\kappa$, since $\langle x, y\rangle$ may be identified with a normally distributed random variable. Consequently, Lemma 1.10.2 is applicable and shows that the indicated restriction of $\phi(ix)$ is essentially selfadjoint. Since the cited algebra is invariant under the Wiener transform, and this carries $\phi(x)$ into $\phi(ix)$, the indicated restriction of $\phi(x)$ is also essentially selfadjoint. □

LEMMA 1.10.4. *T_0 transforms $P_0'(x)$ and $Q_0'(x)$ into $P_0(x)$ and $Q_0(x)$ respectively, for all $x \in \mathbf{H}_\kappa$.*

PROOF. Consider first the case of $Q_0(x)$; it is to be shown that $T_0^{-1}Q_0(x)\,T_0 u = Q_0'(x)u$, for every $x \in \mathbf{H}_\kappa$ and algebraic symmetric tensor u. Choosing a maximal orthonormal set $\{e_\lambda\}$ in $\mathbf{H}_\kappa$ which contains a nonzero multiple of x and such that u is a finite linear combination of symmetric products of the e_λ, it follows that it suffices to show that

$$T_0^{-1}Q_0(x)T_0\, x_{n(\cdot)} = Q_0'(x)\, x_{n(\cdot)}$$

for all $x_{n(\cdot)}$ as involved in the proof of Lemma 1.10.1, where $x = e_\mu$ for some fixed μ. It is readily computed that

$$2^{1/2}\, i\, Q_0'(x)\, x_{n(\cdot)} = (n(\mu) + 1)^{1/2}\, x_{n'(\cdot)} - (n(\mu))^{1/2}\, x_{n''(\cdot)},$$

where $n'(\lambda) = n(\lambda) + \delta_{\lambda\mu}$ and $n''(\lambda) = n(\lambda) - \delta_{\lambda\mu}$, with the convention that $x_{n(\cdot)} = 0$ if $n(\lambda) < 0$ for any λ. Thus

$$\begin{aligned} &2^{1/2}\, i\, T_0\, Q_0'(x)\, x_{n(\cdot)} \\ &= (n(\mu) + 1)^{1/2}\, T_0\, x_{n'(\cdot)} - (n(\mu))^{1/2}\, T_0\, x_{n''(\cdot)} \\ &= (n(\mu) + 1)^{1/2}\, \Pi_\lambda[r_{n'(\lambda)}(c)\, \mathbf{F}_c(\langle\cdot, e_\lambda\rangle^{n'(\lambda)})] \\ &\quad - (n(\mu))^{1/2}\, \Pi_\lambda[r_{n''(\lambda)}(c)\, \mathbf{F}_c(\langle\cdot, e_\lambda\rangle^{n''(\lambda)})]. \end{aligned}$$

On the other hand, the formula

$$\sum_{n=0}^{\infty} (s^n/n!)\, \mathbf{F}_c(\langle \cdot, x'\rangle^n) = \exp[(c/2)\, s^2 + is\, \langle \cdot, x'\rangle] \qquad (1.7)$$

is obtainable by dominated convergence and the evaluation of a Gaussian integral, x' being arbitrary in $\mathbf{H}_\varkappa$. Differentiation with respect to s and comparison of coefficients yields the equation

$$\langle \cdot, x'\rangle\, \mathbf{F}_c(\langle \cdot, x'\rangle^n) = icn\, \mathbf{F}_c(\langle \cdot, x'\rangle^{n-1}) - i\mathbf{F}_c(\langle \cdot, x'\rangle^{n+1})$$

($n = 0, 1, 2, \ldots$; as earlier a negative index denotes a vanishing quantity). Setting $x' = e_\mu$, it follows by a simple computation that

$$2^{1/2}\, i\, \langle \cdot, e_\mu\rangle\, T_0 x_{n(\cdot)} = (n(\mu) + 1)^{1/2}\, T_0 x_{n'(\cdot)} - (n(\mu))^{1/2}\, T_0 x_{n''(\cdot)}.$$

Thus $T_0^{-1} Q_0(x) T_0$ and $Q_0'(x)$ agree on the algebraic symmetric tensors.

In the case of $P_0(x)$, it suffices similarly to show that

$$T_0^{-1} P_0(x) T_0 x_{n(\cdot)} = P_0'(x)\, x_{n(\cdot)}.$$

For any $x' \in \mathbf{H}$, let $\partial_{x'}$ denote Frechet differentiation in the direction x', applied to functions F on $\mathbf{H}_\varkappa$: $\partial_{x'} F = \lim_{\varepsilon \to 0} \varepsilon^{-1}[F(y + \varepsilon x') - F(y)]$. The application of $\partial_{x'}$ to both sides of equation 1.7 and comparison of coefficients show that

$$\partial_{x'}\, \mathbf{F}_c(\langle \cdot, x'\rangle^n) = in\, \mathbf{F}_c(\langle \cdot, x'\rangle^{n-1})\, \langle x', x'\rangle.$$

From this it can be inferred, as in the case of $Q_0(x)$, that

$$2^{1/2}\, P_0(x_\mu) T_0 x_{n(\cdot)} = (n(\mu) + 1)^{1/2}\, T_0 x_{n'(\cdot)} + (n(\mu))^{1/2}\, T_0 x_{n''(\cdot)}.$$

A simple computation shows that this is the same as $2^{1/2} T_0 P_0'(x_\mu) x_{n(\cdot)}$. □

LEMMA 1.10.5. *T_0 transforms $C_0(z)$ into $C_0'(z)$, for all $z \in \mathbf{H}_\varkappa$.*

PROOF. This is an immediate consequence of the more detailed Lemma 1.10.4 and the complex-linearity of $C(z)$ as a function of z. □

COMPLETION OF PROOF OF THEOREM 1.10. Note first that for any unitary operator U on $\mathbf{H}$, $\Gamma'(U) C_0'(z) \Gamma'(U)^{-1} = C_0'(Uz)$, by straightforward application of the definitions involved. Now setting T for the closure of T_0^{-1}, to show that the two unitary operators $T\Gamma(U)T^{-1}$ and $\Gamma'(U)$ agree, it suffices to show that they agree on a set of vectors spanning a dense domain in $\mathbf{K}'$. The vectors of the form $C_0'(z_1) \cdots C_0'(z_n)\, v'$, where $z_1, \ldots, z_n$ is an arbitrary finite ordered set of vectors in $\mathbf{H}_\varkappa$, form such a set; indeed, the $x_{n(\cdot)}$ have this form, within a constant factor. That is, it suffices to show that

$$\Gamma(U) T^{-1}\, C_0'(z_1) \cdots C_0'(z_n)\, v' = T^{-1} \Gamma'(U)\, C_0'(z_1) \cdots C_0'(z_n)\, v'.$$

Now

$$T^{-1}\Gamma'(U)\,C_0'(z_1)\cdots C_0'(z_n)\,v'$$
$$= T^{-1}\Gamma'(U)\,C_0'(z_1)\,\Gamma'(U)^{-1}\cdots\Gamma'(U)\,C_0'(z_n)\,\Gamma'(U)^{-1}\Gamma'(U)\,v'$$
$$= T^{-1}C_0'(Uz_1)\cdots C_0'(Uz_n)v';$$

applying Lemma 1.10.5, the latter vector is $C_0(Uz_1)\cdots C_0(Uz_n)v$. On the other hand, $T^{-1}C_0'(z_1)\cdots C_0'(z_n)\,v'$ is, by Lemma 1.10.5, equal to $C_0(z_1)\cdots C_0(z_n)\,v$. Applying $\Gamma(U)$ to this vector gives

$$\Gamma(U)C(z_1)\Gamma(U)^{-1}\cdots\Gamma(U)C(z_n)\Gamma(U)^{-1}v.$$

Since $\Gamma(U)W(z)\,\Gamma(U)^{-1} = W(Uz)$ for arbitrary U and z, $\Gamma(U)\phi(z)\Gamma(U)^{-1} = \phi(Uz)$, hence $\Gamma(U)C(z)\Gamma(U)^{-1} = C(Uz)$. The vector in question is therefore $C(Uz_1)\cdots C(Uz_n)v$, as required. □

Problems

1. Show that Γ_n (that is, the representation Γ restricted to $\mathbf{K}_n$) is continuous in the uniform, strong, and weak operator topologies, while Γ itself is continuous in the strong and weak topologies but not the uniform topology.

2. Prove that multiplication of tensors of finite rank over a Hilbert space $\mathbf{H}$ is not jointly continuous (relative to the Hilbert space topology on $\bigoplus_0^\infty \mathbf{H}^n$) if neither factor is restricted to have bounded rank.

3. Given a selfadjoint operator A in $\mathbf{H}$, show that $\partial\Gamma(A)$ restricts to a densely defined selfadjoint operator in $\mathbf{K}_n$ with range contained in $\mathbf{K}_n$, and describe this operator (as well as its domain) explicitly.

4. Show, using Problem 3, that if A is a nonnegative selfadjoint operator in $\mathbf{H}$ then $\partial\Gamma(A)^n \geq \partial\Gamma(A^n)$ for $n \geq 1$.

5. Show that, up to normalization, the Wiener transform of the function x^n on $(\mathbf{R}^1, g)$ is the nth Hermite polynomial.

6. Let A be a selfadjoint operator in $\mathbf{H}$ such that there is an orthonormal basis $\{e_i\}$ of $\mathbf{H}$ with $Ae_i = \lambda_i e_i$. Show that $\partial\Gamma(A)$ equals the closure of $\Sigma_i\,\lambda_i\,C(e_i)C(e_i)^*$.

7. a) Show that the restriction of Γ_n to the subgroup $U_\varkappa(\mathbf{H})$ of all real unitaries on $\mathbf{H}$ that commute with the (arbitrary) given conjugation $\varkappa$ on $\mathbf{H}$ is irreducible if and only if $\mathbf{H}$ is infinite-dimensional.

b) Generalize (a) to the other irreducible components of the n-fold tensor product of $\mathbf{H}$ with itself, for the case when $\mathbf{H}$ is infinite-dimensional.

(Results a) and b) are due independently to K. Okamoto et al. and J. Pedersen.)

1.9. Implications of wave-particle duality

COROLLARY 1.10.1. *For each $x \in \mathbf{H}_\varkappa$, $P_0'(x)$ and $Q_0'(x)$ are essentially self-adjoint, and their closures $P'(x)$ and $Q'(x)$ are such that $(\mathbf{K}, P', Q')$ is a Heisenberg pair over $\mathbf{H}_\varkappa$.*

PROOF. Since $P_0'(x)$ and $Q_0'(x)$ are unitarily equivalent via T to essentially selfadjoint operators having the indicated properties, they themselves have these (unitarily-invariant) properties. □

COROLLARY 1.10.2. *T transforms $\Gamma(iI_{\mathbf{H}})$ into the Wiener transform.*

PROOF. It has earlier been shown that the Wiener transform carries $P(x)$ into $-Q(x)$, and $Q(x)$ into $P(x)$, for all x. On the other hand, from the relation $\Gamma'(U)C_0'(z)\Gamma'(U)^{-1} = C_0'(Uz)$ it follows on taking $U = iI$ that $\Gamma'(iI)$ transforms $P_0'(x)$ into $-Q_0'(x)$ and $Q_0'(x)$ into $P_0'(x)$, and hence transforms their closures similarly. By the irreducibility of the totality of the $e^{iP(x)}$ and $e^{iQ(x)}$, the Wiener transform is uniquely determined within a constant factor by this property; the constant factor is determined by the property that 1 is carried into 1. It follows by unitary equivalence that $\Gamma'(iI)$ is similarly characterized within a constant factor by the manner in which it transforms $P'(x)$ and $Q'(x)$, and the constant factor may be fixed by the requirement that v' be carried into v'. Since T carries v into v', it must transform $\Gamma(iI)$ in the indicated fashion. □

COROLLARY 1.10.3. *Let $\{e_\lambda\}$ be a maximal orthonormal set in $\mathbf{H}_\varkappa$. The Hilbert space $\mathbf{K}'$ of all symmetric tensors over $\mathbf{H}$ is unitarily equivalent to the space $L_2(M)$, where M is the tensor product of* dim$\mathbf{H}_\varkappa$ *copies of $(\mathbf{R}, g_c)$, where $dg_c = (2\pi c)^{-1/2}e^{-x^2/2c}\,dx$, in such a way that $Q'(e_\lambda)$ is the operation of multiplication by $(2c)^{-1/2}\,a_\lambda$, where a_λ is the function on M mapping the generic element $a \in M$ into its λ-th coordinate. $P'(e_\lambda)$ is represented in its action on the representative in $L_2(M)$ of an algebraic symmetric tensor as follows:*

$$f \mapsto -i(2c)^{1/2}\,\partial f/\partial a_\lambda + i(2c)^{-1/2}\,a_\lambda f$$

(any such representative f being equal a.e. to a polynomial in a finite number of the a_λ).

PROOF. The tensor products of the members of an orthonormal basis in $L_2(\mathbf{R}, g)$, including 1, form an orthonormal basis in $L_2(M)$. Setting F_n for $(n!)^{-1/2}\,F_{1/2}(x^n)$ on $\mathbf{R}$ (where x is a dummy variable) provides such an orthonormal basis in $L_2(\mathbf{R}, g)$. The proof of Lemma 1.10.4 shows that $P'(x)$ and

$Q'(x)$ correspond in the representation of Corollary 1.10.3 to the indicated operators.

COROLLARY 1.10.4. *If U is a real unitary on* **H** *(i.e., one leaving* $\mathbf{H}_\kappa$ *invariant), then* $\Gamma(U)$ *acts on* **P**, *the algebra of all polynomials over* $\mathbf{H}_\kappa$, *as follows:* $f(x) \mapsto f(U^{-1}x)$, *for* $f \in \mathbf{P}$.

PROOF. It is easily seen that the mapping $f(x) \to f(U^{-1}x)$ extends uniquely to a unitary transformation V on **K**. To show that $V = \Gamma(U)$, it suffices to show that these two unitaries transform $P(x)$ and $Q(x)$ in identical fashions, and leave 1 invariant (the latter fact is already established). To show that V transforms $P(x)$ and $Q(x)$ in the same fashion as $\Gamma(U)$, it suffices to treat—instead of $P(x)$ and $Q(x)$—their restrictions to **P**, in view of the essential self-adjointness of these restrictions. This computation is straightforward.

LEXICON. The unitary equivalence of the *particle* representation on the one hand, involving the symmetrized tensor products of the Hilbert space **H** with itself, and of the *wave* representation in the space of square-integrable functionals over a real part $\mathbf{H}_\kappa$ of **H**, is the mathematical manifestation of the so-called "wave-particle duality" which has evolved during several centuries of the study of physical light. From the point of view of an experimentalist, particularly one concerned with "scattering" experiments in which incoming and outgoing free particles are compared, the chief physical observables are the occupation numbers. Physically, these specify the numbers of particles of designated species and parameters in a given state. For example, the statement that the incoming state in a certain physical system consists of so many mesons of such and such energy and momenta can be formulated in terms of statements that certain selfadjoint operators have the state vector in question as an eigenvector, and that the corresponding eigenvalues are the nonnegative integers corresponding to the designated composition of the state. It is quite analogous to the description of the population of a given country at a given time in, say, economic terms. Significant indices of the economic state might be the numbers of individuals of specified income, capital, investment and saving rates, etc.

To deal theoretically with this key experimental object, one makes the following entry in the basic physical-mathematical lexicon. *The "number of particles" in the free boson field over* **H** *in the state represented by the vector x in* **H**, *is* $\partial\Gamma(P_x)$ *where* P_x *is the projection of* **H** *onto the one-dimensional subspace spanned by x, and* $\partial\Gamma(A)$ *denotes for any selfadjoint operator A the selfadjoint generator of the one-parameter unitary group* $\Gamma(e^{itA})$. More generally, the number of particles in states represented by vectors in the given subspace **M** of **H** is $\partial\Gamma(P_{\mathbf{M}})$, where $P_{\mathbf{M}}$ is the projection with range **M**. In particular, the (total) number of particles is $\partial\Gamma(I)$. This physical-mathematical correspondence is justified by a variety of considerations involving mathematical consequences that are in agreement with physical observation. The main qualitative mathematical considerations are these:

1) $\partial\Gamma(P)$ has nonnegative integral eigenvalues, as is physically to be expected of an occupation number. Moreover, the only state vector in **K** in which the expected total number of particles is zero is the vacuum vector v.

2) The occupation number has the natural additivity properties: $\partial\Gamma(P) + \partial\Gamma(P')$ has closure equal to $\partial\Gamma(P + P')$ for any two orthogonal projections P and P' in **H**. Indeed, $\partial\Gamma$ is not only additive in this sense, but countably additive.

3) The occupation number has the natural connection with total attributes. Just as the total income of a population should be the sum of the products of the various incomes with the corresponding occupation numbers for these incomes, so one has for any selfadjoint operator A in **H**, with spectral resolution $A \sim \int \lambda dE(\lambda)$ the relation $\partial\Gamma(A) \sim \int \lambda\partial\Gamma(E(\lambda))$. In particular, if the "single-particle energy" A has the eigenvalues $a_1, a_2, \ldots$, then the "field energy" $\partial\Gamma(A)$ has the eigenvalue $n_1a_1 + n_2a_2 + \cdots$ in the state in which n_j particles each of energy a_j are present.

4) The occupation number has the appropriate transformation properties under automorphisms of the system, in particular under Lorentz transformations in a relativistic field. This is the relation:

$$\Gamma(U)\partial\Gamma(A)\Gamma(U)^{-1} = \partial\Gamma(UAU^{-1}).$$

The proofs (and in the case of problem 3), the detailed mathematical formulation) of these results are left as exercises.

On the other hand, just as the description of the economic state of a country in terms of occupation numbers may not be the best basis for the consideration of economic dynamics, i.e., the temporal development of the state, so in a quantum process the occupation numbers give no specific indication of the dynamics (apart from what may be corollary to kinematical considerations such as Lorentz-invariance). The treatment of the dynamics of quantum systems turns out to be naturally undertaken in terms of field rather than particle concepts, by virtue of the local character of relativistic interactions. In mathematical terms, the *field* is diagonalized in the functional integration representation, just as the *particle numbers* are diagonalized in the tensor product representation.

Chapter 6 will treat quantization in the simple but basic case in which the dynamics is linear. In essence, this is the study of the free boson field over a Hilbert space on which is given a distinguished unitary (or, more generally, symplectic) one-parameter group. The classical linear relativistic wave equations of "even spin" (i.e., involving single-valued representations of the

Poincaré group, rather than of its covering group) can be quantized as a very special case of the general theory. But the quantization of nonlinear wave equations, treated in Chapter 8, naturally involves qualitatively new considerations.

There is a third representation for the free boson field that has remarkable mathematical properties, although physically it is less readily interpreted than the two that have already been treated. Recall that one of these, the Fock-Cook, or particle representation, diagonalizes the occupation numbers; the other—the "renormalized Schrödinger," the "functional integration representation," or the "real wave representation"—can serve to diagonalize the "field variables" at a fixed time throughout space.

The third, or complex wave, representation gives a kind of diagonalization of the creation operators. Since the latter are nonnormal, a literal diagonalization is impossible. What is achieved instead is a simple representation of the creation operators as multiplication operators, not in L_2 over a measure space (this would contradict the nonnormality of the operators), but in a closely related space, consisting precisely of the holomorphic (or more conveniently, for present purposes, antiholomorphic) members of L_2 over a space having both a measure and a complex structure.

In the rough terminology employed by physicists one might say that we make an expansion of the state vectors into the (virtual) eigenfunctions of the annihilation operators; these are the so-called "coherent" states.

Before entering into the complex wave representation itself, we discuss and treat Bose-Einstein quantum fields from historical and axiomatic points of view, from which the relation to the other two representations may be better appreciated.

The concept of a free quantum field of particles satisfying Bose-Einstein statistics is commonly introduced in a highly structured manner involving space-time, invariant wave equations thereon, etc.; but in algebraic essence, all that is involved is a given complex Hilbert space **H**, as earlier indicated. The concrete structure of this space may vary from application to application, but many of the central results of the theory are independent of this special structure.

Mathematically, the *free boson field* over a given complex Hilbert space **H**, to be denoted as **B(H)**, may be defined as a quadruple $(\mathbf{K}, W, \Gamma, v)$ consisting of

1) a complex Hilbert space **K**;
2) a Weyl system W on **H** with values in $U(\mathbf{K})$; i.e. a (strongly, as normally understood) continuous mapping $z \mapsto W(z)$ from **H** to the unitary operators on **K** satisfying the Weyl relations

$$W(z)W(z') = e^{\frac{1}{2}i\,\mathrm{Im}\langle z, z'\rangle} W(z + z')$$

for arbitrary z and z' in **H**;

3) a continuous representation Γ from $U(\mathbf{H})$ into $U(\mathbf{K})$ satisfying the relation

$$\Gamma(U)W(z)\,\Gamma(U)^{-1} = W(Uz)$$

for arbitrary $U \in U(\mathbf{H})$ and $z \in \mathbf{H}$; and

4) a unit vector v in **K** having the properties that $\Gamma(U)v = v$ for all $U \in U(\mathbf{H})$, and that the $W(z)v$, $z \in \mathbf{H}$, span **K** topologically. (For brevity, we cite the latter condition as cyclicity of v for W.)

The constraints 1–4 do not uniquely determine the free boson field, but serve to do so with the addition of

5) Γ is "positive" in the sense that if A is any nonnegative selfadjoint operator in **H**, then $\partial\Gamma(A)$ is likewise nonnegative—where for any selfadjoint A in **H**, $\partial\Gamma(A)$ denotes the *selfadjoint* generator of the one-parameter unitary group $\{\Gamma(e^{itA}): t \in \mathbf{R}\}$ provided by Stone's theorem.

We may take conditions 1)–5) as the definition of the free boson field. This axiomatic characterization must, of course, be supplemented by suitable existence considerations, which are supplied by the concrete representation of **B(H)** given earlier, or by that given later.

Historically, the free boson field was introduced quite heuristically by Dirac in 1926 in purely formal analogy with the Heisenberg quantization of a single-particle system. The Hilbert spaces **H** and **K** and aspects of Γ are rather implicit in this treatment, and the Weyl system W appears in the form of an infinite sequence of putative operators: $p_1, q_1; p_2, q_2; \ldots$, satisfying the Heisenberg commutation relations, whose existence Dirac merely hypothesized. An explicit representation of this early version of the free boson field was first given by Fock (1932) in a nonrelativistic and mathematically heuristic format. A rigorous construction, which assumes and exploits the Hilbertian character of the so-called single-particle space **H** and is adaptable to the relativistic context, was first given by Cook (1953). The Fock-Cook representation satisfied conditions 1)–5), except that 2) was clear only in its infinitesimal form. The global form of 2) in the Fock-Cook, or particle, representation follows from the unitary equivalence of this representation with the functional integration representation. This "wave" representation was implicit in theoretical physical practice which, however, dealt with functional integrals only at a quite formal level.

In a general scientific way, the Fock-Cook representation expresses the particle properties of the quantum field by providing explicit diagonalizations for the so-called occupation numbers $\partial\Gamma(P)$, where P ranges over a maximal commuting set of projections on **H**. These are the properties typically most directly observable in high-energy experiments. The functional integration representation expresses the wave properties, by virtue of its explicit diagonalization of the values of the field at different points of space, at a fixed time. The wave properties are conceptually fundamental, and technically the functional integration representation has been the primary basis for progress in constructive quantum field theory in recent decades.

However, in each of the two classic representations—particle and wave—for the free boson field, some of the important operators have highly complicated actions. On the other hand, in the complex wave representation, virtually all of the important operators have a remarkably simple formal appearance, despite the greater complexity of the physical interpretation of the representation.

The relation with heuristic developments due to theoretical physicists may be amplified as follows. In the case of wave functions in one dimension, the harmonic oscillator hamitonian $N = \frac{1}{2}(p^2 + q^2 - 1)$, the position operator q, and the creation operator $C = 2^{-\frac{1}{2}}(p + iq)$, all have simple spectra in the sense that they admit a cyclic vector. According to the general ideas of quantum mechanics, any one of them may be used as a complete state labeling operator. In Fock (1928) the use of C was originated in this connection, and it was noted that this led to the consideration of holomorphic wave functions; but no Hilbert space of such functions was formulated, and the only specific domain of definition cited was the unit disc. The more comprehensive treatment of Dirac (1949) remained formal, and the case of a quantum field is reduced to the one-dimensional case via a representation as a direct product that is not invariant. In this work there is again no actual Hilbert space $\mathbf{K}$, or representation Γ of $U(\mathbf{H})$, and only formally defined field operators. Rigorous mathematical correlatives to some of the theoretical physics initiatives toward the representation of boson fields by analytic functions are given at the end of this chapter, but precise formal correlation is somewhat elusive.

1.10. Characterization of the free boson field

Constraint 5) in the preceding section, whose notation is used here, is equivalent to the apparently much weaker condition that its conclusion holds for just one positive operator A. This is a consequence of the following abstract characterization of the free boson field over a Hilbert space.

THEOREM 1.11. *Let* $\mathbf{H}$ *be a given complex Hilbert space, and let* W *be a Weyl system over* $\mathbf{H}$ *with representation space* $\mathbf{K}$ *and cyclic vector* v *in* $\mathbf{K}$.

Suppose there exists a positive selfadjoint operator A *in* $\mathbf{H}$ *that annihilates no nonzero vector, and a one-parameter unitary group* Γ' *on* $\mathbf{K}$ *with the properties*

a) $\Gamma'(t)W(z)\Gamma'(-t) = W(e^{itA}z)$ *for all* $t \in \mathbf{R}$ and $z \in \mathbf{H}$;
b) $\Gamma'(t)v = v$ *for all* $t \in \mathbf{R}$;

c) $\Gamma'(t) = e^{itH}$, *where* H *is selfadjoint and nonnegative.*

Then there exists a unique representation Γ *of* $U(\mathbf{H})$ *into* $U(\mathbf{K})$ *which extends* Γ' *in the sense that* $\Gamma(e^{itA}) = \Gamma'(t)$ *for all* $t \in \mathbf{R}$, *and such that* $(\mathbf{K}, W, \Gamma, v)$ *satisfies conditions* 1)–5) *of Section* 1.9.

PROOF. Set

$$f(u) = \langle e^{-uH}W(z)v, W(z)v\rangle,$$

where $u = s + it$ with $s \geq 0$, and z is arbitrary in $\mathbf{H}$. Then f is bounded and continuous in the half-plane $s \geq 0$, and holomorphic in the interior $s > 0$; we denote the totality of all such functions as $\mathfrak{B}$. By virtue of the Weyl relations, the boundary values $f(it)$ may be expressed as

$$f(it) = e^{\frac{1}{2}i\,\mathrm{Im}\langle z_t, z\rangle}\langle W(z_t - z)v, v\rangle; \qquad z_t = e^{-itA}z.$$

Now setting

$$g(u) = \exp\{-\tfrac{1}{2}\langle e^{-uA}z, z\rangle\},$$

then $g \in \mathfrak{B}$ also. Accordingly, $fg \in \mathfrak{B}$, and fg has the boundary values

$$fg(it) = \langle W(z_t - z)v, v\rangle\, e^{-\frac{1}{2}\mathrm{Re}\langle z_t, z\rangle}.$$

Replacing z by $-z$, it follows that $\langle W(-z_t + z)v, v\rangle\, e^{-\frac{1}{2}\mathrm{Re}\langle z_t, z\rangle}$ is also the boundary value function of an element of $\mathfrak{B}$. But this function is the complex conjugate of the function $(fg)(it)$. Accordingly, fg must be a constant, which may be evaluated as $e^{-\langle z,z\rangle/2}$ by setting $t = 0$, yielding the equation

$$\langle W(z_t - z)v, v\rangle = \exp(-\|z_t - z\|^2/4).$$

By virtue of the triviality of the null space of A, the $z_t - z$ are dense in $\mathbf{H}$ as z and t vary, as follows from the spectral theorem; and it follows in turn by continuity that

$$\langle W(z)v, v\rangle = \exp(-\|z\|^2/4)$$

for all $z \in \mathbf{H}$.

This result implies that the inner product $\langle W(Uz)v, W(Uz')v\rangle$, where z and z' are arbitrary in $\mathbf{H}$, is independent of $U \in U(\mathbf{H})$; and this implies in turn that the linear transformation T_0:

$$\sum_i a_i W(z_i)v \mapsto \sum_i a_i W(Uz_i)v$$

is well defined and isometric on the domain $\mathbf{D}$ of all finite linear combinations of the $W(z)v$, $z \in \mathbf{H}$. Since $\mathbf{D}$ is dense in $\mathbf{K}$, T_0 extends uniquely to a unitary operator on all of $\mathbf{K}$, which we denote as $\Gamma(U)$; by construction, $\Gamma(U)W(z)\Gamma(U)^{-1} = W(Uz)$ and $\Gamma(U)v = v$ for all $U \in U(\mathbf{H})$. That Γ is a representation of $U(\mathbf{H})$ follows from its intertwining relation with W, noting also that $\Gamma(U)$ is the unique unitary operator on $\mathbf{K}$ that transforms $W(z)$ into $W(Uz)$ and leaves v invariant, by virtue of the cyclicity of v. In order to show that Γ is continuous from $U(\mathbf{H})$ into $U(\mathbf{K})$, it suffices to show that $\Gamma(U)W(z)v$ is a continuous function of U, for any fixed $z \in \mathbf{H}$, at the element $U = I$ of $U(\mathbf{H})$. To this end it suffices in turn to note that

$$\langle\Gamma(U)W(z)v, W(z)v\rangle = \exp(\tfrac{1}{2}i \operatorname{Im}\langle Uz, z\rangle) \exp(-\|Uz - z\|^2/4),$$

which is continuous in U.

To show the positivity of the representation Γ, suppose now that A is any nonnegative selfadjoint operator in $\mathbf{H}$; it suffices to show that $\Gamma(e^{itA})$ is a positive-frequency function of t, i.e., that $\int \langle\Gamma(e^{itA})w, w'\rangle g(t)dt = 0$ for functions $g \in L_2(\mathbf{R}^1)$, whose Fourier transforms vanish on the negative half-axis, and for arbitrary vectors w and w' in $\mathbf{K}$. To this end it suffices to show that $\langle\Gamma(e^{itA})W(z)v, W(z')v\rangle$ is a positive-frequency function of t for arbitrary z and z' in $\mathbf{H}$. But, by the earlier evaluation of $\langle W(z)v, v\rangle$ and the Weyl relations, this function is

$$\exp \tfrac{1}{4}(-\|z\|^2 - \|z'\|^2 + 2\langle z_{-t}, z'\rangle);$$

since $\langle z_{-t}, z'\rangle$ is a positive-frequency function of t, so also is $\exp \tfrac{1}{2}\langle z_{-t}, z'\rangle$.

Finally, the unicity of the representation Γ follows from the cyclicity of v under W, which implies that any unitary operator that commutes with all $W(z)$ and leave v invariant must be the identity. □

Corollary 1.11.1. *Any two systems* $(\mathbf{K}, W, \Gamma, v)$ *satisfying conditions* 1)–5) *are unitarily equivalent.*

Proof. Given a second such system, denoted by primes, there exists a unique unitary transformation T from $\mathbf{K}$ onto $\mathbf{K}'$ that carries $W(z)v$ into $W'(z)v'$, for all $z \in \mathbf{H}$, by an argument employed in the preceding proof. The respective Weyl systems are then unitarily equivalent via T, and the definition of $\Gamma(U)$ as the unique unitary on $\mathbf{K}$ that transforms $W(z)$ into $W(Uz)$ and leaves v invariant shows that it is transformed by T into $\Gamma'(U)$. □

1.11. The complex wave representation

The existence of the free boson field has been established, but it will be illuminating to give a new construction, which will serve at the same time to define the complex wave representation. This construction is based on functional integration, in connection with which the following notation will be employed: if $\mathbf{H}$ is a real Hilbert space, $\mathbf{S}$ will denote the space of "L_2" or "square-integrable" tame functions on **H**, i.e., those tame functions f for which $E(|f|^2) < \infty$, where E denotes the expectation relative to the isonormal distribution g on **H**. If f is a tame function on **H**, the corresponding random variable in $(\mathbf{H}, g)$ will be denoted as $\theta(f)$ and also called tame. In particular, if $f \in \mathbf{S}$, $\theta(f) \in L_2(\mathbf{H}, g)$.

Now if **H** is a given complex Hilbert space, it has also the structure of a real Hilbert space $\mathbf{H}^{\#}$, with inner product equal to the real part of the complex inner product given in **H**. The isonormal distribution g on **H** is defined as that on $\mathbf{H}^{\#}$, and $L_2(\mathbf{H}, g)$ is defined as $L_2(\mathbf{H}^{\#}, g)$. In the case of a real space **H**, the space $L_2(\mathbf{H}, g)$ is the completion of the algebra of all polynomials on **H**, with respect to the inner product

$$\langle\theta(f), \theta(f')\rangle = \int_{\mathbf{L}} f(x)\overline{f'(x)}\, dg(x),$$

L being any finite-dimensional subspace of **H** on which the polynomials f and f' are based. Recall that a polynomial on **H** is defined as a function of the form $f(x) = p(\langle x, e_1\rangle, \ldots, \langle x, e_n\rangle)$, where p is a polynomial function on $\mathbf{R}^n$, and the e_j are arbitrary vectors in **H**, finite in number; it is no essential loss of generality, and it will henceforth be assumed, that the e_j are orthonormal in this representation. The possibility of defining $L_2(\mathbf{H}, g)$ as the completion of the polynomials rather than of the bounded tame functionals may, as earlier noted, be regarded as a generalization of the completeness of the Hermite functions in $L_2(\mathbf{R})$.

This means that if **H** is a complex Hilbert space, then $L_2(\mathbf{H}, g)$ consists of the completion of the algebra $\mathbf{P}_r(\mathbf{H})$ of the functions of the form

$$f(x) = p(\mathrm{Re}\langle x, e_1\rangle, \ldots, \mathrm{Re}\langle x, e_n\rangle);$$

such a function we call a *real-analytic polynomial*. But in addition to the algebra $\mathbf{P}_r(\mathbf{H})$, there are two other simple unitarily-invariant algebras of polynomials: the *complex-analytic*, defined as those of the form

$$f(x) = p(\langle x, e_1\rangle, \ldots, \langle x, e_n\rangle),$$

where p is a polynomial function on $\mathbf{C}^n$; and the *complex-antianalytic*, i.e., the complex conjugates of those just indicated. The totality of complex-antianalytic polynomials on **H** will be denoted as $\mathbf{P}(\mathbf{H})$; the representation space **K** for the complex wave representation will consist of the closure of $\mathbf{P}(\mathbf{H})$ in $L_2(\mathbf{H}, g)$.

To begin with, a system $(\mathbf{K}', W', \Gamma', v')$ will be defined that fails to represent the free boson field only in that condition 5 is violated and that v' is not cyclic for the $W'(z)$. Let g denote the isonormal distribution on **H** with variance parameter $c = \sigma^2$, and let $\mathbf{K}' = L_2(\mathbf{H}, g)$. For any vector $z \in \mathbf{H}$, define the operator $W_0'(z)$ on the subspace $\theta(\mathbf{S})$ of $\mathbf{K}'$, where **S** denotes the totality of all square-integrable tame functions on **H**, as follows: for any function f on **H**, let f_z denote the function given by the equation

$$f_z(w) = f(w - \sigma z)\, e^{\langle z,w\rangle/2\sigma - \langle z,z\rangle/4}.$$

If f is tame, so also is f_z; and if $\theta(f) = 0$, $\theta(f_z) = 0$ also, since straightforward computation show that $\|\theta(f)\|^2 = \|\theta(f_z)\|^2$ for arbitrary tame functions f and vectors z in **H**. The mapping $\theta(f) \mapsto \theta(f_z)$ is therefore a well-defined isometry of $\theta(\mathbf{S})$ into itself, and defines an operator $W_0'(z)$ on $\theta(\mathbf{S})$.

It is straightforward to verify that

$$W_0'(z)\, W_0'(z') = e^{\frac{1}{2}i\,\mathrm{Im}\langle z,z'\rangle}\, W_0'(z + z')$$

for arbitrary z and z' in **H**. Setting $z' = -z$, it follows that $W_0'(z)$ is invertible, and so extends to a unique unitary transformation, to be denoted as $W'(z)$, on all of $\mathbf{K}'$. By continuity, the Weyl relations remain valid for W'. It is not difficult to show in addition that the mapping $z \mapsto W'(z)$ is continuous from **H** to $U(\mathbf{K}')$, and it follows that W' is a Weyl system on **H** with representation space $\mathbf{K}'$.

A representation Γ_0' of $U(\mathbf{H})$ on $\theta(\mathbf{S})$ will now be defined as follows: for arbitrary $f \in \mathbf{S}$ and $U \in U(\mathbf{H})$, $\Gamma_0'(U)$ sends $\theta(f)$ into $\theta(f_U)$, where $f_U(z) = f(U^{-1}z)$. Since it is readily verified that $f_U \in \mathbf{S}$ when $f \in \mathbf{S}$, and that the map $f \mapsto f_U$ preserves the expectation functional E, $\Gamma_0'(U)$ is well defined and is isometric from $\theta(\mathbf{S})$ into itself. It is straightforward to verify also that $\Gamma_0'(UU') = \Gamma_0'(U)\, \Gamma_0'(U')$ for arbitrary U and U' in $U(\mathbf{H})$, and that $\Gamma_0'(I_{\mathbf{H}}) = I$, so Γ_0' is a representation of $U(\mathbf{H})$ on $\theta(\mathbf{S})$. In particular, $\Gamma_0'(U)$ is invertible for all $U \in U(\mathbf{H})$, and so extends uniquely to a unitary operator $\Gamma'(U)$ defined on all of $\mathbf{K}'$. It follows that Γ' is a representation of $U(\mathbf{H})$ into $U(\mathbf{K}')$, and it is not difficult to verify that Γ' is strongly continuous from $U(\mathbf{H})$ into $U(\mathbf{K}')$ by checking continuity on the spanning subset of $\mathbf{K}'$ consisting of the finite products of real-linear functionals.

Now let v' denote $\theta(1) \in \mathbf{K}'$; then $\Gamma'(U)v' = v'$ for all $U \in U(\mathbf{H})$, and the required intertwining relation between Γ' and W' is readily verified on the dense subset $\theta(\mathbf{S})$, and hence by continuity holds on all of $\mathbf{K}'$. Thus the quadruple $\mathbf{B}' = (K', W', \Gamma', v')$ satisfies all of the conditions on the free boson field over **H** except the positivity condition 5 and the cyclicity for v'. We call $\mathbf{B}'$ the *regular boson field over* **H** in analogy with the term "regular representation" for groups.

The complex wave representation may now be defined as that given by restriction to the subspace spanned by the complex-antianalytic (or *antiholomorphic*) polynomials.

THEOREM 1.12. *Let* $\mathbf{B}' = (\mathbf{K}', W', \Gamma', v')$ *denote the regular boson field over the given complex Hilbert space* **H**. *Let* **K** *denote the cyclic subspace under the action of* W' *generated by* v'. *Then*

1) *the complex antianalytic polynomials* **P** *on* **H** *are dense in* **K**;
2) *setting* $W(z) = W'(z)|\mathbf{K}$ *and* $\Gamma(U) = \Gamma'(U)|\mathbf{K}$ *for all* $z \in \mathbf{H}$ *and* $U \in U(\mathbf{H})$, *and* $v = v'$, *then the quadruple* $\mathbf{B} = (\mathbf{K}, W, \Gamma, v)$ *is a representation of the free boson field over* **H**.

PROOF. By direct computation $W(z)v = \theta(f)$, where f is the tame function on $\mathbf{H}$, $f(x) = e^{\langle z,x\rangle/2\sigma - \langle z,z\rangle/4}$. Thus the functionals $\theta(b_z)$ are all in the subspace spanned by the $W(z)v$, where $b_z(x) = e^{\langle z,x\rangle}$, $z \in \mathbf{H}$. Since $b_{z+z'} = b_z b_{z'}$, the set of all finite combinations of the b_z forms a ring, which contains, for any $\varepsilon > 0$ and finite ordered orthonormal set of vectors $e_1, \ldots, e_m \in \mathbf{H}$ and positive integers $n_1, \ldots, n_m$, the vector

$$c_\varepsilon = \prod_{j=1}^{m} [(\exp\varepsilon\langle e_j,u\rangle - 1)/\varepsilon]^{n_j};$$

It follows by dominated convergence that

$$\theta(c_\varepsilon) \to \theta\left(\prod_{j=1}^{m} \langle e_j, u\rangle^{n_j}\right),$$

showing that the image under θ of any complex antianalytic polynomial is in **K**. Conversely, the application of dominated convergence to the power series expansion of b_z in terms of $\langle z, x\rangle$ shows that $W(z)v$ is in the closure of $\theta(\mathbf{P})$.

Thus v is a cyclic vector for W, and to conclude that **B** is the free boson field over **H** it is only necessary to show that the representation Γ is positive (condition 5). It suffices to show that the number of particles operator $\partial\Gamma(\mathrm{I})$ has a nonnegative spectrum. In fact, $\Gamma(e^{it})$ sends the monomial $\theta(w)$, where $w(u) = \langle e_1, u\rangle^{n_1} \cdots \langle e_m, u\rangle^{n_m}$, into $e^{it(n_1 + \cdots + n_m)}\,\theta(w)$, showing that $\partial\Gamma(\mathrm{I})$ has spectrum consisting of the nonnegative integers (and, incidentally, is diagonalized by the totality of the $\theta(w)$ relative to a fixed orthonormal basis). □

DEFINITION. *An entire function on* **H** is a function F with the property that for every finite-dimensional subspace **M** of **H**, the restriction $F|\mathbf{M}$ is an entire function in the usual sense of dim **M** complex variables. An *antientire function* is one which is the complex conjugate of an entire function.

THEOREM 1.13. *Given* $z \in \mathbf{H}$, *let* b_z *denote the square-integrable tame function* $e^{\langle z/2\sigma^2, \cdot\rangle}$ *on* **H**. *If* u *is any vector in* **K**, *then*

1) *the function* F *on* **H** *given by the equation*

$$F(z) = \langle u, \theta(b_z)\rangle$$

is an antientire function on $\mathbf{H}$*;*

2) $\sup_{\mathbf{L}} \int_{\mathbf{L}} |F(z)|^2 \, dg(z) = \|u\|^2$, *the supremum being taken over the set of all finite-dimensional (complex) subspaces* $\mathbf{L}$ *of* $\mathbf{H}$*;*

3) *if* $u = \theta(f)$, f *being a square-integrable tame antientire function, then* $F(z) = f(z)$, $z \in \mathbf{H}$.

Conversely, if F *is an antientire function such that*

$$\sup_{\mathbf{L}} \int_{\mathbf{L}} |F(z)|^2 \, dg(z) < \infty,$$

then there exists a unique vector $u \in \mathbf{K}$ *such that the foregoing holds.*

Proof. The term *antimonomial* will be used for a vector in $\mathbf{K}$ (or, when indicated by the context, a function on $\mathbf{H}$) that is a constant multiple of one of the form $\theta(w)$ (respectively, w), where $w(x) = \langle e_1, x\rangle^{n_1} \cdots \langle e_m, x\rangle^{n_m}$.

Observe next that the antimonomials $\Psi_{\mathbf{n}} = 2^{-n/2}(\mathbf{n}!)^{-1/2}\sigma^{-n}\theta(w)$, where $\mathbf{n}$ stands for the multi-index $n_1, n_2, \ldots, w$ being as earlier and n being $\Sigma_j n_j$, form an orthonormal basis in $\mathbf{K}$, relative to any given orthonormal basis $e_1, e_2, \ldots$ in $\mathbf{H}$. Orthogonality is a consequence of the vanishing of the integrals $\int_{\mathbf{C}} |z|^s z^t \, dg(z)$ when s and t are nonnegative integers with $t \neq 0$, as follows from rotational invariance; normalization follows by evaluation of the integral when $t = 0$. It follows that for any vector $u \in \mathbf{K}$, $\|u\|^2 = \Sigma_{\mathbf{n}} |a_{\mathbf{n}}|^2$ where $a_{\mathbf{n}} = \langle u, \Psi_{\mathbf{n}}\rangle$.

Suppose that $u = \theta(f)$, where f is a square-integrable antientire tame function based on the finite-dimensional complex subspace $\mathbf{L}$ of $\mathbf{H}$. Then the coefficients $a_{\mathbf{n}}$ vanish unless $\mathbf{n} \in \mathbf{n}(\mathbf{L})$, where $\mathbf{n}(\mathbf{L})$ denotes the set of all multi-indices $\mathbf{n}$ such that $n_j = 0$ when e_j is not in $\mathbf{L}$. Thus $f = \Sigma_{\mathbf{n}\in\mathbf{n}(\mathbf{L})} a_{\mathbf{n}} \Psi_{\mathbf{n}}$, where as usual we omit reference to θ in the finite-dimensional context. A calculation shows that

$$\int_{\mathbf{L}} \Psi_{\mathbf{n}}(x) \, \bar{b}_z(x) \, dg(x) = \Psi_{\mathbf{n}}(z)$$

for $\mathbf{n} \in \mathbf{n}(\mathbf{L})$; this is known as the "reproducing" property of the kernel $\bar{b}_z(x)$. It follows that

$$F(z) = \sum_{\mathbf{n}\in\mathbf{n}(\mathbf{L})} a_{\mathbf{n}} \langle \Psi_{\mathbf{n}}, b_z\rangle = \sum_{\mathbf{n}\in\mathbf{n}(\mathbf{L})} a_{\mathbf{n}} \Psi_{\mathbf{n}}(z) = f(z),$$

where the sums converge in $L_2(\mathbf{L}, g)$. As a consequence we have

$$\|u\|^2 = \int_{\mathbf{L}} |F(z)|^2 \, dg(z).$$

Statement 1 now follows from the expansion of $\exp[\langle z/2\sigma^2, \cdot\rangle]$.

Concerning 2), let u be arbitrary in $\mathbf{K}$. Given a finite-dimensional complex subspace $\mathbf{L}$ spanned by the $\{e_j\}$, let $u_{\mathbf{L}} = \Sigma_{\mathbf{n}\in\mathbf{n}(\mathbf{L})} a_{\mathbf{n}} \Psi_{\mathbf{n}}$. Clearly $u_{\mathbf{L}}$ is a square-integrable tame function based on $\mathbf{L}$, and the supremum over $\mathbf{L}$ of $\|u_{\mathbf{L}}\|$ is $\|u\|$.

Moreover, if $F_{\mathbf{L}}(z) = \langle u_{\mathbf{L}}, \theta(b_z)\rangle$, it is easily seen that F and $F_{\mathbf{L}}$ agree on $\mathbf{L}$. Thus we have

$$\sup_{\mathbf{L}} \int_{\mathbf{L}} |F(z)|^2\, dg(z) = \sup_{\mathbf{L}} \|u_{\mathbf{L}}\|^2 = \|u\|^2.$$

Taking note of the arbitrariness of the given orthonormal set in $\mathbf{H}$, 2) follows.

For 3), suppose conversely that F is a given antientire function on $\mathbf{H}$ for which $\|F|\mathbf{L}\|_{L_2(\mathbf{L},g)}$ remains bounded as $\mathbf{L}$ ranges over the set of all finite-dimensional complex subspaces of $\mathbf{H}$. Let $f_{\mathbf{L}}$ be the function on $\mathbf{H}$ given by the equation

$$f_{\mathbf{L}}(z) = \sum_{\mathbf{n}\in n(\mathbf{L})} a_{\mathbf{n}} \Psi_{\mathbf{n}}(z),$$

where

$$\Psi_{\mathbf{n}}(z) = 2^{-n/2}(n!)^{-1/2}\sigma^{-n}\langle e_1, z\rangle^{n_1}\cdots\langle e_m, z\rangle^{n_m}$$

and $a_{\mathbf{n}} = \langle F|\mathbf{L}, \Psi_{\mathbf{n}}\rangle$. Then $f_{\mathbf{L}}$ is a square-integrable tame function on $\mathbf{H}$ that is antientire and based on $\mathbf{L}$. Evidently

$$\|f_{\mathbf{M}} - f_{\mathbf{L}}\|^2 = \sum_{\mathbf{n}\in n(\mathbf{M})\Delta n(\mathbf{L})} |a_{\mathbf{n}}|^2,$$

where Δ denotes the symmetric difference. It follows that as $\mathbf{L} \to \mathbf{H}$, $\{\theta(f_{\mathbf{L}})\}$ is convergent to a vector $u \in \mathbf{K}$. By the same argument as in the preceding paragraph, the antientire function $z \mapsto \langle u, \theta(b_z)\rangle$ corresponding to u is identical with the original antientire function F. Unicity of u is evident from the fact that

$$F(z) = \sum_{\mathbf{n}} a_{\mathbf{n}} 2^{-n/2}(n!)^{-1/2}\sigma^{-n}\langle e_1, z\rangle^{n_1}\cdots\langle e_m, z\rangle^{n_m},$$

as follows by a limiting argument from the finite-dimensional case, the series being convergent by virtue of the Schwarz inequality. □

It follows that the Weyl system operators $W(z)$ and the representation Γ of $U(\mathbf{H})$ for the free boson field may be defined in a pointwise fashion, and not merely as Hilbert space limits, in accordance with

COROLLARY 1.13.1. *The free boson field* $\mathbf{B}(\mathbf{H})$ *over* $\mathbf{H}$ *may be represented as follows:*

1) $\mathbf{K}$ *is the Hilbert space of all antientire functions* F *on* $\mathbf{H}$ *for which the norm*

$$\|F\| = \sup_{\mathbf{L}} \|F|\mathbf{L}\|_{L_2(\mathbf{L},g)}$$

is finite, the supremum being taken over all finite-dimensional subspaces **L** *of* **H**, *with the inner product*

$$\langle F, F' \rangle = \lim_{\mathbf{L} \to \mathbf{H}} \langle F|\mathbf{L}, F'|\mathbf{L} \rangle_{L_2(\mathbf{L}, g)}.$$

2) For any $z \in \mathbf{H}$, $W(z)$ *is the operator*

$$F(x) \mapsto F(x - \sigma z)\, e^{\langle z,x \rangle / 2\sigma - \langle z,z \rangle /4}, \qquad F \in \mathbf{K}.$$

3) For any $U \in U(\mathbf{H})$, $\Gamma(U)$ *is the operator*

$$F(z) \mapsto F(U^{-1}z), \qquad F \in \mathbf{K}.$$

4) v *is the function identically* 1 *on* **H**.

Proof. This involves straightforward limiting arguments of the type already employed, and the formalities are omitted. □

Problems

1. Show that in part 2 of Theorem 1.13 the integrals $\int_{\mathbf{L}} |F(z)|^2\, dg(z)$ over *real* subspaces may be unbounded, as shown by the following example due to M. Vergne: $F(z) = \exp(-a_n \langle e_n, z \rangle^2)$, where $|\Sigma\, a_n| < \sigma$, $\Sigma |a_n|^2 < \infty$, and $\Sigma |a_n| = \infty$.

2. Show that a linear differential operator with polynomial coefficients on $L_2(\mathbf{R})$ is carried by the unitary equivalence of the real- and complex-wave representations into a linear differential operator with polynomial coefficients on the present space **K** of antianalytic functions in $L_2(\mathbf{C}, g)$.

1.12. Analytic features of the complex wave representation

This representation is less familiar than the other two treated in Chapter 1, but has probably the richest mathematical theory. In this section a number of useful analytic features that emerge in the course of the proof of Theorem 1.13 will be made more explicit.

A natural orthonormal basis in the complex wave representation is given by

Corollary 1.13.2. *If* $e_1, e_2, \ldots$ *is any orthonormal basis for* **H**, *then an orthonormal basis for the preceding space* **K** *is provided by the antianalytic polynomials*

$$p(z) = 2^{-n/2}\,(n_1!\,n_2!\ldots)^{-1/2}\,\sigma^{-n}\,\langle e_1, z \rangle^{n_1} \langle e_2, z \rangle^{n_2} \ldots,$$

as the n_i range independently over the nonnegative integers subject to the constraint that $n = \Sigma_i n_i$ be finite.

The corresponding expansion for any vector $F \in \mathbf{K}$ is convergent pointwise, and

$$|F(z)| \leq \|F\| \exp(\langle z, z\rangle/(4\sigma^2)), \qquad z \in \mathbf{H}.$$

PROOF. This is a straightforward consequence of the theorem. □

We now rationalize the term "complex wave representation" by making explicit the sense in which this representation pseudo-diagonalizes the creation operators. We first recall the

DEFINITION. For any representation $\mathbf{B} = (\mathbf{K}, W, \Gamma, \nu)$ of the free boson field over the given Hilbert space $\mathbf{H}$, and vector $z \in \mathbf{H}$, the *creation operator for the vector z*, to be denoted as $C(z)$, is defined as the operator $2^{-1/2}(\phi(z) - i\phi(iz))$, where $\phi(z)$ denotes the selfadjoint generator of the one-parameter group $\{W(tz): t \in \mathbf{R}\}$. The *annihilation operator for the vector z*, to be denoted as $C^*(z)$, is defined as the operator $2^{-1/2}(\phi(z) + i\phi(iz))$; this is in fact identical to $C(z)^*$, by virtue of

COROLLARY 1.13.3. *The operators $C(z)$ and $C^*(z)$ are closed, densely defined, and mutually adjoint. In the complex-wave representation, $C(z)$ has domain consisting of all $F \in \mathbf{K}$ such that $\langle z, \cdot\rangle F(\cdot) \in \mathbf{K}$, and sends any such function into $a^{-1}\langle z, \cdot\rangle F(\cdot)$, where $a = i2^{1/2}\sigma$. $C(z)^*$ has domain consisting of all $F \in \mathbf{K}$ such that $\partial_z F(\cdot) \in \mathbf{K}$, where $\partial_z F(u) = \lim_{\varepsilon\to 0} \varepsilon^{-1}(F(u + \varepsilon z) - F(u))$, and sends any such function into $-a\, \partial_z F$.*

PROOF. That $C(z)$ and $C^*(z)$ are closed, densely defined, and mutually adjoint is valid for an arbitrary Weyl system, by familiar smoothing arguments, as in Theorem 1.1, and we omit the proof. Since convergence in the Hilbert space $\mathbf{K}$ implies pointwise convergence of the corresponding antientire functions on $\mathbf{H}$, the action of $C(z)$ and $C(z)^*$ may be computed by taking pointwise limits. This is straightforward and has the stated results.

It remains to show that the domains of $C(z)$ and $C(z)^*$ include all vectors to which the corresponding pointwise-defined operators are applicable as operators in $\mathbf{K}$. To this end let $\mathbf{K}_n$ denote the subspace of $\mathbf{K}$ consisting of the closure of homogeneous polynomials in $\mathbf{K}$ of degree n, and let F denote any element of $\mathbf{K}$ such that $\langle z, \cdot\rangle F(\cdot)$ is again in $\mathbf{K}$. It is easily seen that multiplication by $\langle z, \cdot\rangle$ is bounded on $\mathbf{K}_n$ and maps it into $\mathbf{K}_{n+1}$; consequently, $\|\langle z, \cdot\rangle F(\cdot)\|^2 = \sum_{n=0}^{\infty} \|\langle z, \cdot\rangle F_n(\cdot)\|^2$, where F_n denotes the component of F in $\mathbf{K}_n$. On the other hand, it is readily verified that $\mathbf{K}_n$ is in the domains of all the $C(z)$ and $C(z)^*$. Setting $F^N = \Sigma_{n<N} F_n$, it follows that $F^N \to F$ and that $C(z)F^N \to a^{-1}\langle z, \cdot\rangle F(\cdot)$, and, since $C(z)$ is closed, it follows in turn that F is in the domain of $C(z)$. The same argument applies to $C(z)^*$ with $n + 1$ replaced by $n - 1$. □

Lastly, the intertwining operator between the complex and real wave representations is given as follows: let $\mathbf{H}$ be a complex Hilbert space, let a be an arbitrary positive number, and let g_a denote the isonormal distribution of variance a on $\mathbf{H}$. For arbitrary $z \in \mathbf{H}$, let $W_c'(z, a)$ denote the following operator on functions over $\mathbf{H}$, where $\sigma = a^{1/2}$:

$$W_c'(z, a): f(u) \to f(u - \sigma z)\, e^{\langle z, u\rangle/2\sigma - \langle z, z\rangle/4}.$$

Let $W_c(z, a)$ denote the unitary operator on the subspace $\mathbf{K}$ of $L_2(\mathbf{H}, g_a)$ that carries $\theta(p)$ into $\theta(W_c'(z, a)p)$ for any antiholomorphic polynomial p on $\mathbf{H}$ (as in Theorem 1.13). Now let b be an arbitrary positive number, and let $\mathbf{H}'$ be an arbitrary real part of $\mathbf{H}$. The real wave representation may be defined as follows: let $\mathbf{K}' = L_2(\mathbf{H}', g_b)$, and for arbitrary $z = x + iy \in \mathbf{H}$, where $x, y \in \mathbf{H}'$, let $W_r'(z, b)$ denote the following operator on functions over $\mathbf{H}'$, where $\nu = b^{1/2}$:

$$W_r'(z, b): f(u) \to f(u - \nu 2^{1/2} x) \exp[\langle z, u\rangle/\nu 2^{1/2} - \tfrac{1}{2}\langle z, x\rangle].$$

Let $W_r(z, b)$ denote the unitary operator on $\mathbf{K}'$ that carries $\theta(f)$ into $\theta(W_r'(z, b)f)$ for any tame square-integrable function f. The vacuum vector is represented by the unit function $1_{\mathbf{H}}$ in $\mathbf{K}$ and $1_{\mathbf{H}'}$ in $\mathbf{K}'$. (The respective Γ's are determined by the invariance of the vacuum vectors and their intertwining with the Weyl systems, and need not be involved here.)

By Theorem 1.12, for any given a and b, there exists a unique unitary transformation T from $\mathbf{K}$ onto $\mathbf{K}'$ such that $W_c(z, a) = T^{-1}W_r(z, b)T$ and $T(\theta(1_{\mathbf{H}})) = \theta(1_{\mathbf{H}'})$. The explicit expressions for T and T^{-1} on sufficiently regular functions are simplest when $b = 2a$ (which in fact exemplifies the impossibility of entirely transforming away the $2^{1/2}$ factor sometimes involved in boson field analysis). Because of the simplicity of scaling considerations, it will suffice here to treat the case $a = 1/2$, $b = 1$.

THEOREM 1.14. *The (unitary) intertwining operator T from the complex onto the real wave representation for the free boson field over* $\mathbf{H}$ *has the following action on antientire functions F on* $\mathbf{H}$: $F \to \theta(f)$, *where*

$$f(u) = \int_{\mathbf{H}} \exp[\langle z, u\rangle - \tfrac{1}{2}\langle z, \bar{z}\rangle] F(z)\, dg_{1/2}(z). \qquad (1.8)$$

The integral exists in the Lebesgue sense for tame antientire functions of order < 2, *or as a limit in mean otherwise.*

T^{-1} *carries any tame functional* $\theta(f) \in L_2(\mathbf{H}', g_1)$ *into the antientire function* F *on* $\mathbf{H}$ *given by the equation*

$$F(z) = \exp(-\langle \bar{z}, z\rangle/2) \int_{\mathbf{H}'} \exp(\langle u, z\rangle) f(u)\, dg_1(u). \qquad (1.9)$$

More generally, for arbitrary $f \in \mathbf{K}'$, $T^{-1}f = F$, *where*

$$F(z) = \exp(-\langle \bar{z}, z\rangle/4)\, \langle f, \theta(e^{\langle z, \cdot\rangle})\rangle.$$

PROOF. Since $TW_c(z, \tfrac{1}{2})v = W_r(z, 1)Tv'$, where v and v' are the respective vacua, it follows that T carries $\exp(2^{-1/2}\langle z, w\rangle - \tfrac{1}{4}\|z\|^2)$ into $\exp(2^{-1/2}\langle z, u\rangle - \tfrac{1}{2}\langle z, x\rangle)$, where w is variable in $\mathbf{H}$ and u is variable in $\mathbf{H}'$, for any fixed $z \in \mathbf{H}$. Thus T sends $e^{\langle z, \cdot\rangle}$ into $e^{\langle z, \cdot\rangle - \frac{1}{2}\langle z, \bar{z}\rangle}$. The reproducing kernel property (which by complex conjugation applies to holomorphic as well as antiholomorphic entire functions) shows that the integral expression given in the theorem is correct for functions $F(w)$ of the form $F(w) = e^{\langle z, w\rangle}$ for some fixed $z \in \mathbf{H}$. It follows from the unitarity of T and Lebesgue convergence theory that equation 1.8 is valid also for arbitrary antiholomorphic polynomials F; it then follows, in turn, that equation 1.8 holds with an absolutely convergent Lebesgue integral on the right for arbitrary tame antientire functions of order less than 2. The general case follows by formation of a limit in mean, as justified in part by the functional integration theory earlier developed.

To establish the form given for T^{-1}, note that in case $f(u) = e^{\langle z, u\rangle}$ for some fixed $z \in \mathbf{H}$, the integral expression may be evaluated explicitly as an elementary Gaussian integral and found to agree with the expression for T^{-1} given above on functions of this type. More generally, equation 1.9 holds for arbitrary tame vectors in $L_2(\mathbf{H}', g_1)$ by virtue of their approximability in this space by finite linear combinations of the $e^{\langle w, u\rangle}$. It follows for general f from the unitarity of T, the density of tame functionals in $\mathbf{K}'$, and the continuity of $F(z)$ as a function of F for fixed z. □

It is interesting to note that, as a result of the standardizations and choice of respective variance parameters, the projective mapping corresponding to T (i.e., from $L_2(\mathbf{H}, g_{1/2})$ modulo equivalence via constant factors to $L_2(\mathbf{H}', g_1)$ similarly projectified) is simple restriction from $\mathbf{H}$ to $\mathbf{H}'$ for the vectors of the form $e^{\langle z, \cdot\rangle}$; these have been interpreted as "coherent states" in physical applications.

We also note that the intertwining operators between the particle and wave representations are of a different nature from those between the two wave representations, since the particle representation is in a space of tensors of varying rank, rather than a function space. In the case of the real wave representation, the intertwining operator is given earlier in this chapter; for the complex wave representation it is simpler, and readily derived from that for the real wave representation.

Bibliographical Notes on Chapter 1

The Hilbert-space particle representation for the universal boson field is due to Cook (1953). The problem of integration in Hilbert space and its role in the representation of boson fields was discussed by Friedrichs (1953). The isonormal distribution was introduced by Segal (1954) and developed as an instance of functional integration with respect to a nonstrict distribution in Segal (1956a), which made applications to boson fields. This included the formulation of the real wave representation and its unitary equivalence to the particle representation. The complex wave representation was introduced by Segal (1962; 1978); the former paper includes the positive-energy characterization of the free boson field.

Functional integration for strict distributions was introduced in 1921 by Wiener in the case of Brownian motion and generalized by Kolmogoroff (1933). The "homogeneous chaos" was defined and treated by Wiener (1938). Further development of analysis in Wiener space was made by Cameron (1945) Cameron and Martin (1944; 1945a,b; 1947) and by Kakutani (1950).

2

The Free Fermion Field

2.1. Clifford systems

The theory of the free fermion field is essentially distinct from but nevertheless parallel to that of the free boson field, and the underlying formal analogy is close and useful.

DEFINITION. An *orthogonal space* is a pair $(\mathbf{L}, S)$ consisting of a real topological vector space $\mathbf{L}$ and a given continuous symmetric nondegenerate bilinear form S on $\mathbf{L}$.

Concerning the notions of nondegeneracy, symmetry, and of the algebraic topology on a real vector space, see Chapter 1.

EXAMPLE 2.1. Let $\mathbf{M}$ be finite-dimensional real vector space and $\mathbf{M}^*$ its dual. Let $\mathbf{L}$ denote the direct sum $\mathbf{M}\oplus\mathbf{M}^*$, and let S denote the form

$$S(x\oplus\lambda, x'\oplus\lambda') = \lambda'(x) + \lambda(x')$$

for arbitrary $x\oplus\lambda$ and $x'\oplus\lambda'$ in $\mathbf{L}$. Then $(\mathbf{L}, S)$ is an orthogonal space, and will be called the *orthogonal space built from* $\mathbf{M}$. Similarly for the case of an arbitrary given dual pair of topological vector spaces.

DEFINITION. Let $(\mathbf{L}, S)$ be a given orthogonal space. A *Clifford system* over $(\mathbf{L}, S)$ is a pair $(\mathbf{K}, \phi)$ consisting of a complex Hilbert space $\mathbf{K}$ and a continuous linear mapping ϕ from $\mathbf{L}$ to the bounded selfadjoint operators on $\mathbf{K}$ (in their strong operator topology) such that

$$\phi(z)\phi(z') + \phi(z')\phi(z) = S(z, z')\mathrm{I}. \tag{2.1}$$

The equations 2.1 are called the *Clifford relations* (later, the *real Clifford relations*).

If **H** is a given complex pre-Hilbert space, and S is given by $S(z, z') = \mathrm{Re}\langle z, z'\rangle$, then the pair $(\mathbf{H}^\#, S)$, where $\mathbf{H}^\#$ is **H** as a real vector space, is an orthogonal space, a Clifford system over which is called simply a *Clifford system over* **H**; if $S(z, z') = c\,\mathrm{Re}(\langle z, z'\rangle)$, where c is a positive constant, we speak of a *Clifford system with variance (parameter) c*.

EXAMPLE 2.2. Let **H** denote an n-dimensional complex Hilbert space, $n < \infty$, and let $\mathbf{H}^\#$ and S be as just indicated. Let **C** denote the Clifford algebra over $\mathbf{H}^\#$, i.e., the unique associative algebra over the complex number field generated by $\mathbf{H}^\#$ together with a unit e, with the following relations for the elements of $\mathbf{H}^\#$: $xy + yx = c\,\mathrm{Re}\langle x, y\rangle$ (x and y being arbitrary in $\mathbf{H}^\#$). Let E denote the unique linear functional in **C** such that $E(AB) = E(BA)$ for arbitrary A and B in **C**, and with $E(e) = 1$. Let * denote the unique adjunction operator on **C** such that $x^* = x$ for all $x \in \mathbf{H}^\#$. (An *adjunction operator* on an algebra is characterized by the properties $(A+B)^* = A^* + B^*$, $(AB)^* = B^*A^*$, $(\alpha A)^* = \bar{\alpha}A^*$ if $\alpha \in \mathbf{C}$, and $A^{**} = A$.) Let **K** denote the Hilbert space consisting of **C** completed with respect to the inner product $\langle A, B\rangle = E(B^*A)$.

For arbitrary $A \in \mathbf{C}$, let L_A denote the operation $B \mapsto AB$ on **C**, and R_A the operation $B \mapsto BA$ on **C**. For any $z \in \mathbf{H}^\#$, it follows from the relations $\langle L_zA, B\rangle = E(B^*zA)$ together with $\langle A, L_zB\rangle = E((zB)^*A) = E(B^*zA)$ that L_z may be identified with a densely defined hermitian operator on **K**. This operator extends uniquely to a bounded operator on **K** because $L_z^2 = c\|z\|^2/2$. Define $\phi(z)$ to be the unique extension to a bounded selfadjoint operator on **K** of L_z; thus one has an example of a Clifford system of variance parameter c (similarly if $\phi(z)$ is defined as R_z). It is evident also that if V is any orthogonal transformation on **H**, i.e., one preserving the form S, and ϕ is a Clifford system, then so is ϕ_V where $\phi_V(x) = \phi(Vx)$. Again, if ϕ and ψ are Clifford systems of the same variance that anticommute, in the sense that

$$\phi(x)\psi(y) + \psi(y)\phi(x) = 0$$

for all $x, y \in \mathbf{H}$, and if a and b are any real numbers such that $|a|^2 + |b|^2 = 1$, then $\zeta(x) = a\phi(x) + b\psi(x)$ defines a Clifford system of the same variance.

The last remark can be applied to yield an interesting class of Clifford systems as follows: let Ω denote the unique automorphism of **C** that carries z into $-z$ for all $z \in \mathbf{H}$. Then Ω anticommutes with all L_z and R_z and $\Omega^2 = 1$; hence

$$z \mapsto iL_z\Omega \qquad \text{and} \qquad z \mapsto iR_z\Omega$$

are Clifford systms. Moreover, since L_z and $R_{z'}$ commute for all z and z', L_z and $R_{z'}\Omega$ anticommute for all z and z'. It follows that

$$z \mapsto aL_z + biR_z\Omega$$

is a Clifford system of variance c, if a and b are as earlier.

A Clifford system $(\mathbf{K}, \phi)$ is called irreducible in case the only closed linear subspaces that are invariant under every $\phi(z)$ are $\mathbf{K}$ and $\{0\}$. None of the examples given is irreducible. This is clear from the fact that a Clifford system extends by general algebra to a representation of the Clifford algebra $\mathbf{C}$ by operators on $\mathbf{K}$. As is well known (see the proof of Lemma 2.3.1), the Clifford algebra over an n-dimensional complex Hilbert space is isomorphic to an $N \times N$ complete matrix algebra, where $N = 2^n$, and the only irreducible representation of an $N \times N$ complete matrix algebra is its natural action on an N-dimensional vector space, whereas in the examples $\mathbf{K}$ has dimension N^2. It will later be seen that irreducibly invariant subspaces can be picked out by various side conditions, including an analogue to holomorphy.

The analogue of the Stone–von Neumann theorem here is the result that every *-representation of a complete complex matrix algebra is a direct integral of copies of its natural, unique, irreducible representation cited. The argument just given shows that this representation occurs with multiplicity N in the reducible representation given above.

Additional concrete examples of Clifford systems are afforded by analogues of the particle and wave representations already given for the free boson field, which are treated below.

The fermion field is simpler than the boson field in having bounded rather than unbounded operators for the field quantities, here $\phi(x)$. This facilitates the treatment of the creation and annihilation operators, which can be defined in formally the same way as in the boson case:

$$C(z) = 2^{-1/2}(\phi(z) - i\phi(iz)), \qquad C(z)^* = 2^{-1/2}(\phi(z) + i\phi(iz));$$

$\phi(\cdot)$ can then be recovered from $C(\cdot)$ by the equation $\phi(z) = 2^{-1/2}(C(z) + C(z)^*)$, while $C(\cdot)$ and $C(\cdot)^*$ are respectively complex linear and antilinear, as functions of their argument.

SCHOLIUM 2.1. *For any Clifford system of variance c over the complex Hilbert space* $\mathbf{H}$,

$$\begin{aligned} &C(z)C(z')^* + C(z')^*C(z) = c\langle z, z'\rangle, \qquad C(iz) = iC(z), \qquad (2.2)\\ &C(z)C(z') + C(z')C(z) = 0 \end{aligned}$$

for arbitrary z and z' in $\mathbf{H}$.

Conversely, if C is any continuous complex-linear mapping from $\mathbf{H}$ *into the bounded linear operators on a Hilbert space* $\mathbf{K}$ *satisfying the relations* 2.2 *and if* $\phi(z)$ *is defined as* $2^{-1/2}(C(z) + C(z)^*)$, *then* $(\mathbf{K}, \phi)$ *is a Clifford system over* $\mathbf{H}$ *of variance* c.

PROOF. Straightforward computation. □

DEFINITION. A pair $(\mathbf{K}, C)$ consisting of a complex Hilbert space $\mathbf{K}$ together with a linear mapping C from a complex Hilbert space $\mathbf{H}$ to the bounded linear operators on $\mathbf{K}$, satisfying relation 2.2, is called a *complex Clifford system over* $\mathbf{H}$ (of variance c). The earlier-defined Clifford system is designated as *real*, when the context requires this distinction. When $c = 1$, we omit the specification "of variance c."

The "free fermion field" is defined in the same way as the free boson field, except for the substitution of anticommutators for commutators, and for the elimination of the analytical considerations required to deal with the intervention of unbounded operators in the boson case. That is to say, in the fermion case there is no need to introduce an analogue to the Weyl relations; the analogue to the Heisenberg relations is fully viable, unlike the Heisenberg relations themselves.

DEFINITION. Let $\mathbf{H}$ be a given complex Hilbert space. A *free fermion field over* $\mathbf{H}$ is a system $(\mathbf{K}, C, \Gamma, v)$ consisting of

1) a complex Hilbert space $\mathbf{K}$;
2) a linear mapping C from $\mathbf{H}$ to the bounded linear operators on $\mathbf{K}$, satisfying the relations
$$C(z)C(z')^* + C(z')^*C(z) = \langle z, z'\rangle,$$
$$C(z)C(z') + C(z')C(z) = 0;$$
3) a continuous representation Γ of the unitary group on $\mathbf{H}$ into the unitary group on $\mathbf{K}$, such that $\Gamma(U)C(z)\Gamma(U)^{-1} = C(Uz)$ for all $z \in \mathbf{H}$ and unitary U on $\mathbf{H}$;
4) a unit vector v in $\mathbf{K}$ that is cyclic for the $C(z)$, $z \in \mathbf{H}$, and such that $\Gamma(U)v = v$ for all unitaries U on $\mathbf{H}$; and satisfies the condition that
5) for any nonnegative selfadjoint operator A in $\mathbf{H}$, $\partial\Gamma(A)$ (defined as the selfadjoint generator of the one-parameter unitary group $\Gamma(e^{itA})$) is also nonnegative.

It seems natural to begin with the question of unicity. Here as in the boson case a much stronger result holds and is physically relevant. Specifically, the positivity postulated in condition 5) need not be assumed for *all* nonnegative selfadjoint operators A on $\mathbf{H}$, but only for any *one* nontrivial such operator. In physical practice, the relevant latter such operator is usually the Hamiltonian for the single-particle system under consideration; but it may also be the generator I of the phase transformations $z \rightarrow e^{it}z$, $\partial\Gamma(\mathrm{I})$ being the number of particles.

THEOREM 2.1. *Let $U(\cdot)$ be a continuous one-parameter unitary group on the given Hilbert space* $\mathbf{H}$, *the selfadjoint generator of which is positive.*

Let $(\mathbf{K}, C, \Gamma_0, v)$ *be a system consisting of a) a complex Hilbert space* $\mathbf{K}$*; b) a linear mapping* C *from* $\mathbf{H}$ *to the bounded linear operators on* $\mathbf{K}$ *such that* $C(z)C(z')^* + C(z')^*C(z) = \langle z, z'\rangle$ *and* $C(z)C(z') + C(z')C(z) = 0$ *for all* z, $z' \in \mathbf{H}$*; c) a continuous one-parameter unitary group* Γ_0 *on* $\mathbf{K}$*, with nonnegative generator, such that*

$$\Gamma_0(t)C(z)\Gamma_0(t)^{-1} = C(U(t)z),$$

for all $z \in \mathbf{H}$ *and* $t \in \mathbf{R}$*; and d) a unit vector* v *in* $\mathbf{K}$ *such that* $\Gamma_0(t)v = v$ *for all* $t \in \mathbf{R}$*, and that is cyclic for the* $C(z)$.

Then if $(\mathbf{K}', C', \Gamma_0', v')$ *is another system satisfying (a) − (d), there exists a unique unitary operator from* $\mathbf{K}$ *onto* $\mathbf{K}'$ *that carries* (C, v, Γ_0) *into* (C', v', Γ_0').

Proof. In essence, the proof parallels that for the boson case by utilizing the positive energy condition in the form of its implications regarding bounded holomorphic functions in a half-plane, which arise from temporal evolution of vacuum expectation values.

Let z be arbitrary in $\mathbf{H}$ and w be arbitrary in $\mathbf{K}$; let H denote the generator of Γ_0. Since $H \geq 0$, the function

$$f(\lambda) = \langle e^{-\lambda H}\phi(z)v, v\rangle,$$

where $\phi(z) = 2^{-1/2}(C(z) + C(z)^*)$, is holomorphic and bounded in the open half-plane $\operatorname{Re}\lambda > 0$ and continuous on the closed half-plane $\operatorname{Re}\lambda \geq 0$. Setting $\lambda = s + it$ with s and t real, then

$$f(it) = \langle e^{-itH}\phi(z)v, v\rangle = \langle \phi(z_t)v, v\rangle,$$

where $z_t = U(-t)z$, showing that as a function of t, $\langle\phi(z_t)v, v\rangle$ is the boundary value function of a bounded holomorphic function in the half-plane $s > 0$. By moving e^{-itH} to the right side of the inner product, it follows that $f(it)$ is a constant (independent of t); but more parallel to the later induction argument is the observation of the selfadjointness of $\phi(z_t)$, from which this result follows by basic complex variable theory. Setting $\mu(z)$ for the constant $f(it)$, then $|\mu(z)| \leq \|z\|$, so that μ is a continuous real-linear functional on $\mathbf{H}$. It has consequently the form

$$\mu(z) = \operatorname{Re}\langle z, u\rangle$$

for a unique element $u \in \mathbf{H}$. But $\mu(z_t) = \mu(z)$ for all t, which implies that $U(-t)u = u$ for all t. Since, however, the generator A has positive spectrum, U leaves no nonzero vector fixed. Hence $u = 0$, implying that $\mu = 0$, which shows that $\langle\phi(z)v, v\rangle = 0$ for all $z \in \mathbf{H}$.

An induction argument may now be used to show that for arbitrary $u_1, \ldots, u_k$ in $\mathbf{H}$, $\langle\phi(u_1)\cdots\phi(u_k)v, v\rangle$ is uniquely determined by the Clifford relations and the positive-energy condition, i.e., is the same for the corresponding primed quantities. Note first that it follows from the Clifford relations that

$$\phi(u_1)\phi(u_2)\cdots\phi(u_k) = \pm\ \phi(u_k)\cdots\phi(u_2)\phi(u_1) + G,$$

where G is in the linear space of products of not more than $k - 2$ of the $\phi(u_j)$. Now suppose that the $\langle\phi(u_1)\cdots\phi(u_k)v, v\rangle$ are uniquely determined when $k < n$ in the sense indicated; this has been shown to be the case for $n = 2$. To establish it for the indicated value of n, replace u_n by z_t, and note that

$$\langle\phi(u_1)\cdots\phi(u_{n-1})\phi(z_t)v, v\rangle = \langle\phi(z_t)v, \phi(u_{n-1})\cdots\phi(u_2)\phi(u_1)v\rangle.$$

This expression differs from its complex conjugate

$$\langle\phi(u_{n-1})\cdots\phi(u_2)\phi(u_1)v,\phi(z_t)v\rangle = \langle\phi(z_t)\phi(u_{n-1})\cdots\phi(u_2)\phi(u_1)v, v\rangle$$

by vacuum expectation values of products of fewer than n of the field operators $\phi(u_j)$, by the observation made above, and a possible sign depending on the parity of n. Applying the induction hypothesis, it follows that either the real or the imaginary part of $g(t) = \langle\phi(z_t)\phi(u_{n-1})\cdots\phi(u_2)\phi(u_1)v, v\rangle$ is determined uniquely, i.e., is the same as that of $h(t) = \langle\phi'(z_t)\phi'(u_{n-1})\cdots\phi'(u_2)\ \phi'(u_1)v, v\rangle$. But the same argument as earlier shows that $g(t) - h(t)$ is the boundary value function of a bounded holomorphic function f on the right half-plane. Since the generator of U has positive spectrum, $\lim_{t\to\infty} f(t) = 0$. Consequently $g(t) = h(t)$, and the induction argument is complete.

A similar argument to that in the boson field case now concludes the proof. □

2.2. Existence of the free fermion field

Having shown the unicity of the free fermion field, consider now the existence question. The construction that is closest to that in the boson case is via the particle representation, which is described in

THEOREM 2.2. *Let* **H** *be a given complex Hilbert space. Let* **K** *denote the Hilbert space direct sum* $\oplus_{n\geq 0}\mathbf{K}_n$ *of all (covariant) antisymmetric tensors over* **H** *of rank n. For any unitary or antiunitary operator U on* **H**, *let* $\Gamma(U)$ *denote the direct sum of the operators* $\Gamma_n(U)$ *on* $\mathbf{K}_n$, *where* $\Gamma_n(U)$ *is the restriction to* $\mathbf{K}_n$ *of the n-fold tensor product* $U\otimes U\otimes\cdots\otimes U$ $(n = 0, 1, \ldots)$. *Let v denote the 0-tensor* 1. *Let* $C(z)$ *denote the closure of the linear operator on the algebraic direct sum of the* $\mathbf{K}_n$ *whose action on* $\mathbf{K}_n$ *consists of antisymmetrized tensor multiplication by z, followed by multiplication by* $(n + 1)^{1/2}$.

Then $(\mathbf{K}, C, \Gamma, v)$ *is the free fermion field over* **H**.

PROOF. Recall from Section 1.8 that antisymmetrization A takes the form

$$A(x_1 \otimes \cdots \otimes x_k) = 1/k! \sum_{\lambda \epsilon S_k} (sgn\, \lambda)\, x_{\lambda(1)} \otimes \cdots \otimes x_{\lambda(n)}$$

where the summation is over the group S_k of all permutations of $1, \ldots, k$, and that tensor multiplication followed by antisymmetrization is denoted $\wedge$. In these terms, $C(z)u = (n+1)^{1/2} z \wedge u$ for $u \in \mathbf{K}_n$. Let $\mathbf{P}$ denote the subset of $\mathbf{K}$ consisting of the algebraic linear span of the tensors of the form $x_1 \wedge \cdots \wedge x_n$, the x_j being arbitrary in $\mathbf{H}$. Then $\mathbf{P}$ is dense in $\mathbf{K}$, and includes an orthonormal basis $\{x_{n(\cdot)}\}$ for $\mathbf{K}$ of the following form: let e_μ be an arbitrary orthonormal basis for $\mathbf{H}$, μ ranging over a well-ordered index set M. Let $n(\cdot)$ be any function from M to the set $\{0, 1\}$ that vanishes except for finitely many, and possibly all, values, of μ. If $n(\mu) = 0$ for all μ, define $x_{n(\cdot)}$ as v; otherwise, let $x_{n(\cdot)}$ be defined as $(k!)^{1/2} e_{\mu(1)} \wedge \cdots \wedge e_{\mu(k)}$, where $\mu(1) < \cdots < \mu(k)$ and these are the values of μ for which $n(\mu) = 1$. The $x_{n(\cdot)}$ are evidently mutually orthogonal, and $\langle x_{n(\cdot)}, x_{n(\cdot)} \rangle = (k!) \langle y, y \rangle$, where

$$y = 1/k! \sum_{\lambda \epsilon S_k} (\text{sgn}\, \lambda)\, e_{\lambda(1)} \otimes \cdots \otimes e_{\lambda(k)},$$

the indices $\mu(1), \ldots, \mu(k)$ being redesignated as $1, \ldots, k$ for simplicity. Now

$$\begin{aligned}\langle y, y \rangle &= (k!)^{-2} \sum_{\lambda, \lambda'} (\text{sgn}\, \lambda\lambda') \langle e_{\lambda(1)} \otimes \cdots \otimes e_{\lambda(k)}, e_{\lambda'(1)} \otimes \cdots \otimes e_{\lambda'(k)} \rangle \\ &= (k!)^{-2} \sum_{\lambda} \sum_{\lambda'} \delta_{\lambda\lambda'} = (k!)^{-2} \sum_{\lambda} 1 = 1/k!\end{aligned}$$

Thus the $x_{n(\cdot)}$ form an orthonormal basis for $\mathbf{K}$.

Observe next that the indicated operation $C(z)$: $u \to (n+1)^{1/2} z \wedge u$ for $u \in \mathbf{K}_n$ is bounded by $\|z\|$. To show this it suffices to consider the case in which z has unit norm; it may then be included in some orthonormal basis for $\mathbf{H}$, which may be taken as the given one $\{e_\mu\}$. But for any basis vector $x_{n(\cdot)}$ of the indicated form,

$$C(e_0) x_{n(\cdot)} = (n+1)^{1/2} (n!)^{1/2} e_0 \wedge (e_1 \wedge \cdots \wedge e_n) = x_{n'(\cdot)}$$

where $n'(\mu) = n(\mu) + \delta_{\mu,0}$; while $C(e_k) x_{n(\cdot)} = 0$ if $1 \le k \le n$.

Since $C(z)$ as given is uniformly bounded from $\mathbf{K}_n$ to $\mathbf{K}_{n+1}$, it has a unique extension to a bounded linear operator from all of $\mathbf{K}$ to itself, also denoted as $C(z)$. Consider now the verification of the complex Clifford relations. In order to show that $C(z)C(z') + C(z')C(z) = 0$, it suffices to check the action on the $x_{n(\cdot)}$, employing an orthonormal basis $\{e_\mu\}$ in $\mathbf{H}$ for which z and z' are linear combinations of e_1 and e_2. To verify that $C(z)C(z')^* + C(z')^*C(z) = \langle z, z' \rangle I$, a similar procedure may be followed (alternatively, direct computation in an arbitrary basis gives the result). The remaining defining properties of the free fermion field follow straightforwardly. □

It follows from Scholium 1.14 that the space **P** of algebraic antisymmetric tensors forms an algebra relative to tensor multiplication followed by antisymmetrization. This algebra, generated by **H**, is isomorphic to the "Grassmann" algebra over **H**, which is generated by **H** and a unit, with the relation $x^2 = 0$ for all x in **H**. An essentially distinctive property of the symmetric and antisymmetric tensors, relative to other symmetry classes of tensors, is that the subsets annihilated by symmetrization or antisymmetrization form *ideals* in the algebra of all tensors (Segal, 1956a).

2.3. The real wave representation

In order to treat the analogue of the wave representation of the free boson field, in which the canonical Q's are simultaneously diagonalized, it is convenient to introduce a formulation of Clifford systems in terms of P's and Q's analogous to that for boson fields.

DEFINITION: Let $\varkappa$ be a conjugation on the complex Hilbert space **H**. A *dual Clifford system over* $(\mathbf{H}, \varkappa)$ (of variance c) is a system $(\mathbf{K}, P, Q)$ consisting of a complex Hilbert space **K**, together with linear mappings P and Q from $\mathbf{H}_\varkappa$ into the bounded selfadjoint operators on **K** with the properties

$$P(x)P(y) + P(y)P(x) = c\langle x, y\rangle = Q(x)Q(y) + Q(y)Q(x);$$
$$P(x)Q(y) + Q(y)P(x) = 0$$

for arbitrary x and y in $\mathbf{H}_\varkappa$.

If $(\mathbf{K}, \phi)$ is a given real Clifford system over **H**, the *dual Clifford system over* $(\mathbf{H}, \varkappa)$ is defined as $(\mathbf{K}, P, Q)$ with $P(x) = \phi(x)$ and $Q(x) = \phi(ix)$ for $x \in \mathbf{H}_\varkappa$. Conversely, if $(\mathbf{K}, P, Q)$ is a given dual Clifford system, the *corresponding real Clifford system* over $(\mathbf{H}, \varkappa)$ is defined as $(\mathbf{K}, \phi)$ with $\phi(z) = P(x) + Q(y)$ if $z = x + iy$ with x and y in $\mathbf{H}_\varkappa$.

Where it is a matter of indifference, or the context makes clear, the type of Clifford system under consideration—real, complex, or dual—the simple term *Clifford system* will be used.

EXAMPLE 2.3. The dual Clifford system associated with the free fermion field transforms less simply under the representation Γ of $U(\mathbf{H})$ than do the corresponding real and complex systems, but enjoys the following properties:

1) If $U \in U(\mathbf{H})$ is such that $\varkappa U \varkappa = U$, then

$$\Gamma(U)P(x)\Gamma(U)^{-1} = P(Ux), \qquad \Gamma(U)Q(x)\Gamma(U)^{-1} = Q(Ux)$$

for all $x \in \mathbf{H}_\kappa$.

2) If $\mathbf{M}$ is any closed (complex-linear) subspace of $\mathbf{H}$, and $i(\mathbf{M})$ is the unitary operator on $\mathbf{H}$ such that $i(\mathbf{M})x = ix$ for all $x \in \mathbf{M}$ while $i(\mathbf{M})x = x$ for all $x \in \mathbf{M}^\perp$, then for $x \in \mathbf{M}$,

$$\Gamma(i(\mathbf{M}))P(x)\Gamma(i(\mathbf{M}))^{-1} = Q(x), \qquad \Gamma(i(\mathbf{M}))Q(x)\Gamma(i(\mathbf{M}))^{-1} = -P(x),$$

while for $x \in \mathbf{M}^\perp$, $\Gamma(i(\mathbf{M}))$ commutes with both $P(x)$ and $Q(x)$.

The wave representation for the fermion field cannot simultaneously diagonalize the Q's, since they do not commute with each other, but is analogous to the wave representation for the boson field in representing them simultaneously as left multiplications acting on an algebra, which has also the structure of a pre-Hilbert space. In the boson case this algebra happens to be commutative: it can be taken to consist of all sufficiently regular elements of $L_2(M)$ for a suitable measure space M. In the fermion case, the relevant algebra is noncommutative, and some preliminaries are required to set it up. Later it will be seen that both the fermion and boson cases can be viewed as representations associated with an operator ring $\mathbf{A}$ in Hilbert space and a particular vector v which may be called a "trace vector" (in physical applications, it is the free vacuum state vector). For any such pair $(\mathbf{A}, v)$, there is a corresponding integration and L_p-theory analogous to the Lebesgue theory, as applied to a probability measure space M in which $\mathbf{A}$ is all multiplications on $L_2(M)$ by bounded measurable functions and v is the function identically 1.

DEFINITION. Let $\mathbf{H}_\kappa$ be a real pre-Hilbert space of infinite or even dimension. The *Clifford algebra over* $\mathbf{H}_\kappa$ (of variance c) is the algebra $\mathbf{C}$ (over the complex field) generated by a unit e together with the elements of $\mathbf{H}_\kappa$, with the relations $xy + yx = c\langle x, y\rangle e$, for arbitrary x and y in $\mathbf{H}_\kappa$. The unique linear functional E on $\mathbf{C}$ such that $E(ab) = E(ba)$ for arbitrary a and b in $\mathbf{C}$, and such that $E(e) = 1$, is called the *trace* on $\mathbf{C}$. The canonical injection of $\mathbf{H}_\kappa$ into $\mathbf{C}$ will be denoted as η.

Before continuing, we explain why the functional E exists and is unique. We use the fact that when $\mathbf{H}_\kappa$ is of finite even dimension n, $\mathbf{C}$ is *-algebraically isomorphic to the algebra of all complex matrices of order $2^{n/2}$, where the * on $\mathbf{C}$ is the unique adjunction operator relative to which every element of $\mathbf{H}_\kappa$ is invariant, and the usual hermitian conjugate on the matrices. Only multiples of the usual matrix trace have the indicated *centrality* property that $E(ab) = E(ba)$ for all a and b, among linear functionals on the matrices, and its multiple by $2^{-n/2}$ is then the present trace. It follows that even when $\mathbf{H}_\kappa$ is infinite-dimensional (the case of primary concern), the trace is unique if it exists at all; for any element lies in the Clifford algebra over some finite and even-dimensional subspace of $\mathbf{H}_\kappa$, and the restrictions to this algebra of any two putative traces defined everywhere on **C** must agree. On the other hand, E may be defined by essentially the same argument. More specifically, if u is a given element of **C**, which is generated as an algebra by $\mathbf{H}_\kappa$, then u is contained in the subalgebra **C**(**M**) generated by some subspace of finite even dimension; and $E(u)$ may be defined as the trace (normalized in the indicated fashion) of the matrix representing u in the isomorphism of **C**(**M**) with a matrix algebra. If **N** is another subspace with the same properties as **M**, the same value of $E(U)$ will be obtained because **N** and **M** may be simultaneously imbedded in a larger subspace **L** of $\mathbf{H}_\kappa$ of finite even dimension. The restriction of the trace on **C**(**L**) to **C**(**M**) has the characteristic properties of the trace on **C**(**M**), hence agrees with it; the same is true with **M** replaced by **N**. Therefore, the traces on **C**(**M**) and **C**(**N**) agree on the common part of these subalgebras of **C**(**L**).

THEOREM 2.3. *Let $\mathbf{H}_\kappa$ be a real part of the given complex Hilbert space* **H** *(assumed of infinite or finite even dimension). Let* (**C**, E) *denote the Clifford algebra* **C** *over $\mathbf{H}_\kappa$ together with its trace E. Let* **K**$'$ *denote the Hilbert space obtained by completion of* **C** *with respect to the inner product $\langle A,B\rangle = E(B^*A)$. For arbitrary $x \in \mathbf{H}_\kappa$, let $P'(x)$ and $Q'(x)$ denote the operators on* **K**$'$ *given by the equations*

$$P'(x)u = i(\Omega u)x; \qquad Q'(x)u = xu$$

where Ω is as earlier. Let v' denote the element e of **K**$'$.

Then (**K**$'$, P', Q') *is a dual Clifford system over* **H** *of unit variance, with respect to $\mathbf{H}_\kappa$, with cyclic vector v'; and* (**K**$'$, C', v'), *where C' is the corresponding complex system, is unitarily equivalent to the subsystem* (**K**, C, v) *of the free fermion field via a unique unitary transformation.*

This unitary transformation from **K** *to* **K**$'$ *may be characterized as that which, for every finite orthonormal subset $e_1,\ldots,e_n$ of $\mathbf{H}_\kappa$, carries $(n!)^{1/2}$ $e_1\wedge\cdots\wedge e_n$ into $(-i2^{1/2})^n\, e_1\cdots e_n$ and v into v'.*

As in the proof for the boson case, an explicit unitary equivalence is set up between a basis for the Hilbert space **K** of the particle representation treated earlier and a basis of the wave representation Hilbert space **K**$'$. This unitary equivalence is in fact independent of the choice of basis.

The essentially new point of the proof is the following remarkable property of the Clifford algebra as a noncommutative probability space; a similar property is well known to be valid for the conventional isonormal probability distribution and in fact characterizes it among commutative probability distributions (cf. Kac, 1939).

LEMMA 2.3.1. *Let $\mathbf{H}_\kappa$ be a real Hilbert space and let $\mathbf{M}_j$ $(j = 1,2,\ldots,n)$ be mutually orthogonal linear subspaces (all of infinite or finite even dimen-*

sion). Let E denote the trace on the Clifford algebra $\mathbf{C}(\mathbf{H}_\kappa)$. *Then, if* $u_j \in \mathbf{C}(\mathbf{M}_j)$, $E(u_1 u_2 \cdots u_n) = E(u_1)E(u_2)\cdots E(u_n)$.

PROOF OF LEMMA 2.3.1. Note first that it suffices to show that for arbitrary orthonormal $e_1, \ldots, e_m$ in $\mathbf{H}_\kappa$, $E(e_1 \cdots e_m) = E(e_1)\cdots E(e_m)$. For if u_j is arbitrary in $\mathbf{C}(\mathbf{M}_j)$, it may be expressed in the form

$$u_j = \sum_k u_{jk},$$

where each summand u_{jk} is of the form $ce_1 \cdots e_m$ for some constant c and orthonormal $e_1, \ldots, e_m$ in $\mathbf{M}_j$. Therefore

$$\begin{aligned} E(u_1 \cdots u_n) &= E(\sum_{k_1, \ldots, k_n} u_{1,k_1} \cdots u_{n,k_n}) \\ &= \sum_{k_1, \ldots, k_n} E(u_{1,k_1} \cdots u_{n,k_n}); \end{aligned}$$

on assuming that $E(e_1 \cdots e_m) = E(e_1)\cdots E(e_m)$, it follows that $E(u_{1,k_1} \cdots u_{n,k_n}) = E(u_{1,k_1})\cdots E(u_{n,k_n})$, hence

$$E(u_1 \cdots u_n) = E(\sum_{k_1} u_{1,k_1})\cdots E(\sum_{k_n} u_{n,k_n}) = E(u_1)E(u_2)\cdots E(u_n)$$

In order to verify that $E(e_1 \cdots e_m) = E(e_1)\cdots E(e_m)$, it is appropriate to make use of an explicit representation for the Clifford algebra. One that is often used is as follows: we take $n \geq m$ with n even, and represent the basis vector e_j on the tensor product of n copies of the two-dimensional vector space $\mathbf{C}^2$ by the matrix

$$e_j' = 2^{-1/2}(1' \otimes 1' \otimes \cdots \otimes 1' \otimes A \otimes 1 \otimes \cdots \otimes 1)$$

where

$$1' = \begin{bmatrix} 1 & 0 \\ 0 & -1 \end{bmatrix} \quad A = \begin{bmatrix} 0 & 1 \\ 1 & 0 \end{bmatrix} \quad 1 = \begin{bmatrix} 1 & 0 \\ 0 & 1 \end{bmatrix}$$

and the jth factor is A. Then

$$e_j'e_k' + e_k'e_j' = \delta_{jk} \qquad (j, k = 1, \ldots, n),$$

and mapping linear combinations of the e_j into corresponding combinations of the e_j', a representation of the Clifford algebra over n-space into the complete matrix algebra in a space of dimension 2^n is obtained. The restriction of the normalized trace on the latter algebra to the subalgebra representing the Clifford algebra must agree with the functional on this subalgebra corresponding to the trace on the Clifford algebra, by the unicity of the latter. Therefore, the trace of any product of the e_j, $1 \leq j \leq m$, is the same as the normalized trace of the corresponding product of the e'_j. Observing that A has vanishing trace, it follows that this trace vanishes in all cases ($m \geq 1$).

PROOF OF THEOREM. It is straightforward to verify that $(\mathbf{K}', P', Q')$ is indeed a dual Clifford system over $\mathbf{H}$ with respect to $\mathbf{H}_\times$ (cf. Example 2.2). To show its unitary equivalence with the system described in the preceding theorem, we define explicitly a unitary transformation D from $\mathbf{K}$ to $\mathbf{K}'$ which implements this equivalence. Let $\{e_\mu\}$ be any well ordered orthonormal basis for $\mathbf{H}_\times$, let $\{x_{n(\cdot)}\}$ be the corresponding orthonormal basis for $\mathbf{K}$ (with $x_{n(\cdot)} = v$ if $n(\mu) = 0$ for all μ), and let $y_{n(\cdot)}$ denote the element of $\mathbf{C}$ given by the equation

$$y_{n(\cdot)} = (-i2^{1/2})^k e_{\mu(1)} \cdots e_{\mu(k)}$$

where $\mu(1) < \cdots < \mu(k)$ and these are the indices μ for which $n(\mu) = 1$ (with $y_{n(\cdot)} = v'$ if $n(\mu) = 0$ for all μ). Then the $y_{n(\cdot)}$ form an orthonormal basis for $\mathbf{K}'$, and there exists a unique unitary transformation D from $\mathbf{K}$ to $\mathbf{K}'$ which carries $x_{n(\cdot)}$ into $y_{n(\cdot)}$.

Note next that if $(\mathbf{K}', C')$ is the complex Clifford system corresponding to the dual system $(\mathbf{K}', P', Q')$, then

$$DC(z)x_{n(\cdot)} = C'(z)y_{n(\cdot)} \tag{2.3}$$

for all $z \in \mathbf{H}$ and basis vectors $x_{n(\cdot)}$. Because of the complex-linearity of $C(\cdot)$ and $C'(\cdot)$, it suffices to establish equation 2.3 in the case in which z is real. Moreover, since $\|C(z)\| \leq c\,\|z\|$, it suffices to check equation 2.3 on a dense subset, and ultimately for a basis of $\mathbf{H}_\times$, say the $\{e_\mu\}$. We may evidently suppose that the notation is such that the e_μ involved in $x_{n(\cdot)}$ are $e_1, \ldots, e_k$, and that z is either an element e_0 of the basis which is orthogonal to the $e_1, \ldots, e_k$, or that it is one of them. In the latter event, both sides vanish, noting that $C'(e_j)y_{n(\cdot)} = 0$ if $1 \leq j \leq n$. In the former event, the left side is

$$D(k+1)^{1/2}(k!)^{1/2} A(e_0 \otimes A(e_1 \otimes \cdots \otimes e_k)) = Dx_{n'(\cdot)} = y_{n'(\cdot)}$$

where $n'(\mu) = n(\mu) + \delta_{\mu,0}$; and the right side is

$$-i2^{-1/2}((\Omega y_{n(\cdot)})e_0 + e_0 y_{n(\cdot)}) = y_{n'(\cdot)};$$

thus the required agreement holds.

Since $Dv = v'$, it follows that $(\mathbf{K}, C, v)$ is unitarily equivalent to $(\mathbf{K}', C', v')$ via the unitary transformation D. If D were not unique, say if D_j $(j = 1, 2)$ had the same property, then $D_1^{-1} D_2$ would be a unitary opertor on $\mathbf{K}$ which commutes with all $C(z)$ and leaves v invariant. Since v is a cyclic vector for the $C(z)$, such a unitary operator must be the identity. It follows also that D not only carries $x_{n(\cdot)}$ into $y_{n(\cdot)}$ for one particular orthonormal basis of $\mathbf{H}_\times$, but does so for all such bases; for any such D must implement the unitary equivalence between $(\mathbf{K}, C, v)$ and $(\mathbf{K}', C', v')$, and this unitary equivalence is unique. □

COROLLARY 2.3.1. *There exists a unique representation* Γ' *of* $U(\mathbf{H})$ *on* $\mathbf{K}'$ *with the properties*

$$\Gamma'(U)C'(z)\,\Gamma'(U)^{-1} = C'(Uz), \qquad \Gamma'(U)v' = v'$$

for all $z \in \mathbf{H}$.

Moreover, $\partial\Gamma'(A) \geq 0$ *for all nonnegative selfadjoint operators* A *on* $\mathbf{H}$.

PROOF. Taking $\Gamma'(U) = D^{-1}\Gamma(U)D$ shows the existence of Γ'. The proof of unicity is similar to that given for D. Recalling that $\partial\Gamma'(A)$ is defined as the selfadjoint generator of the one-parameter subgroup $\{\Gamma'(e^{itA}): t \in \mathbf{R}\}$, it is evident in the particle representation (with the use of spectral theory) that $\partial\Gamma(A) \geq 0$ for all $A \geq 0$, and hence the same is true of $\partial\Gamma'(A) \geq 0$. □

COROLLARY 2.3.2. Γ' *is the unique continuous representation of* $U(\mathbf{H})$ *on* $\mathbf{K}'$ *enjoying the transformation properties 1 and 2 given in Example 2.3.*

PROOF. It suffices to show that the transformations considered in Example 2.3 generate a subgroup $\mathbf{U}'$ that is strongly dense in $U(\mathbf{H})$. Since those unitaries on $\mathbf{H}$ that are the identity on a cofinite-dimensional subspace form a strongly dense subset of $U(\mathbf{H})$, it suffices to show that any such unitary is in $\mathbf{U}'$. Actually, it suffices to show that every unitary that is the identity on the orthocomplement of a two-dimensional subspace of $\mathbf{H}$ is in $\mathbf{U}'$, for such unitaries generate $U(n)$ for every n, as follows from the diagonalizability of unitary transformations. Thus, it suffices to show that $U(2)$ is generated by $O(2)$ together with the element

$$\begin{bmatrix} i & 0 \\ 0 & 1 \end{bmatrix}$$

in $U(2)$. On conjugating the generator

$$\begin{bmatrix} 0 & 1 \\ -1 & 0 \end{bmatrix}$$

of the $O(2)$ subgroup of $U(2)$ by the cited element, forming commutators, and taking the linear span, a four-dimensional space, and hence the totality of the Lie algebra of $U(2)$, is obtained. □

COROLLARY 2.3.3. *If* $(\mathbf{K}, P, Q, \Gamma, v)$ *is the free fermion field over the complex Hilbert space* $\mathbf{H}$ *(in terms of a dual Clifford system, with respect to the real part* $\mathbf{H}_\kappa$*), and if* $\mathbf{R}_Q$ *denotes the W*-algebra generated by the* $Q(x)$, $x \in \mathbf{H}_\kappa$, *then*

1) v is a separating, cyclic, and trace vector for $\mathbf{R}_Q$*;*
2) the center of $\mathbf{R}_Q$ *consists only of scalars; and*
3) $\mathbf{R}_Q$ *is an approximately finite* II_1 *factor in the terminology of Murray and von Neumann.*

PROOF. The Clifford algebra $\mathbf{C}$ over $\mathbf{H}_\kappa$ together with the inner product given on it have the properties

$$\langle ab, c\rangle = \langle b, a^*c\rangle = \langle a, cb^*\rangle; \qquad \langle ab, ab\rangle \le c_a\langle b, b\rangle \qquad (c_a = \text{const.})$$

The first property is immediate from the definition; to derive the second, note that $\langle ab, ab\rangle = E(b^*a^*ab)$—but E is a positive linear functional, and $a^*a \le c_a$ for a suitable constant c_a (in the usual ordering on hermitian operators), so $E(b^*a^*ab) \le c_a\langle b, b\rangle$. Moreover, $\mathbf{C}$ has a unit e. This means that the system $(\mathbf{C}, \langle\cdot,\cdot\rangle)$ forms a "Hilbert algebra" (cf. SK). While a proof could be given that is largely independent of the general theory of such algebras—the present algebra is one of the simplest—in part it would reproduce an argument used in the general theory, and we shall, instead, simply appeal to one of the basic results about Hilbert algebras: if $\mathbf{K}'$ is the Hilbert space formed by completion of $\mathbf{C}$ with respect to the inner product $\langle\cdot,\cdot\rangle$; if for any $a \in \mathbf{C}$, L_a and R_a denote the operations of left and right multiplication by a, acting on $\mathbf{C}$; and if $\mathbf{L}$ and $\mathbf{R}$ denote the W^*-algebras generated by the bounded extensions $\overline{L}_a$ and $\overline{R}_a$ of L_a and R_a to all of $\mathbf{K}'$; then the commutor $\mathbf{L}'$ of $\mathbf{L}$ (i.e., the set of all bounded operators T such that $TL = LT$ for all $L \in \mathbf{L}$) is $\mathbf{R}$, vice versa.

In the context of Corollary 2.3.3, $\mathbf{L}$ is $\mathbf{R}_Q$ and $\mathbf{R}$ is the W^*-algebra generated by the $\overline{R}_a$, $a \in \mathbf{H}_\kappa$. Consider now assertion 1. That v is cyclic for $\mathbf{R}_Q$ is immediate since $\mathbf{R}_Q v$ certainly contains $\mathbf{C}$, which is dense. It is also cyclic for the W^*-algebra generated by the $\overline{R}_a$, $a \in \mathbf{H}_\kappa$, by the same token; hence it is separating for the commutor of this algebra, i.e., for $\mathbf{R}_Q$. Now if A and B are any operators in the algebra $\mathbf{L}_\mathbf{C}$ generated (purely algebraically) by the $\overline{L}_a$, $a \in \mathbf{C}$, then it is an entirely algebraic fact about the trace that $\langle ABv, v\rangle = \langle BAv, v\rangle$. On the other hand, every operator M in $\mathbf{R}_Q$ is a strong limit of a uniformly bounded net of operators in $\mathbf{L}_\mathbf{C}$, by virtue of the fact that if a *-algebra is strongly dense in a W^*-algebra, then its unit ball is strongly dense in the unit ball of the W^*-algebra (cf. Dixmier, 1981). It follows by a simple approximation that $\langle MNv, v\rangle = \langle NMv, v\rangle$ for all operators M and N in $\mathbf{R}_Q$—i.e., v is a trace vector.

Now let A be an element of the center of $\mathbf{R}_Q$, i.e., $AX = XA$ for all $X \in \mathbf{R}_Q$. For any linear subspace $\mathbf{M}$ of $\mathbf{H}_\kappa$ of finite even dimension, let $\mathbf{R}^\mathbf{M}$ denote the W^*-algebra generated by the $Q(x)$ for $x \in \mathbf{M}$; as an algebra, $\mathbf{R}^\mathbf{M}$ is finite-dimensional, being isomorphic to the Clifford algebra over $\mathbf{M}$. Denoting as $E(T)$ the extension $\langle Tv, v\rangle$ of the trace to all of $\mathbf{R}_Q$, the functional $F(X) = E(AX)$ defined for $X \in \mathbf{R}_Q$ is linear and central, i.e., $F(XY) = F(YX)$ for arbitrary $Y \in$

$\mathbf{R}_Q$. The restriction $F_\mathbf{M}$ of F to $\mathbf{R}^\mathbf{M}$ has consequently the same property, and must therefore conincide with a scalar multiple of the trace on $\mathbf{R}^\mathbf{M}$. Thus, there exists a scalar operator $A_\mathbf{M}$ such that $E(AX) = E(A_\mathbf{M}X)$ for all $X \in \mathbf{R}^\mathbf{M}$.

It will follow that A is a scalar operator if it can be shown that $A_\mathbf{M} \to A$ strongly as $\mathbf{M} \to \mathbf{H}_\kappa$. To this end, note first that $\|A_\mathbf{M}\| \leq \|A\|$. Again, this could be proved directly in the present case but we shall rather appeal to a simple general result, to the effect that if $\mathbf{S}$ is a W^*-subalgebra of the W^*-algebra $\mathbf{R}$, if E is a trace (central state) defined on $\mathbf{R}$, and if A is any element of $\mathbf{R}$, then the operator A' in $\mathbf{S}$ such that $E(XA) = E(XA')$ for all $X \in \mathbf{S}$ (the conditional expectation of A with respect to $\mathbf{S}$) has the property that $\|A'\| \leq \|A\|$. Second, if $u \in \mathbf{C}$, then $A_\mathbf{M}u \to Au$ as $\mathbf{M} \to \mathbf{H}_\kappa$, indeed $A_\mathbf{M}u = Au$ for sufficiently large $\mathbf{M}$ (depending on u), by a simple argument (or familiar property of conditional expectation). It follows by approximation that $A_\mathbf{M}u \to Au$ for all $u \in \mathrm{K}'$.

That $\mathbf{R}_Q$ is an approximately finite II_1 factor simply summarizes the facts that it has a trace, is a "factor" (i.e., has trivial center), and is generated by the finite-dimensional subalgebras $\mathbf{R}^\mathbf{M}$. □

2.4. The complex wave representation

The development of an analog to the boson field complex wave representation for the fermion case depends naturally on the interplay between the real and complex structures in the underlying Hilbert spaces. It is necessary to make this more explicit.

In the following there will be no occasion to consider the Clifford algebra over $\mathbf{H}$ as a complex space with a complex quadratic form. (There is in fact no unitarily invariant such form, just as there is no unitarily invariant complex conjugation.) All Clifford algebras considered here will be over real spaces with real forms and complex coefficients in the algebra itself. (They are thus somewhat analogous to the algebra of complex polynomials over a real linear space.) In the case of a given complex space $\mathbf{H}$, the following are unitarily-invariant notions.

DEFINITION. Let $\mathbf{H}$ be a given complex Hilbert space. The *Clifford algebra over* $\mathbf{H}$, denoted $\mathbf{C}(\mathbf{H})$, is defined as the Clifford algebra $\mathbf{C}(\mathbf{H}^\#)$ over the real Hilbert space $\mathbf{H}^\#$. The canonical injection of $\mathbf{H}$ (or $\mathbf{H}^\#$) into $\mathbf{C}(\mathbf{H})$ will be denoted as η (note that η is *real-linear* but not *complex*-linear.) An element u of $\mathbf{C}(\mathbf{H})$ (resp. its Hilbert space completion, denoted $L_2(\mathbf{C}(\mathbf{H}))$ is called *holomorphic/antiholomorphic* if it is in the subalgebra generated by the elements of $\mathbf{C}(\mathbf{H})$ of the form $\eta(x) + i\eta(ix)$ / $\eta(x) - i\eta(ix)$, $x \in \mathbf{H}$ (resp. its Hilbert space completion).

The question may arise of whether a holomorphic/antiholomorphic element of $L_2(\mathbf{C}(\mathbf{H}))$ that lies in $\mathbf{C}(\mathbf{H})$ is also holomorphic/antiholomorphic as an element of the latter. That this is the case will be a consequence of Theorem 2.4. This involves the following analog to differentiation.

DEFINITION. A *pseudo-derivation* on a Clifford algebra $\mathbf{C}$ over a space $\mathbf{L}$ is defined as a linear map ∂ such that $\partial(uv) = (\partial u)v + (\Omega u)\partial v$ for all $u, v \in \mathbf{C}$.

THEOREM 2.4. *Let* $\mathbf{H}$ *be a given complex Hilbert space, and* $\mathbf{C}(\mathbf{H})$ *the Clifford algebra over* $\mathbf{H}$. *Let* $\tilde{\mathbf{H}}$ *denote the subspace* $\eta(\mathbf{H}) + i\eta(\mathbf{H})$ *of* $\mathbf{C}(\mathbf{H})$, *consisting of all elements u of* $\mathbf{C}(\mathbf{H})$ *of the form* $\eta(x) + i\eta(x')$, *with* $x, x' \in \mathbf{H}$. *Let* $w \mapsto \bar{\psi}$ *denote the operation of complex conjugation on* $\mathbf{C}(\mathbf{H})$, *i.e., the unique antilinear automorphism of* $\mathbf{C}(\mathbf{H})$ *as a ring that extends the mapping* $\eta(x) + i\eta(x') \mapsto \eta(x) - i\eta(x')$, *with x and x' in* $\mathbf{H}$. *Then*

i) for every nonzero $z \in \tilde{\mathbf{H}}$ *of the form* $z = \eta(x) \pm i\eta(ix)$, $x \in \mathbf{H}$, *there is a unique pseudo-derivation* ∂ *on* $\mathbf{C}(\mathbf{H})$ *such that* $\partial z = e$ *and* $\partial y = 0$ *for all y in* $\tilde{\mathbf{H}}$ *that are orthogonal to z; and*

ii) ∂ *admits a bounded linear extension* ∂_z *to all of* $\mathrm{L}_2(\mathbf{C}(\mathbf{H}))$, *and an element u of this space is holomorphic if and only if* $\partial_{\bar{z}} u = 0$ *for all* $z \in \tilde{\mathbf{H}}$ *of the form* $z = \eta(x) + i\eta(ix)$, $x \in \mathbf{H}$ *(resp. antiholomorphic if* $\partial_z u = 0$ *for all such z).*

LEMMA 2.4.1. *Let* $\mathbf{H}'$ *be a given real Hilbert space, and let* $\mathbf{H}$ *denote its complexification* $\mathbf{H}' + i\mathbf{H}'$. *The mapping* $x + iy \mapsto 2^{1/2}(\eta(x) + i\eta(y))$ *from* $\mathbf{H}$ *into the Clifford algebra* $\mathbf{C}(\mathbf{H}')$ *is an isometry.*

PROOF. It is immediate that the mapping is linear and one-to-one, so what is to be proved is that $\langle x + iy, x' + iy'\rangle = 2\langle\eta(x) + i\eta(y), \eta(x') + i\eta(y')\rangle$ for arbitrary x, y, x', y' in $\mathbf{H}'$. In fact, the right side of this putative equality is

$$\begin{aligned}&2\,tr\,\{(\eta(x) + i\eta(y))\,(\eta(x') - i\eta(y'))\}\\ &= 2\,tr\,\{\eta(x)\eta(x') + \eta(y)\eta(y') + i\eta(y)\eta(x') - i\eta(x)\eta(y')\}\\ &= \langle x, x'\rangle + \langle y, y'\rangle + i\langle y, x'\rangle - i\langle x, y'\rangle = \langle x + iy, x' + iy'\rangle.\end{aligned}$$ □

LEMMA 2.4.2. *Let* $\mathbf{H}'$ *be real Hilbert space, and let* $\mathbf{M}$ *denote any complex-linear subspace of* $\eta(\mathbf{H}') + i\eta(\mathbf{H}')$. *Then the subalgebra of* $\mathbf{C}(\mathbf{H}')$ *generated by* $\mathbf{M}$ *and the identity is *-isomorphic (automatically in a trace-preserving fashion) to the abstract Clifford algebra* $\mathbf{C}(\mathbf{M})$ *over* $\mathbf{M}$ *(relative to the restriction of the inner product* $E(A^*B)$ *on* $\mathbf{C}(\mathbf{H}')$ *to an inner product on* $\mathbf{M}$*).*

PROOF. Since $\mathbf{M}$ is either even- or infinite-dimensional, the Clifford algebra over $\mathbf{M}$ relative to the given inner product is simple; the nonzero *-homomorphism from **C(M)** to **C(H′)** given above is thus an isomorphism. Since the trace is unique, the isomorphism is necessarily trace-preserving. □

LEMMA 2.4.3. *Let* **H′** *be a given real Hilbert space, and* **H** *its complexification, identified with* $\eta(\mathbf{H}') + i\eta(\mathbf{H}')$ *as in Lemma 2.4.1. If* $\{e_\mu\}$ *is an orthonormal basis for* **H′**, *then the products* $f_{\mu(1)}f_{\mu(2)}\cdots\bar{f}_{\nu(1)}\bar{f}_{\nu(2)}\cdots$, *where* $f_\mu = \eta(e_\mu) + i\eta(ie_\mu)$, *form an orthonormal basis for* **C(H)**.

PROOF. The f_μ and $\bar{f}_\mu$ form an orthonormal basis for $\mathbf{H}^\#$, by Lemma 2.4.1 and a simple computation. Applying Lemma 2.4.2 and the standard form for the basis of a Clifford algebra over a real Hilbert space, relative to a given orthonormal basis in the latter, the lemma follows.

PROOF OF THEOREM. Note that for arbitrary nonzero x in $\tilde{\mathbf{H}} = \eta(\mathbf{H}) + i\eta(\mathbf{H})$, and a and b in the subalgebra $R(x^\perp)$ generated by the orthocomplement of x in **H**, the a and xb are orthogonal, as a consequence of Lemma 2.4.3. Using Lemma 2.4.3 once again, it follows that every element $u \in \mathbf{C(H)}$ may be expressed uniquely in this form: $u = a + xb$.

Now defining $\partial u = b$, then if $u' = a' + xb'$,

$$uu' = (aa' + xbxb') + (axb' + xba');$$

the term $aa' + xbxb'$ is in $\mathbf{R}(x^\perp)$, while $axb' = x(\Omega a)b'$. Thus

$$\begin{aligned}
&\partial(uu') = ba' + (\Omega a)b', \\
&(\partial u)u' + (\Omega u)\partial u' = b(a' + xb') + \Omega(a + xb)b';
\end{aligned}$$

so ∂ is a pseudo-derivation. The remainder of assertion i) follows from the observation that the difference of two pseudo-derivations with the given property is a pseudo-derivation which vanishes both on x and $\mathbf{R}(x^\perp)$, and hence vanishes identically.

To see that ∂ is bounded in the $L_2(\mathbf{C(H)})$ norm, note that with u as earlier,

$$\|u\|^2 = \|a\|^2 + \|xb\|^2 \geq \|xb\|^2 = \|x\|^2\|b\|^2$$

To show that if u is holomorphic, then $\partial_{\bar{z}}u = 0$ for all $z = \eta(x) + i\eta(ix)$, $x \in \mathbf{H}$, it suffices to treat the case in which $u \in \tilde{\mathbf{H}}$, since the general case follows by induction, using the fact that $\partial_{\bar{z}}$ is a bounded pseudo-derivation. Now if $u \in \tilde{\mathbf{H}}$, and u is holomorphic, then $u = \eta(y) + i\eta(iy)$ for some $y \in \mathbf{H}$. To show that $\partial_{\bar{z}}u = 0$, it suffices to show that $\langle u, \bar{z}\rangle_{\tilde{\mathbf{H}}} = 0$, as follows:

$$\begin{aligned}
&2\langle \eta(y) + i\eta(iy), \eta(x) - i\eta(ix)\rangle \\
&= 2\langle \eta(y), \eta(x)\rangle - 2\langle \eta(y), i\eta(ix)\rangle + 2\langle i\eta(iy), \eta(x)\rangle - 2\langle i\eta(iy), i\eta(ix)\rangle \\
&= \mathrm{Re}\langle y, x\rangle + i\mathrm{Re}\langle y, ix\rangle + i\mathrm{Re}\langle iy, x\rangle - \mathrm{Re}\langle ix, iy\rangle = 0
\end{aligned}$$

Suppose conversely that $u \in L_2(\mathbf{C}(\mathbf{H}))$ and that $\partial_{\bar z} u = 0$ for all $z = \eta(x) + i\eta(ix)$, $x \in \mathbf{H}$. Let $\mathbf{M}$ denote any finite-dimensional complex-linear subspace of $\tilde{\mathbf{H}}$ that contains x and ix. Then $\partial_{\bar z}$ commutes with the operation P of projection of $L_2(\mathbf{C}(\mathbf{H}))$ onto $\mathbf{C}(\mathbf{M})$, as a subspace of $\mathbf{C}(\mathbf{H})$ (i.e., "conditional expectation with respect to $\mathbf{C}(\mathbf{M})$"). For if $u = a + \bar{z}b$ as earlier, with $z \in \mathbf{M}$, then $Pu = Pa + \bar{z}Pb$, by the property of conditional expectation with respect to a subalgebra $\mathbf{B}$ that if $q \in \mathbf{B}$ and if r is arbitrary in $L_2(\mathbf{C}(\mathbf{H}))$, then $P(qr) = qP(r)$. (This property follows from the characterization of Pr as the unique element s in the closure of $\mathbf{B}$ such that $E(rt) = E(st)$ for all $t \in \mathbf{B}$.)

To conclude the proof for the holomorphic case, it suffices now to assume that $\mathbf{H}$ is finite-dimensional. For, from the commutativity of P with $\partial_{\bar z}$, it follows that if $\partial_{\bar z} u = 0$ for all z as assumed, then $\partial_{\bar z}(Pu) = 0$ for all z. Assuming the finite-dimensional case, it then follows that Pu is holomorphic as an element of $\mathbf{C}(\mathbf{H})$, but $Pu \to u$ as $\mathbf{M} \to \mathbf{H}$.

Consider now the finite-dimensional case. Let $\{e_\mu\}$ be an orthonormal basis for $\mathbf{H}$, and set $f_\mu = \eta(e_\mu) + i\eta(ie_\mu)$. Then, by Lemma 2.4.3, every element $u \in \mathbf{C}(\mathbf{H})$ is a linear combination of the $f_{\mu(1)}f_{\mu(2)} \cdots \bar{f}_{\nu(1)}\bar{f}_{\nu(2)} \cdots$. If $\partial_{\bar f_{\mu(i)}} u = 0$, then those members of the cited orthonormal basis of $\mathbf{C}(\mathbf{H})$ in which $\bar{f}_{\mu(i)}$ occurs have vanishing coefficient in the representation of u in terms of this basis. It follows that if $\partial_{\bar f_\mu} u = 0$ for all μ, then none of the basis vectors involving one or more of the $\bar{f}_\mu$ occurs, which means that u is holomorphic according to the earlier definition.

The antiholomorphic case now follows by entirely parallel arguments. □

We are now in a position to show how the free fermion field over a given complex Hilbert space $\mathbf{H}$ can be simply represented in terms of the antiholomorphic spinors, as the space $\mathbf{K}'$ of antiholomorphic elements of $L_2(\mathbf{C}(\mathbf{H}))$ may be termed, since the so-called spin representation of the Lie algebra of the orthogonal group $O(\mathbf{H}^\#)$ takes place in a Hilbert space which may be naturally represented as $\mathbf{K}'$.

THEOREM 2.5. *Let* $\mathbf{H}$ *be a given Hilbert space, let* $\mathbf{K}'$ *denote the space of antiholomorphic spinors over* $\mathbf{H}$, *and for* $x \in \mathbf{H}$, *let* $\phi(x)$ *denote the operator on* $L_2(\mathbf{C}(\mathbf{H}))$ *given by* $\phi(x) = 2^{-1/2}(L_x + iR_{ix}\Omega)$. *For any unitary* U *on* $\mathbf{H}$, *let* $\Gamma_0(U)$ *denote the unitary operator on* $L_2(\mathbf{C}(\mathbf{H}))$ *that extends the automorphism of* $\mathbf{C}(\mathbf{H})$ *that in turn extends the operator* U *on* $\mathbf{H}$. *Let* v' *denote the identity* e_0 *in* $L_2(\mathbf{C}(\mathbf{H}))$.

Then $\mathbf{K}'$ *is invariant under the* $\phi(x)$ *and* $\Gamma_0(U)$, *and if* $\phi'(x)$ *and* $\Gamma'(U)$ *denote the restrictions of these operators to* $\mathbf{K}'$, *the quadruple* $(\mathbf{K}', \phi', \Gamma, v')$ *is unitarily equivalent to the free fermion field over* $\mathbf{H}$.

PROOF. It is easy to check that $\phi(x)$ is selfadjoint, and that the Clifford relations

$$\phi(x)\phi(y) + \phi(y)\phi(x) = \mathrm{Re}\langle x, y\rangle \qquad (x, y \in \mathbf{H})$$

are satisfied. Now if $z \in \tilde{\mathbf{H}}$ has the form $z = \eta(x) - i\eta(ix)$, $x \in \mathbf{H}$, then denoting as ϕ^e the extension of $\phi(\cdot)$ to a homomorphism from $C(\mathbf{H})$ into the bounded linear operators on $L_2(\mathbf{C}(\mathbf{H}))$, it is readily checked that the following relations hold:

$$\phi^e(z) = 2^{-1/2}(L_z - R_z\Omega); \qquad \phi^e(\bar{z}) = 2^{-1/2}(L_{\bar{z}} + R_{\bar{z}}\Omega).$$

Supposing that $u \in \mathbf{K}'$, then the following relations hold:

$$\phi^e(z)\, u = 2^{1/2}L_z u, \qquad \phi^e(\bar{z})u = 2^{1/2}\partial_z u.$$

To check these relations, it suffices to consider the case in which u is a finite product of elements of the form $\eta(e_\mu) - i\eta(ie_\mu)$, where the $\{e_\mu\}$ form an orthonormal basis for $\mathbf{H}$, and in which x is itself a member of this basis. The relations then follow straightforwardly. Hence $\mathbf{K}'$ is invariant under the $\phi^e(z)$ and $\phi^e(\bar{z})$, which implies in turn that it is invariant under the $\phi(x)$.

Since $\Gamma'(U)$ carries the product

$$(\eta(e_{\mu(1)}) - i\eta(ie_{\mu(1)}))\,(\eta(e_{\mu(2)}) - i\eta(ie_{\mu(2)}))\cdots$$

into the product

$$(\eta(Ue_{\mu(1)}) - i\eta(iUe_{\mu(1)}))\,(\eta(Ue_{\mu(2)}) - i\eta(iUe_{\mu(2)}))\cdots,$$

it leaves $\mathbf{K}'$ invariant. It is immediate that $\Gamma'(U)$ satisfies the relations

$$\Gamma'(U)\phi'(x)\Gamma'(U)^{-1} = \phi'(Ux); \qquad \Gamma'(U)v = v \text{ for all } U.$$

To complete the proof, it suffices now to show that $\Gamma'(U(t))$ has a nonnegative generator for some strictly positive energy one-parameter unitary group on $\mathbf{H}$, and that v' is cyclic in $\mathbf{K}'$ for the $\phi(x)$. The latter is immediate from the relation $\phi^e(z)u = 2^{1/2}L_z u$. To verify the former, let $U(t) = e^{it}$; then $(\eta(U(t)x) - i\eta(iU(t)x) = e^{it}(\eta(x) - i\eta(ix))$, as is easily verified, from which it follows that $\mathbf{K}'$ is spanned by eigenvectors of $\Gamma'(U(t))$ on which the generator has nonnegative eigenvalues and so is itself nonnegative. □

COROLLARY 2.5.1. *The (unique) unitary equivalence T between the complex wave and particle representations of the free fermion field over a given Hilbert space $\mathbf{H}$ carries $\bar{f}_1\bar{f}_2\cdots\bar{f}_k$, where $\bar{f}_\mu = \eta(e_\mu) - i\eta(ie_\mu)$, $\{e_\mu\}$ being any orthonormal basis of $\mathbf{H}$, into $(n!)^{1/2}A(e_1\otimes e_2\otimes\cdots\otimes e_k)$, A being the antisymmetrization projection.*

PROOF. Using the stochastic independence of the e_i and ie_i in the Clifford algebra, it follows that the $\bar{f}_1\bar{f}_2\cdots\bar{f}_k$ form an orthonormal basis in $\mathbf{K}'$, as the e_μ vary over the given basis of $\mathbf{H}$. It has already been determined that the putative corresponding vectors in the particle representation space, say $\mathbf{K}_p$, form an orthonormal basis there. Consequently there exists a unique unitary operator carrying the one basis in $\mathbf{K}'$ into that in $\mathbf{K}_p$. Denoting this operator as T, it is immediate from its definition that it makes the respective vacuum state representatives correspond. Since $\Gamma(U)$ is uniquely determined, for any unitary U on $\mathbf{H}$, by the property that $\Gamma(U)v = v$ and that $\Gamma(U)C(z)\Gamma(U)^{-1} = C(Uz)$ for all z in some spanning subset of $\mathbf{H}$, it suffices, in order to conclude the proof, to show that

$$T^{-1}C(y)\bar{f}_1\bar{f}_2\cdots\bar{f}_k = C_p(y)T^{-1}\bar{f}_1\bar{f}_2\cdots\bar{f}_k,$$

for all y that are one of the e_j. If $1 \le j \le k$, then both sides vanish, so it may be supposed that y is another basis vector, say e_0. Then

$$C(e_0)\bar{f}_1\bar{f}_2\cdots\bar{f}_k = 2^{-1/2}(\phi(e_0) - i\phi(ie_0))\bar{f}_1\bar{f}_2\cdots\bar{f}_k = 2^{-1/2}\phi^c(z)\bar{f}_1\bar{f}_2\cdots\bar{f}_k$$

where $z = (\eta(e_0) - i\eta(ie_0))$. Now applying an equation obtained in the proof of Theorem 2.5, it results that

$$C(e_0)\bar{f}_1\bar{f}_2\cdots\bar{f}_k = \bar{f}_0\bar{f}_1\bar{f}_2\cdots\bar{f}_k,$$

which is carried by T^{-1} into $((n+1)!)^{1/2}A(e_0\otimes e_1\otimes\cdots\otimes e_k)$. On the other hand,

$$\begin{aligned}C_p(e_0)T^{-1}\bar{f}_1\bar{f}_2\cdots\bar{f}_k &= C_p(e_0)(n!)^{1/2}A(e_1\otimes e_2\otimes\cdots\otimes e_k)\\ &= (n+1)^{1/2}(n!)^{1/2}A(e_0\otimes e_1\otimes\cdots\otimes e_k),\end{aligned}$$

since $C_p(e_0)$ consists of antisymmetrized tensor multiplication by e_0 followed by multiplication by $(n+1)^{1/2}$. The observation that the operaton of antisymmetrized tensor multiplication commutes with antisymmetrization with respect to a subset of the variables involved completes the proof. □

The use of the unicity theorem in the proof of Theorem 2.5 is avoidable, being readily replaced by considerations in the proof of Corollary 2.5.1, which independently establish the equivalence of the structure given in the theorem with the particle representation of the free fermion field.

The explicit form of the intertwining operator between the complex and real wave representations for the free fermion field is obtained straightforwardly by combining the equivalences of the wave representations with the particle representation which have been obtained. Specifically, with the same notation as earlier, the $(-i2^{-1/2})^n e_1e_2\cdots e_n$ in the real wave representation corresponds to $\bar{f}_1\bar{f}_2\cdots\bar{f}_n$ in the complex wave representation, relative to an arbitrary orthonormal basis in $\mathbf{H}$.

Problems

1. Let $\{e_n\}$ be an orthonormal basis for the Hilbert space $\mathbf{H}$, and let $C(z)$ denote the creation operator for the vector $z \in \mathbf{H}$ in the free fermion field. Let

$a_n = C(e_n)^*$, i.e., the annihilation operator for the nth "mode." Show that the number operator $N = \partial\Gamma(I)$ equals $\Sigma_n a_n a_n^*$, where the sum is convergent in the strong topology.

2. Let $(\mathbf{K}, \phi, \Gamma, \nu)$ denote the free fermion field over the Hilbert space $\mathbf{H}$. Show that the representation Γ_n of the unitary group on $\mathbf{H}$ given by $U \rightarrow \Gamma(U)|\mathbf{K}_n$ is irreducible. (Cf. Segal, 1956a.)

3. With the notation of Problem 2, let $\varkappa$ be a conjugation on $\mathbf{H}$, and let $U_\varkappa(\mathbf{H})$ be the subgroup of $U(\mathbf{H})$ consisting of transformations that commute with $\varkappa$ (i.e., the orthogonal group on $\mathbf{H}_\varkappa$). Show that the restriction of Γ_n to $U_\varkappa(\mathbf{H})$ is irreducible if and only if $\mathbf{H}$ is infinite-dimensional. (As in the boson case, this result is due independently to K. Okamoto et al. and J. Pedersen.)

4. Let $\mathbf{S}$ be the space of all tensors in the algebraic direct sum $\mathbf{T} = \oplus_n \mathbf{H}^n$ that are annihilated by antisymmetrization. Show that $\mathbf{S}$ is an ideal in the tensor algebra $\mathbf{T}$.

Bibliographical Notes on Chapter 2

The particle representation for the universal fermion field (in its invariant Hilbert space form) is due to Cook (1953). See Segal (1956b) for the real wave representation and its unitary equivalence to the particle representation. The complex wave representation was developed by Shale and Stinespring (1964).

3

Properties of the Free Fields

3.1. Introduction

The free fields appear as a natural mathematical extension of classical lines of investigation. Notable among these are those of Schur, Weyl, and Brauer on the decomposition of tensor representations of the classical groups; Wiener's work on functional integration and its further developments by Cameron and Martin, Kac, and others; and a variety of developments in number theory concerned with the symplectic group and theta-functions, along lines treated, e.g., by Cartier, Igusa, and Weil. Ideas connected with free field algebra are involved in works of Bernstein, Leray, Quillen, and Vergne, among others.

In this chapter we give mathematical characterizations of the free fields on the basis of simple physical properties. It will be seen that, in essence, the fundamental desideratum of positive energy, together with the assumption of "canonical" commutation or anticommutation relations is sufficient to pick out the free fields from an enormous class of other fields. Before getting to this point, however, it would seem interesting to display some of the simpler general properties of the free fields. In part, these are analogous to familiar finite-dimensional results. For example, the infinite unitary group $U(\mathbf{H})$ continues to act irreducibly on the space of symmetric tensors over $\mathbf{H}$ of a given rank, etc. But there are some surprising differences, in part in the nature of simplifications in the infinite-dimensional cases; e.g., the orthogonal group $O(\mathbf{H}')$ on a real Hilbert space acts ergodically on $L_2(\mathbf{H}', n)$ when $\mathbf{H}'$ is infinite-dimensional, quite unlike the finite-dimensional case.

Familiarity with such properties of the free fields will convey some feeling for the relations between the finite and infinite-dimensional situations, further prepare the ground for the consideration of general statistics, and help to consolidate the analogy between the boson and fermion cases.

3.2. The exponential laws

How are the free fields over the direct sum $\mathbf{H}_1 \oplus \mathbf{H}_2$ of two Hilbert spaces related to the free fields over each of $\mathbf{H}_1$ and $\mathbf{H}_2$? At first glance this might appear as a dry formal question, but the answer is surprisingly simple and useful.

THEOREM 3.1. *The free boson field over a direct sum of Hilbert spaces is the tensor product of the corresponding free boson fields, in the following sense:*

If $\mathbf{H}_j$ $(j = 1, 2)$ *are given complex Hilbert spaces, and if* $(\mathbf{K}, W, \Gamma, v)$ *is the free boson field over their direct sum* $\mathbf{H} = \mathbf{H}_1 \oplus \mathbf{H}_2$, *then* $\mathbf{K}$ *is unitarily equivalent to* $\mathbf{K}_1 \otimes \mathbf{K}_2$ *in such a way that*

$$W(z_1 \oplus z_2) \cong W_1(z_1) \otimes W_2(z_2), \qquad \Gamma(U_1 \oplus U_2) \cong \Gamma_1(U_1) \otimes \Gamma_2(U_2), \qquad \text{and } v \cong v_1 \otimes v_2$$

where $(\mathbf{K}_j, W_j, \Gamma_j, v_j)$ *denotes the free boson field over* $\mathbf{H}_j$, z_j *is arbitrary in* $\mathbf{H}_j$, *and* U_j *is arbitrary in* $U(\mathbf{H}_j)$.

Symbolically, we may write $\mathbf{B}(\mathbf{H} \oplus \mathbf{H}') \cong \mathbf{B}(\mathbf{H}) \otimes \mathbf{B}(\mathbf{H}')$, where $\mathbf{B}(\mathbf{H})$ denotes the free boson field over $\mathbf{H}$. This result, suitably formulated (in terms of von Neumann's concept of infinite products of Hilbert spaces with distinguished unit vectors), is equally applicable to infinite direct sums of Hilbert spaces, as a corollary to the case of a direct sum of two spaces and the von Neumann theory.

PROOF OF THEOREM. Note first that in any free boson field, the vacuum vector v is a cyclic vector for the Weyl operators $W(z)$. To set up a unitary equivalence between $\mathbf{K}$ and $\mathbf{K}_1 \otimes \mathbf{K}_2$, we make $W(z_1 \oplus z_2)v$ correspond to $(W_1(z_1) \otimes W_2(z_2))(v_1 \otimes v_2)$, z_j being arbitrary in $\mathbf{H}_j$. Since these vectors span $\mathbf{K}$ and $\mathbf{K}_1 \otimes \mathbf{K}_2$ respectively, there exists a unique unitary transformation T from $\mathbf{K}$ onto $\mathbf{K}_1 \otimes \mathbf{K}_2$ such that $TW(z_1 \oplus z_2)v = (W_1(z_1') \otimes W_2(z_2))(v_1 \otimes v_2)$ provided that the respective inner products are equal; i.e.,

$$\begin{aligned} &\langle W(z_1 \oplus z_2)v, W(z_1' \oplus z_2')v \rangle \\ &= \langle (W_1(z_1) \otimes W_2(z_2))(v_1 \otimes v_2), (W_1(z_1') \otimes W_2(z_2'))(v_1 \otimes v_2) \rangle \end{aligned}$$

Using the fact that for any free boson field

$$\langle W(z)v, v \rangle = \exp(-\tfrac{1}{4} \|z\|^2),$$

it is easily seen that the above equation holds.

It is a direct consequence of this construction for T that $TW(z_1 \oplus z_2)T^{-1} = W_1(z_1) \otimes W_2(z_2)$ and $Tv = v_1 \otimes v_2$. To show that $T\Gamma(U_1 \oplus U_2)T^{-1} =$

$\Gamma_1(U_1)\otimes\Gamma_2(U_2)$, it suffices to note that the latter operator leaves $v_1\otimes v_2$ fixed and that it transforms $W_1(z_1)\otimes W_2(z_2)$ by conjugation in a fashion corresponding to the transformation of $W(z_1\oplus z_2)$ by $\Gamma(U_1\oplus U_2)$. □

In the fermion case, the exponential law represented by Theorem 3.1 cannot hold without some modification, because the Clifford operators $\phi(z)$ and $\phi(z')$ for orthogonal z and z' anticommute, rather than commute; thus it is impossible that $\phi(z + z') \cong \phi(z)\otimes\phi(z')$ in a simple formal sense. But with the insertion of a suitable twist, the boson results generalize appropriately. The twist in question is provided by the automorphism Ω which played a considerable role in the preceding chapter.

THEOREM 3.2. *The free fermion field over the direct sum of Hilbert spaces is the skew product of the corresponding free fermion fields in the following sense:*

If $\mathbf{H}_j(j = 1, 2)$ *are given complex Hilbert spaces, and if* $(\mathbf{K}, \phi, \Gamma, v)$ *is the free fermion field over their direct sum* $\mathbf{H} = \mathbf{H}_1\oplus\mathbf{H}_2$, *then* $\mathbf{K}$ *is unitarily equivalent to* $\mathbf{K}_1\otimes\mathbf{K}_2$ *in such a way that*

$$\phi(z_1\oplus z_2) \cong \phi_1(z_1)\otimes \mathrm{I}_2 + \Omega_1\otimes\phi_2(z_2),$$
$$\Gamma(U_1\oplus U_2) \cong \Gamma_1(U_1)\otimes\Gamma_2(U_2),$$
$$v \cong v_1\otimes v_2;$$

where $\mathrm{I}_2 = \mathrm{I}_{\mathbf{K}_2}$, $\Omega_1 = \Gamma(-\mathrm{I}_{\mathbf{H}_1})$, $(\mathbf{K}_j, \phi_j, \Gamma_j, v_j)$ *denotes the free fermion field over* $\mathbf{H}_j$, z_j *is arbitrary in* $\mathbf{H}_j$, *and* U_j *is arbitrary in* $U(\mathbf{H}_j)$.

Symbolically, we may write $\mathbf{F}(\mathbf{H}\oplus\mathbf{H}') \cong \mathbf{F}(\mathbf{H})\underline{\otimes}\mathbf{F}(\mathbf{H}')$, where $\mathbf{F}(\mathbf{H})$ denotes the free fermion field over $\mathbf{H}$, and $\underline{\otimes}$ is called the *skew product*. As in the case of boson fields, this result also can be extended to infinite direct sums.

PROOF OF THEOREM. We define a unitary transformation T from $\mathbf{K}_1\otimes\mathbf{K}_2$ onto $\mathbf{K}$ by choosing orthonormal bases $\{e_\mu\}$ and $\{f_\nu\}$ in $\mathbf{H}_1$ and $\mathbf{H}_2$, and mapping $x_{m(\cdot)}\otimes x_{n(\cdot)}$ into $x_{m(\cdot)+n(\cdot)}$ (in the notation of Section 2.2). In order to show that $T(C_1(z)\otimes I_2 + \Omega_1\otimes C_2(w))T^{-1} = C(z\oplus w)$, it suffices by continuity to prove this for the cases when z and w are finite linear combinations of the e_μ and f_ν, and by linearity is suffices in turn to prove this when z and w are among the e_μ and f_ν. It then suffices to establish the equality of the operators in question in their respective actions on the basis vectors $x_{m(\cdot)}\otimes x_{n(\cdot)}$ or $x_{m(\cdot)+n(\cdot)}$ respectively. It may then be assumed that either z is among the e_μ with $m(\mu) = 0$, and $w = 0$, or $z = 0$ and w is among the f_ν with $n(\nu) = 0$; for if $m(\mu) = 1$ or $n(\nu) = 1$, then both sides vanish. In the case when $z \neq 0$ while $w = 0$, the equality to be checked is

$$T(C_1(e_0)\otimes I_2)T^{-1}x_{m(\cdot)+n(\cdot)} = C(e_0)\,x_{m(\cdot)+n(\cdot)}$$

In the real wave representation, relabeling indices as in Chapter 2 and assuming $x_{m(\cdot)} = e_1 \cdots e_m$ and $x_{n(\cdot)} = f_1 \cdots f_n$, the left-hand side takes the form $T(e_0e_1 \cdots e_m \otimes f_1 \cdots f_n) = x_{m'(\cdot)+n(\cdot)}$ where $m'(\mu) = m(\mu) + \delta_{\mu,0}$. By the definition of T this agrees with the right-hand side, with the convention that all μ precede all ν. In the case $z = 0$ and $w = f_0$, the left-hand side takes the form

$$\begin{aligned} &T(\Omega_1\otimes C_2(f_0))T^{-1}(e_1 \cdots e_m f_1 \cdots f_n) \\ &= T(\Omega_1\otimes C_2(f_0))(x_{m(\cdot)}\otimes x_{n(\cdot)}) = (-1)^m x_{m(\cdot)+n'(\cdot)}, \end{aligned}$$

where $m = \Sigma_\mu\, m(\mu)$ and $n'(\nu) = n(\nu) + \delta_{\nu,0}$. The right-hand side is $f_0(e_1 \cdots e_m)(f_1 \cdots f_n) = (-1)^m(e_1 \cdots e_m)(f_0f_1 \cdots f_n)$, and so agrees with the left.

By construction, T carries $v_1\otimes v_2$ into v. To show that $\Gamma(U_1\oplus U_2)$ is as stated, it suffices, noting that it leaves v invariant, to show that it transforms the $C(z\oplus w)$ in the same way as does $T(\Gamma_1(U_1)\otimes\Gamma_2(U_2))T^{-1}$. By the definition, $C(z\oplus w)$ is transformed into $C(U_1z\oplus U_2w)$. On the other hand, under transformation by $\Gamma(U_1)\otimes\Gamma(U_2)$, $C_1(z)\otimes \mathrm{I} + \Omega_1\otimes C_2(w)$ is carried into $C_1(U_1z)\otimes \mathrm{I} + \Omega_1\otimes C_2(U_2w)$, since Ω_1 commutes with all $\Gamma(U_1)$, and the required equality follows. □

3.3. Irreducibility

The irreducibility of the free boson and fermion fields over a given Hilbert space has already been seen. In this section a different proof is given that uses positivity in a way that is often effective. We recall that a selfadjoint set **R** of operators is one such that if $A \in \mathbf{R}$, then $A^* \in \mathbf{R}$.

THEOREM 3.3. *Let H be a nonnegative selfadjoint operator in a Hilbert space **K** such that a unique vector v (within proportionality) is annihilated by H. Let **R** denote a selfadjoint set of bounded linear operators on **K** for which v is cyclic, and suppose that $e^{itH}\mathbf{R}e^{-itH} \subseteq \mathbf{R}$ for all $t \in \mathbf{R}$. Then **R** is irreducible on **K**.*

PROOF. It is no essential loss of generality to assume that **R** is an algebra, since it may otherwise be replaced by the algebra it generates. Assuming this, $\mathbf{R}v$ is dense in **K**. Letting P denote the projection of **K** onto an **R**-invariant subspace, then $PR = RP$ for all $R \in \mathbf{R}$, hence $\langle PRv, v\rangle = \langle RPv, v\rangle$. Setting $\Gamma(t) = e^{itH}$, then $\Gamma(t)v = v$ and $\Gamma(t)R\Gamma(t)^{-1} \in \mathbf{R}$ for all $t \in \mathbf{R}$ and $R \in \mathbf{R}$, from which it follows that $\langle P\Gamma(t)Rv, v\rangle = \langle R\Gamma(t)^{-1}Pv, v\rangle$. Since the right side of the equation can be extended to a bounded analytic function in the lower half-plane, and the left side can be extended to a bounded analytic function in the upper half-plane, $\langle P\Gamma(t)Rv, v\rangle$ is independent of t. This implies that $Pv = cv$ by virtue of the unicity of v and cyclicity of v for $\mathbf{R}$, where c is a scalar. Now returning to the equality $PR = RP$ and letting u be arbitrary in $\mathbf{K}$, it results that $\langle PRv, u\rangle = \langle RPv, u\rangle$. But since $Pv = cv$, this gives the equation $\langle Rv, Pu\rangle = c\langle Rv, u\rangle$. The cyclicity of v for $\mathbf{R}$ then implies that $\langle x, Pu\rangle = c\langle x, u\rangle$ for all $x \in \mathbf{K}$, whence $Pu = \bar{c}u$, which implies in turn that $P = 0$ or I. □

COROLLARY 3.3.1. *The free boson and fermion fields over a given complex Hilbert space* $\mathbf{H}$ *are irreducible under the respective Weyl or Clifford operators,* $W(z)$ *or* $\phi(z)$.

PROOF. In either case, set H to be $\partial\Gamma(I)$, v to be the vacuum vector, and $\mathbf{R}$ to be the totality of the $W(z)$ or $\phi(z)$, $z \in \mathbf{H}$. The hypotheses of Theorem 3.3 are then satisfied. □

LEXICON. The operator $N = \partial\Gamma(I)$ is called the "number of particles" or "number operator." The subspace $\mathbf{K}_n$, represented in the particle representation of the free fields by the n-fold symmetrized (resp. antisymmetrized) tensor power of $\mathbf{H}$, and definable as the eigenspace of N with eigenvalue n, is called the "n-particle subspace."

3.4. Representation of the orthogonal group by measure-preserving transformations

Let $\mathbf{L}$ be a real Hilbert space, and let $O(\mathbf{L})$ be the group of all orthogonal transformations on $\mathbf{L}$. Since the observation (of Koopman) that a group of measure-preserving transformations induces canonically a unitary representation of the group, and the subsequent application of this idea to classical mechanics (notably by von Neumann), there has been considerable development of the theory of such "flows." We shall not require a highly structured definition, but simply define a flow as a system (G, T, M, m), where G is a group, M is a space, T is an action of G on M as a transformation group, and m is a measure on M that is left invariant by the group action (or, on occasion, is such that any two translates under G are mutually absolutely continuous). A kind of converse of Koopman's observation was emphasized by Wiener, who gave a way to represent suitable orthogonal transformations as measure-preserving transformations on "Wiener space," a space of continuous functions interpretable physically as Brownian motion paths. In algebraic essence, however, Wiener's construction (later extended by others) is tantamount to the application of the representation Γ associated with the free boson field.

In this section the canonical "lifting" or transfer of an orthogonal group to a group of measure-preserving transformations is developed, and it is shown how an extensive class of ergodic flows may be derived from this construction. We recall that a flow is ergodic if there exist no invariant measurable sets, modulo null sets, other than the entire space M and the empty set; this is equivalent, as is easily seen, to the nonexistence of invariant measurables, other than constants. Ergodic flows have much the same role in the general theory of flows as irreducible representations have in the theory of group representations: every flow is obtainable as a type of direct sum or integral of ergodic flows.

Koopman's observation applies to the group of transformations $O(\mathbf{L})$ acting on the generalized measure space $(\mathbf{L}, g)$, where g is the isonormal distribution on $\mathbf{L}$, by virtue of the invariance of g under orthogonal transformations on $\mathbf{L}$. We recall that the idea consists in associating with a given measure-preserving transformation T on a measure space (M, m) the unitary transformation $U(T)$ on $L_2(M, m)$ defined by

$$U(T): f(p) \rightarrow f(T^{-1}p).$$

It should be noted that any such transformation T determines a corresponding transformation $\tilde{\mathbf{T}}$ on the measure ring $\tilde{\mathbf{M}}$ of (M, m), consisting of the Boolean ring of all measurable sets in M modulo null sets; $\tilde{\mathbf{T}}$ is a measure-preserving Boolean automorphism of $\tilde{\mathbf{M}}$; if E is a measurable subset of M, $\tilde{\mathbf{T}}$ carries the corresponding residue-class $\tilde{E}$ modulo the ideal of null sets into the residue class of $T(E)$, and the measure $\tilde{m}(\tilde{E})$ of $\tilde{E}$ is defined as that of E.

It is evident that $U(T)$ has the property that if it carries f into f' and g into g', and if $fg \in L_2(M, m)$, then it carries fg into $f'g'$. It is not difficult to see that this property is characteristic of the unitary transformations on $L_2(M, m)$ that arise from measure-preserving automorphisms of the measure ring. It is more complicated to show that any such automorphism may be defined by a measure-preserving point transformation, if not on the space M then on some modification with an isomorphic measure ring to (M, m); but we shall have no occasion to use this result. The essential point is that the notion of a measure-preserving transformation is conceptually equivalent to that of a multiplicative unitary transformation on L_2.

DEFINITION. If T is any linear transformation on the real space $\mathbf{L}$, the corresponding operator $x + iy \mapsto Tx + iTy$ on the complexification $\mathbf{H}$ of $\mathbf{L}$ will be denoted as T^c and called the *complexification* of T.

THEOREM 3.4. *Let* $\mathbf{L}$ *be a real Hilbert space and let* G *be a subgroup of* $O(\mathbf{L})$ *that leaves no finite-dimensional subspace of* $\mathbf{L}$ *invariant. Then* $\Gamma(G^c)$ *acts ergodically on* $L_2(\mathbf{L}, g)$.

LEMMA 3.4.1. *Let U and V be unitary representations of a group G on complex Hilbert spaces $\mathbf{H}_U$ and $\mathbf{H}_V$ whose tensor product has a nonzero invariant vector. Then both U and V leave invariant a nontrivial finite-dimensional subspace.*

PROOF. Let $\varkappa$ denote an arbitrary conjugation on $\mathbf{H}_V$, and let $\bar{V}$ denote the representation $\varkappa V \varkappa$ of G. Then $\bar{V}$ has a nontrivial finite-dimensional subspace if and only if V does. Thus it suffices to show that if the tensor product $U \otimes \bar{V}$ has a nonzero invariant vector, then both U and V have nontrivial finite-dimensional invariant subspaces.

To this end, let w be a nonzero invariant vector for $U \otimes \bar{V}$. Then for each fixed vector x in $\mathbf{H}_U$, the function of y, $\langle x \otimes \varkappa y, w\rangle$, is an antilinear function of y, and so has the form $\langle u, y\rangle$ for some vector u in $\mathbf{H}_V$. Setting $u = Tx$, it is easy to verify that T is bounded and linear, and that $TU(a) = V(a)T$ for all a in G. It follows that $U(a)$ commutes with T^*T, which is a Hilbert-Schmidt operator, since for any finite orthonormal set $e_1, e_2, \dots, e_n$ in $\mathbf{H}_U$, $\Sigma_k \|Te_k\|^2 \leq \|w\|^2$. As the spectral manifolds of T^*T reduce $U(a)$, it follows that U must have an invariant subspace of positive finite dimension. By symmetry, the same is true of V. □

PROOF OF THEOREM. Let $(\mathbf{K}, W, \Gamma, v)$ denote the free boson field over the complexification $\mathbf{H}$ of $\mathbf{L}$. Ergodicity then means that the only invariant vectors in $\mathbf{K}$ for all $\Gamma(g^c)$, $g \in G$, are those of the form λv, $\lambda \in \mathbf{C}$. To show this, let u be any invariant vector, and let u_n denote its component in the n-particle subspace; then u_n is likewise invariant under all $\Gamma(g^c)$, $g \in G$. On the other hand, the n-particle subspace $\mathbf{K}_n$ is a subspace of the n-fold unrestricted tensor product $\mathbf{H}\otimes\cdots\otimes\mathbf{H}$, and the action of $\Gamma(g^c)$ on $\mathbf{K}_n$ is the restriction of the n-fold tensor product $g^c\otimes\cdots\otimes g^c$ from the unrestricted n-fold product to the symmetrized subspace. Thus if $\mathbf{K}_n$ admits a nontrivial invariant vector under the $\Gamma(g)$, so also does $\mathbf{H}\otimes\cdots\otimes\mathbf{H}$ under the $g^c\otimes\cdots\otimes g^c$ with $g \in G$, but this is impossible by Lemma 3.4.1 together with the hypotheses that G leaves no nontrivial finite-dimensional subspace of $\mathbf{L}$ invariant (this condition implying that G^c leaves no finite-dimensional subspace of $\mathbf{H}$ invariant). □

COROLLARY 3.4.1. *Let G be a group of unitary operators on the complex Hilbert space $\mathbf{H}$ leaving invariant no nontrivial finite-dimensional subspace. Then if $(\mathbf{K}, W, \Gamma, v)$ denotes the free boson field over $\mathbf{H}$, the only invariant vectors for $\Gamma(\mathbf{G})$ are the scalar multiples of v.*

PROOF. This is simply a reformulation of what was established in the preceding proof. □

The theorem is applicable to the fermion case with a natural extension of the notion of ergodicity.

DEFINITION. A group of automorphisms of a W^*-algebra is called *ergodic* if the only invariant projections are 0 and I.

EXAMPLE 3.1. In the case of the free boson field over the complexification **H** of **L**, the $\Gamma(O^c)$ for arbitrary $O \in O(\mathbf{L})$ induce automorphisms $\gamma(O)$: $X \mapsto \Gamma(O)X\Gamma(O)^{-1}$ of the W^*-algebra **Q** generated by the $\phi(x)$ with $x \in \mathbf{L}$. (The $\phi(x)$ are unbounded; one defines the W^*-algebra they generate by their bounded functions in the standard sense.) In this way $O(\mathbf{L})$ is represented as an automorphism group of **Q**, and its action is ergodic. For if P is any invariant projection in **Q**, then Pv is an invariant vector in **K**. By the ergodicity in the earlier sense established by the theorem, $Pv = \alpha v$ for some scalar α. But v is a cyclic and separating vector for the maximal abelian algebra **Q**, so $P = \alpha I$.

Conversely, the absence of any nontrivial invariant projection in **Q** implies ergodicity in the original sense. This may be seen most naturally by the unitary equivalence of **K** with $L_2(\mathbf{Q}, v)$, where the latter notation refers to the space of all closed densely defined operators on **K** with domain containing v that are *affiliated with* **Q**, in the sense that they commute with all unitaries in the commutor **Q**′ (or, equivalently, have polar decomposition with partially isometric constituent and bounded functions of their selfadjoint constituent in **Q**). In the present instance, in which **Q** is abelian, all such operators are normal. Endowing $L_2(\mathbf{Q}, v)$ with the inner product $\langle S, T\rangle = \langle Sv, Tv\rangle$, the map Ω: $T \mapsto Tv$ is unitary from $L_2(\mathbf{Q}, v)$ onto **K**, and has the property that $\Omega S'\Omega^{-1} = S$ for arbitrary $S \in \mathbf{Q}$, where S' denotes left multiplication by S acting on $L_2(\mathbf{Q}, v)$.

An invariant element of **K** will consequently have the form Sv for some $S \in L_2(\mathbf{Q}, v)$ that is invariant under the unitary extension of $\gamma(O^c)$ from **Q** to all of $L_2(\mathbf{Q}, v)$. But the spectral projections of S are then also invariant, for an automorphism of a W^*-algebra that leaves invariant a normal element will also leave invariant its spectral projections in **Q**. This implies the lack of nontrivial invariant elements in $L_2(\mathbf{Q}, v)$, and hence the ergodicity in this original sense. (The foregoing argument may seem slightly pedantic, but is designed to substantially subsume the fermion case.)

COROLLARY 3.4.2. *Let* **L** *be a real Hilbert space. Then* $O(\mathbf{L})$ *acts ergodically on the* W^**-algebra* **Q** *generated by the* $Q(x)$, $x \in \mathbf{L}$, *via the automorphisms induced by the* $\Gamma(T^c)$ *for* $T \in O(\mathbf{L})$ *(with the previously used notation regarding the free fermion field).*

PROOF. This follows directly by the argument for Corollary 3.4.1 in the light of Example 3.1 which applies without essential change to the fermion case. □

Theorem 3.4 may be used to characterize the functionals that are based on a given closed linear subspace. If **M** is a closed linear subspace of the complex Hilbert space **H**, an element of the free boson (resp. fermion) field—by which we mean an element of the associated Hilbert space **K**—is said to be *based on* **M** in case it is in the cyclic subspace generated by v under the action of the $W(z)$ (resp. $\phi(z)$) with $z \in$ **M**. It is readily shown from Theorem 3.1 that a vector $u \in$ **K** is based on **M** if and only if it is invariant under all $\Gamma(U)$ with U an arbitrary unitary on **H** that is the identity on **M**. But the following stronger result holds.

COROLLARY 3.4.3. *Let G be a group of unitary operators on the complex Hilbert space* **H** *that leaves invariant the closed linear subspace* **M**, *but leaves no finite-dimensional subspace of* **M** *invariant, and whose restrictions to the orthocomplement of* **M** *are the identity operator. Let* (**K**, W, Γ, v) *denote the free boson (resp. fermion) field over* **H**. *An element* $u \in$ **K** *is based on* $\mathbf{M}^{\perp}$ *if and only if it is invariant under all* $\Gamma(U)$ *with* $U \in G$.

PROOF. The proof is left as an exercise. □

A further strengthened result, applicable to arbitrary measurable functionals that are not necessarily square-integrable, is applicable in terms of the real wave representation. In this connection a measurable functional f on (**L**, n), **L** being a given real Hilbert space, is said to be based on the closed linear subspace **M** of **L** in case any of the following equivalent conditions is satisfied: (i) f is a limit in measure of tame functionals based on subspaces of **M**; (ii) every bounded Borel function f is a limit in L_2(**L**, n) of tame functionals based on subspaces of **M**; or (iii) f is measurable with respect to the σ-ring generated by the coordinate functionals on **M** (within the Boolean ring of all idempotent measurable functionals).

COROLLARY 3.4.4. *Let G be a group of orthogonal transformations on the real Hilbert space* **L** *which leaves invariant the closed linear subspace* **M**, *but leaves invariant no subspace of positive finite dimension, and whose restrictions to the orthocomplement of* **M** *are the identity operator. A measurable functional on* (**L**, n) *is based on* $\mathbf{M}^{\perp}$ *if and only if it is invariant under G*.

PROOF. The proof is again left as an exercise. □

3.5. Bosonic quantization of symplectic dynamics

In the foregoing it has largely been assumed that there is given a priori a complex Hilbert space inhabited by single-particle wave functions, or classical fields. In actuality, it frequently happens that a complex Hilbert space structure is not given to start with, but rather a lesser structure—symplectic or orthogonal, depending on whether bosons or fermions are involved—together with a temporal evolution group (often contained in a larger symmetry group) leaving the structure invariant. The complex Hilbert space structure must be sought on the basis of the desiderata that it should extend the given limited structure and be invariant under temporal evolution; and, in expression of the physical constraint of stability, this evolution should be represented by a one-parameter unitary group whose selfadjoint generator is positive. In terms of this new structure, the original symplectic dynamics becomes unitary, and the quantized boson field associated with this symplectic dynamics is shown to exist and to be equivalent to the free boson field over the derived Hilbert space.

Such a complex structure plays a generally fundamental role in quantum mechanics (in the treatment of fermions as well as bosons, cf. below, e.g., regarding the Dirac "hole" theory). This role is often obscure in the familiar literature, which uses complex structures derived from successful precedent, or somewhat opportunistically on occasion, rather than determined from general principle. The importance of the complex structure in a symplectic (or, in the case of fermions, an orthogonal) context is clear from the dependence of the energy concept on the complex structure. Without a complex structure, in a real space, the eigenvalues of the generator of time evolution will be conjugate complex numbers—pure imaginary ones in the simplest cases. But correlation with the experimental use of the notion of energy represents it as a real number. At the same time, observational convention fixes positivity rather than negativity of the energy as the basic constraint in the description of stable systems, increased stability being observed as the experimentally defined energy decreases. Although in some contexts a complex structure is given or arises very simply (e.g., by Fourier transformation and restriction to the positive-frequency component), it may be inappropriate in its original form (e.g., in hole theory). In any case, i is effectively not simply an absolute entity such as π or $2^{1/2}$, but a conveniently chosen matrix J such that $J^2 = -I$.

Consider for example the differential equation

$$(\Box + m^2)\varphi + V\varphi = 0 \tag{3.1}$$

where V is a given bounded continuous, nonnegative function on euclidean space and $\Box = \partial_t^2 - \Delta$ on $\mathbf{M}_0$ (Minkowski space, represented as $\mathbf{R} \times \mathbf{R}^n$). Let $\mathbf{L}$ denote the class of all real solutions to equation 3.1 that are in $C_0^\infty(\mathbf{R}^n)$

at each time, and let $S(s)$ be the linear operator on **L** that maps the solution $\varphi(t, \cdot)$ into $\varphi(t - s, \cdot)$, where the solution is regarded as a function of time with values that are functions on space. The general theory of evolutionary differential equations shows that there is a unique element of **L** having given *Cauchy data* $\varphi(t_0, \cdot)$ and $\partial_t\varphi(t_0, \cdot)$ that are in $C_0^\infty(\mathbf{R}^n)$, and that $S(t)$ acts continuously on **L** (in the direct sum topology on the Cauchy data at arbitrary times). There is a simple, natural symplectic structure in **L** that is invariant under S, defined by the antisymmetric form A on **L**:

$$\begin{aligned} A(\varphi_1, \varphi_2) &= \langle \varphi_1(t), \partial_t\varphi_2(t)\rangle - \langle \varphi_2(t), \partial_t\varphi_1(t)\rangle \\ &= \int (\varphi_1\partial_t\varphi_2 - \varphi_2\partial_t\varphi_1)\, d^nx; \end{aligned}$$

i.e., the inner product in the first equation is in $L_2(\mathbf{R}^n)$, and the time t is arbitrary, the result being independent of t as a consequence of equation 3.1.

If this form A is substituted for the imaginary part of the inner product in the treatment of the free boson field given earlier, a natural definition of the "quantization" of equation 3.1 emerges. Specifically, one wants a system $(\mathbf{K}, W, \Gamma, \nu)$ as earlier, with, however, Γ limited to be one-parameter unitary group rather than a representation of the full unitary group on a Hilbert space. This may or may not exist; but when it does, it is unique and connected with the introduction of a complex pre-Hilbert space structure in **L**. Whether it exists or not, there is always a C^*-algebraic quantization, in which temporal evolution is represented by a one-parameter group of automorphisms of the algebra (see Chap. 5). But in general there is no mathematical reason for such a group to admit a vacuum state; it is only physical stability that is indicative of its existence. It will be seen that a vacuum exists if and only if an appropriate complex Hilbert space structure can be introduced, which is the case if and only if the symplectic action is "stably unitarizable"—in which case this unitarization (defined below) is unique.

Definition. Let $S(\cdot)$ be a continuous one-parameter group of symplectic transformations on the symplectic vector space $(\mathbf{L}, A)$. The system $(\mathbf{L}, A, S)$ is *unitarizable* if there is a complex pre-Hilbert space structure on **L** such that the imaginary part of the inner product equals A, and such that S extends to a continuous one-parameter unitary group U on the Hilbert space completion **H** of **L**. In this case $(\mathbf{H}, U)$ is said to be a *unitarization* of $(\mathbf{L}, A, S)$. A unitary group U is *stable* (resp. *strictly stable*) if the selfadjoint generator of $U(\cdot)$ is nonnegative (resp. positive). A *quantization* of the system $(\mathbf{L}, A, S)$ is a system $(\mathbf{K}, W, \Gamma, \nu)$ consisting of a complex Hilbert space **K**, a Weyl system over $(\mathbf{L}, A)$ with representation space **K**, a continuous one-parameter unitary group Γ on **K** such that $\Gamma(t)W(z)\Gamma(t)^{-1} = W(S(t)z)$, and a unit vector ν in **K** such that $\Gamma(t)\nu = \nu$ for all t. Such a quantization is said to be *stable* in case the unitary group $\Gamma(\cdot)$ is stable.

Before treating the quantization of symplectic systems, we note some quite general properties of complex structures in temporally evolving linear systems. These clarify the mathematical role of positivity of the energy and will be used later.

SCHOLIUM 3.1. *Let $U(\cdot)$ be a strictly stable unitary group on the complex Hilbert space* **H**, *and let T be a continuous real-linear transformation on* **H** *that commutes with $U(t)$ for all $t \in \mathbf{R}$. Then T is complex linear. In particular, any real-linear closed invariant subspace is complex-linear.*

PROOF. Let $T^{\#}$ denote the unique real-linear operator on **H** such that $\mathrm{Re}(\langle Tx, y\rangle) = \mathrm{Re}(\langle x, T^{\#}y\rangle)$ for all $x, y \in H$. Let A denote the selfadjoint generator of $U(\cdot)$, and set $f(z) = \langle e^{-zA}x, T^{\#}y\rangle$, $g(z) = \langle e^{-zA}Tx, y\rangle$, for z in the half-plane $\mathrm{Re}(z) \geq 0$. Then f and g are bounded and continuous in the half-plane and analytic in its interior, and $\mathrm{Re}\, f(t) = \mathrm{Re}\, g(t)$ for $t \in \mathbf{R}$. Thus the real part of the bounded analytic function $h(z) = f(z) - g(z)$ in the upper half-plane vanishes on the real axis. By the Schwarz reflection principle it can be continued analytically into the entire plane and remains bounded there. By Liouville's theorem, h is identically constant. Since A is strictly positive, $h = 0$.

It follows that $f = g$, and in particular $\langle Tx, y\rangle = \langle x, T^{\#}y\rangle$ for all $x, y \in \mathbf{H}$, whence $\langle Tix, y\rangle = \langle ix, T^{\#}y\rangle = i\langle x, T^{\#}y\rangle = i\langle Tx, y\rangle$, which implies that $Tix = iTx$. Thus T is complex linear. If in particular, **M** is a real-linear closed subspace invariant under the $U(t)$, the real-linear projection P on $\mathbf{H}^{\#}$ having range **M** commutes with the $U(t)$, implying that P is complex-linear, whence so also is **M**. □

The following corollary to the proof of the scholium will be useful in the later treatment of fermions.

SCHOLIUM 3.2. *Let $U(\cdot)$ be a one-parameter orthogonal group on the real Hilbert space* **H** *with inner product $S(\cdot,\cdot)$. Then there exists at most one complex structure J on* **H** *with the following properties:*

i) **H** *acquires the structure of a complex Hilbert space if for arbitrary x, $y \in$* **H** *the inner product $\langle x, y\rangle$ is defined as $S(x, y) - iS(Jx, y)$, and for arbitrary real a, b, $(a + ib)x$ is defined as $ax + bJx$;*
ii) J commutes with the $U(t)$ for all $t \in \mathbf{R}$; and
iii) The selfadjoint generator of the one-parameter unitary group on **H** *represented by $U(\cdot)$ [by virtue of ii)] is positive.*

PROOF. The real part of $\langle U(t)x, y\rangle$ is $S(U(t)x, y)$. By an argument given in the proof of Scholium 3.1, $\langle U(t)x, y\rangle$ is determined by $S(U(\cdot)x, y)$. The complex structure is thereby determined. □

The direct analog of Scholium 3.2 for bosons, which will be used later in this section, involves the introduction of a complex structure into a given linear symplectic space $(\mathbf{L}, A)$. A complex structure J on $\mathbf{L}$ is called *symplectic* if $A(Jx, Jy) = A(x, y)$ for all $x, y \in \mathbf{L}$, and called *positive* in case $A(Jx, x) \geq 0$ for all $x \in \mathbf{L}$. It is *invariant* relative to a given symplectic group if it commutes with all the symplectics in the group. Given a symplectic and positive complex structure in $(\mathbf{L}, A)$, $\mathbf{L}$ can be given the structure of complex pre-Hilbert space by the definition $\langle x, y\rangle = A(Jx, y) + iA(x, y)$. If J is invariant under a given symplectic transformation T on $\mathbf{L}$, T will be unitary relative to this complex pre-Hilbert structure.

SCHOLIUM 3.3. *Let $U(\cdot)$ be a one-parameter group of symplectic transformations on the linear symplectic space $(\mathbf{L}, A)$. Then there exists at most one complex structure J on $\mathbf{L}$ that is invariant, positive, symplectic, and such that the selfadjoint generator of the one-parameter unitary group $U(\cdot)$ in the completion of $\mathbf{L}$ as a complex Hilbert space is positive.*

PROOF. This is the same as for the preceding Scholium with the replacement of the real by the imaginary part. □

THEOREM 3.5. *With the foregoing notation, suppose that there exists a stable quantization of $(\mathbf{L}, A, S)$. Then there exists a complex Hilbert space $\mathbf{H}$ and a symplectic isomorphism T from the linear subset $\mathbf{L}_0$ of $\mathbf{L}$ spanned algebraically by the vectors of the form $S(t)x - x$, $x \in \mathbf{L}$, into a dense subset of $\mathbf{H}$, such that the closure $U(t)$ of $TS(t)T^{-1}$ is a unitary and strictly stable continuous one-parameter group on $\mathbf{H}$.*

PROOF. We remark to begin with that although in practice $\mathbf{L}_0$ will be dense in $\mathbf{L}$, this will obviously not be the case without some restriction to the effect that S acts nontrivially. In case, for example, $S(t)$ is the identity for all t, the stability constraint on the quantization is vacuous, and no nontrivial conclusion can be drawn. It will be seen that the restriction to the subset $\mathbf{L}_0$ is a natural one.

LEMMA 3.5.1. *Let f be a bounded continuous complex-valued function on $\mathbf{R}$ such that $f(0) = 1$ and*

$$f(s + t)f(s - t) = f(s)^2|f(t)|^2 \tag{3.2}$$

for $s \in [-1, 1]$. *Then* $f(s) = \exp(ias - b^2s^2)$ *for suitable real constants a and b*.

PROOF. Let $g(t) = \log|f(t)|$; then g is well defined and continuous in some neighborhood of $t = 0$, and $g(0) = 0$,

$$g(s + t) + g(s - t) = 2\,[g(s) + g(t)]$$

for s and t sufficiently near 0. Setting $s = 0$, it follows that g is even, and by a simple induction on the positive integer n, that $g(nt) = n^2g(t)$ for $nt \in [-\varepsilon, \varepsilon]$, for some fixed positive number ε. Setting $h(t) = g(t)t^{-2}$, then h is continuous for $|t| \in (0, \varepsilon)$ and has the property that $h(nt) = h(t)$ for $n|t| \in (0, \varepsilon)$; or, equivalently, $h(t) = h(t/n)$ for $|t| \in (0, \varepsilon)$. It follows that $\lim_{n\to\infty} h(t/n)$ exists and is the same, say k, for all t with $|t| < \varepsilon$, showing that $h(t) = k$ for such t.

Thus $|f(t)| = \exp(kt^2)$ for sufficiently small $t > 0$. Taking absolute values in equation 3.2, and replacing t by s, the interval on which $|f(t)|$ has the stated form doubles, and by induction it follows that $|f(t)| = \exp(kt^2)$ for all t, implying that $k = -b^2$ for some real b. Now dividing equation 3.2 by the equation resulting from it by taking absolute values, the following functional equation for $f(t)/|f(t)| = p(t)$ results:

$$p(s + t)p(s - t) = p(s)^2.$$

In particular, taking $t = s$, it follows that $p(2s) = p(s)^2$; and setting $x = s + t, y = s - t$,

$$p(x)p(y) = p(\tfrac{1}{2}(x + y))^2 = p(x + y),$$

showing that p is a continuous character on $\mathbf{R}$, and hence of the form $p(t) = e^{iat}$. □

LEMMA 3.5.2. *Let* $\mathbf{M}$ *denote the subspace of all vectors in* $\mathbf{K}$ *that are left invariant by the* $\Gamma(t)$, *and suppose* $v_j \in \mathbf{M}$ *for* $j = 1, 2, 3, 4$. *Denoting* $S(t)x$ *as* x_t, *then for all x and y in* $\mathbf{L}$,

$$\langle W(x_t)v_1, W(-y)v_2\rangle \langle W(x_t)v_3, W(y)v_4\rangle$$

and

$$\langle W(y)v_1, W(-x_t)v_2\rangle \langle W(-y)v_3, W(-x_t)v_4\rangle$$

are independent of t, and equal.

PROOF. By the Weyl relations,

$$\langle W(x_t)v_1, W(-y)v_2\rangle = e^{\iota A(y,x_t)}\langle W(y)v_1, W(-x_t)v_2\rangle;$$

$$\langle W(x_t)v_3, W(y)v_4\rangle = e^{-\iota A(y,x_t)}\langle W(-y)v_3, W(-x_t)v_4\rangle.$$

Multiplying the equations together, it results that

$$\langle W(x_t)v_1, W(-y)v_2\rangle \langle W(x_t)v_3, W(y)v_4\rangle$$
$$= \langle W(y)v_1, W(-x_t)v_2\rangle \langle W(-y)v_3, W(-x_t)v_4\rangle.$$

On the other hand, using the invariance of the v_j under the $\Gamma(t)$, for any $v \in \mathbf{M}$, $W(x_t)v = \Gamma(t)W(x)v$, and substituting in the last equation, it follows that each factor on the left side of the equation can, as a function of t, be extended to a bounded holomorphic function in the upper half-plane. By the same token, each factor on the right side of the equation can be extended to a bounded holomorphic function in the lower half-plane. It follows as earlier that both sides are constant as functions of t. □

LEMMA 3.5.3. *Suppose $z \in \mathbf{L}$ has the property that $\langle W(sz)v, u\rangle = 0$ for all $s \in [-1, 1]$ and all u that are invariant under $\Gamma(t)$ and orthogonal to v. Then there exist real constants a and b such that $\langle W(sz)v, v\rangle = \exp(ias - b^2s^2)$, $s \in \mathbf{R}$.*

PROOF. Apply Lemma 3.5.2 with $v_1 = v_2 = v_3 = v_4 = v$, $x = sz$, and $y = rz$; then

$$\langle W(sz_t)v, W(-rz)v\rangle \langle W(sz_t)v, W(rz)v\rangle$$
$$= \langle W(rz)v, W(-sz_t)v\rangle \langle W(-rz)v, W(-sz_t)v\rangle,$$

both sides being independent of t. Setting $f(s) = \langle W(sz)v, v\rangle$, the right side of this equation takes the form $f(s - r)f(s + r)$, taking $t = 0$. On the other hand, $W(sz_t)v$ on the left side may be expressed as $\Gamma(t)W(sz)v = e^{itH}W(sz)v$, where H denotes the selfadjoint generator of the one-parameter group Γ. The holomorphic continuation of $\langle W(sz_t)v, W(-rz)\rangle$ to the upper half-plane then takes the form, as a function of the complex variable $t + iu$, $u > 0$, $\langle e^{itH-uH}W(sz)v, W(-rz)\rangle$, which for $t = 0$ and $u \to \infty$ tends to $\langle PW(sz)v, W(-rz)v\rangle$, where P is the projection of $\mathbf{K}$ onto $\mathbf{M}$.

Using now the hypothesis on $W(sz)$, $PW(sz)v = P_v W(sz)v$, where P_v is the projection onto the one-dimensional subspace spanned by v, so that $PW(sz)v = \langle W(sz)v, v\rangle v$. It follows that the left side of the equation under consideration may be expressed as $f(s)^2 |f(r)|^2$, and Lemma 3.5.3 follows now directly from Lemma 3.5.1. □

LEMMA 3.5.4. *For arbitrary $x \in \mathbf{L}$, there exists $\varepsilon > 0$ such that for all $\lambda \in [-\varepsilon, \varepsilon]$ and $t \in \mathbf{R}$, $z = \lambda(x_t - x)$ has the property described in Lemma 3.5.3.*

PROOF. Applying Lemma 3.5.2 with $v_1 = v_2 = v_3 = v$, $v_4 = u$, where $u \in \mathbf{M}$ and $\langle u, v\rangle = 0$, and $y = x$, it results that the second expression given there vanishes identically due to the orthogonality of u and v, so that the first expression likewise does so:

$$\langle W(x_t)v, W(-x)v\rangle \langle W(x_t)v, W(x)u\rangle = 0.$$

Since each of the two factors in the last equation is holomorphically extendable, at least one of them vanishes identically. In particular, $\langle W(x_t)v, W(x)u\rangle = 0$ for all t unless $\langle W(2x)v, v\rangle = 0$.

Setting $y = \lambda x$, then for sufficiently small λ, $\langle W(2sy)v, v\rangle \neq 0$ for $s \in [-1, 1]$, by continuity, implying that $\langle W(s(y_t - y)v, u\rangle = 0$ for all t and all $u \in \mathbf{M}$ that are orthogonal to v, as stated by Lemma 3.5.4. □

LEMMA 3.5.5. *Let $\phi(x)$ for $x \in \mathbf{L}$ denote the selfadjoint generator of the one-parameter unitary group $\{W(tx)\colon t \in \mathbf{R}\}$. Then a) v is in the domain of $\phi(x)$ for $x \in \mathbf{L}_0$, b) if T denotes the map*

$$T\colon x \mapsto 2^{1/2}\,[\phi(x)v - \langle\phi(x)v, v\rangle v]$$

from $\mathbf{L}_0$ into $\mathbf{K}$, then T is symplectic from $\mathbf{L}_0$ to a complex linear subset of $\mathbf{K}$; and c) $TS(t)x = \Gamma(t)Tx$ for all $x \in \mathbf{L}_0$ and $t \in \mathbf{R}$.

PROOF. Suppose z is as in Lemma 3.5.4 for some $x \in \mathbf{L}$. Then, by Lemma 3.5.3, $\langle W(sz)v, v\rangle = \exp(ias - b^2s^2)$ for all s and suitable constants a and b (depending on z). Then v is in the domain of $\phi(z)$, since

$$\begin{aligned}&\langle s^{-1}(W(sz) - I)v, s^{-1}(W(sz) - I)v\rangle\\ &= s^{-2}(2 - 2\,\mathrm{Re}\langle W(sz)v,v\rangle) = 2s^{-2}[1 - \exp(-b^2s^2)\cos(as)],\end{aligned}$$

which remains bounded as $s \to 0$.

It follows readily from the Weyl relations (Chapter 1) that if $v \in \mathbf{D}(\phi(x))$ and $v \in \mathbf{D}(\phi(y))$, then $v \in \mathbf{D}(\phi(x + y))$, and $\phi(x + y)v = \phi(x)v + \phi(y)v$. This fact together with the result just obtained establishes (a).

As earlier seen (by differentiation from the Weyl relations), $[\phi(x), \phi(y)] \subset -iA\langle x, y\rangle I$. It follows from this that for any vector v which is in the domain of $\phi(x)\phi(y)$ (and hence in that of $\phi(y)\phi(x)$),

$$\mathrm{Im}\langle\phi(x)v, \phi(y)v\rangle = \tfrac{1}{2}A(x, y). \tag{3.3}$$

On the other hand, equation 3.3 is meaningful if v is only in the domains of $\phi(x)$ and $\phi(y)$; and it remains true, in fact, in this more general situation, as a consequence of the fact that the restriction of $\phi(x)$ (or $\phi(y)$) to the domain of $\phi(x)\phi(y)$ is essentially selfadjoint. This follows from considerations similar to those in the proof of Theorem 1.1, and the details are omitted.

Using the last paragraph, it follows that

$$\begin{aligned}\operatorname{Im}\langle Tx, Ty\rangle &= 2\operatorname{Im}\langle\phi(x)v - \langle\phi(x)v, v\rangle v,\quad \phi(y)v - \langle\phi(y)v, v\rangle v\rangle\\ &= 2\operatorname{Im}\langle\phi(x)v,\quad \phi(y)v\rangle = A(x,y),\end{aligned}$$

showing that T is symplectic. Note that the range of T is contained in the orthogonal complement of $\mathbf{M}$, since it is immediate that $\langle Tx, v\rangle = 0$ from the definition of T, while if $u \in \mathbf{M}$ and $\langle u, v\rangle = 0$, and if z is of the form $y_t - y$ for $y \in \mathbf{L}$, then $\langle W(sz)v, u\rangle = 0$ for sufficiently small s; differentiation with respect to s then implies that $\langle\phi(z)v, u\rangle = 0$, hence $\langle Tz, u\rangle = 0$ for all $z \in \mathbf{L}_0$. Note also that the range of T is invariant under the $\Gamma(t)$, since it is immediate that $TS(t)x = \Gamma(t)Tx$ for all $x \in \mathbf{L}_0$, and that the restriction of $\Gamma(\cdot)$ to the closure $\mathbf{R}$ of the range of T leaves no nonzero vector fixed. Applying Scholium 3.1, it follows that the closure $\mathbf{R}$ of the range of T is a complex-linear subspace of $\mathbf{K}$; it is only necessary to apply this result to the projection whose range is $\mathbf{R}$. This completes the proof of *b*) and *c*). □

LEMMA 3.5.6. *For all $z \in \mathbf{L}_0$ and all $u \in \mathbf{M}$ that are orthogonal to v, $\langle W(z)v, u\rangle = 0$.*

PROOF. Let $\mathbf{L}_1$ denote the set of all elements $z \in \mathbf{L}_0$ such that the conclusion of the lemma holds. From Lemma 3.4.4, we already know that for any $z \in \mathbf{L}_0$, there exists $\varepsilon > 0$ such that $\lambda z \in \mathbf{L}_1$ if $\lambda \in [-\varepsilon, \varepsilon]$. Therefore it suffices to show that $\mathbf{L}_1$ is closed under addition. Suppose that z and z' are in $\mathbf{L}_1$. Define $\langle x, x'\rangle = \langle Tx, Tx'\rangle$ for arbitrary x and x' in $\mathbf{L}_0$. Then by the Weyl relations,

$$e^{\frac{1}{2}\langle z'_s, z\rangle}\langle W(z'_s)v, W(-z)u\rangle = e^{\frac{1}{2}\langle z, z'_s\rangle}\langle W(z)v, W(-z'_s)u\rangle$$

for arbitrary real s. As earlier, the left side of the foregoing equation extends to a bounded holomorphic function in the upper half-plane, while the right side does so in the lower half-plane, implying that each side is constant. It follows that

$$e^{\frac{1}{2}\langle z'_s, z\rangle}\langle W(z'_s)v, W(-z)u\rangle = \langle W(z')v, v\rangle\langle v, W(-z)u\rangle = 0,$$

and substituting $s = 0$, it follows that $\langle W(z + z')v, u\rangle = 0$. □

LEMMA 3.5.7. $\langle W(x)v, v\rangle = k(x)\exp(-\frac{1}{4}\|x\|^2)$, $x \in \mathbf{L}_0$, *where k is a character on $\mathbf{L}_0$.*

PROOF. An argument similar to that for the proof of Lemma 3.5.6 shows that for arbitrary x and y in $\mathbf{L}_0$,

$$\langle W(x + y)v, v\rangle = e^{-\frac{1}{2}\operatorname{Re}\langle x,y\rangle}\langle W(x)v, v\rangle\langle W(y)v, v\rangle.$$

Taking $y = -x$, it results that

$$|\langle W(x)v, v\rangle| = \exp(-\tfrac{1}{4}\|x\|^2),$$

and on utilizing this it follows from the preceding equation that if $k(x)$ is defined as $\langle W(x)v, v\rangle \exp(\frac{1}{4}\|x\|^2)$, then k is a character. □

LEMMA 3.5.8. *For arbitrary $x \in \mathbf{L}$ and real t, $\langle W(x_t - x)v, v\rangle$ is real.*

PROOF. It follows from the Weyl relations that

$$f(t) = \langle W(x_t - x)v, v\rangle \langle W(x_t + x)v, v\rangle = \langle W(x_t)v, W(x)v\rangle \langle W(x_t)v, W(-x)v\rangle.$$

Lemma 3.5.2 implies that $f(t)$ is independent of t. Hence $f(t) = f(0) = \langle W(2x)v, v\rangle$. Similarly we have

$$g(t) = \langle W(x - x_t)v, v\rangle \langle W(x_t + x)v, v\rangle = \langle W(2x)v, v\rangle.$$

Noting that $\langle W(2x)v, v\rangle \neq 0$ for sufficiently small x it follows that

$$1 = f(t)/g(t) = \langle W(x_t - x)v, v\rangle / \langle W(x - x_t)v, v\rangle,$$

which implies that $\langle W(x_t - x)v, v\rangle$ is real since $\langle W(x_t - x)v, v\rangle$ is the complex conjugate of $\langle W(x - x_t)v, v\rangle$. Using Lemma 3.5.7, it follows that $\langle W(x)v, v\rangle \neq 0$ for all x. □

Theorem 3.5 now follows by combining results given in Lemma 3.5.5, taking $\mathbf{H}$ as the closure of the range of T, with Lemmas 3.5.6, 3.5.7, and 3.5.8. □

3.6. Fermionic quantization of orthogonal dynamics

In the last section it was shown that a given linear symplectic dynamics was stably quantizable as a boson field if and only if the given dynamics was stably unitarizable, and that the resulting quantization was essentially unique and equivalent to the boson field over the complex Hilbert space resulting from the unitarization. Although at first glance the fermionic case may look essentially different, nevertheless the parallel theorem holds here—notwithstanding the negative frequencies that occur in physical fermion models, so-called "hole theory," etc. This is one more of innumerable instances of parallels between the boson and fermion cases, and is important for the physical interpretation of fermionic systems exhibiting symmetry between fermions and antifermions (which, in the case of charged particles, have opposite charges).

In the fermionic case there is given a linear orthogonal one-parameter group, rather than a symplectic group. Fermion fields have no direct classical analog, so this group lacks superficial formal resemblance to a classical dynamics. Let $\mathbf{L}$ be a given real topological vector space, S a continuous nondegenerate symmetric form on **L**, and V' a continuous one-parameter orthogonal group on $(\mathbf{L}, S)$. The question is that of the existence and uniqueness of a stable fermionic quantization of the system $(\mathbf{L}, S, V)$, defined analogously to that in the boson case. The basic case is that in which S is positive definite, and the topology in **L** is derived from the corresponding metric, in which case it is no essential loss of generality to assume that **L** is complete.

THEOREM 3.6. *Let* **L** *be a real Hilbert space with corresponding positive definite symmetric form S. Let V' be a continuous one-parameter orthogonal group on* **L**, *leaving no nonzero vector in* **L** *fixed. Suppose that there exists a system* $(\mathbf{K}, \phi, \Gamma, v)$ *such that*

i) **K** *is a complex Hilbert space;*
ii) ϕ *is a (real) linear map from* **L** *into selfadjoint operators on* **K** *such that for all* $x, y \in \mathbf{L}$, $\phi(x)\phi(y) + \phi(y)\phi(x) = S(x, y)$;
iii) $\Gamma(\cdot)$ *is a continuous one-parameter group on* **K** *such that for all* $t \in \mathbf{R}$ *and* $x \in \mathbf{L}$, $\Gamma(t)v = v$ *and* $\phi(V(t)x) = \Gamma(t)\phi(x)\Gamma(t)^{-1}$;
iv) $\Gamma(\cdot)$ *is stable; and*
v) v *is cyclic for the totality of the* $\phi(x)$, $x \in \mathbf{L}$.

Then there exists a unique complex Hilbert space structure on $(\mathbf{L}, S)$ *such that* $S(x, y) = \mathrm{Re}(x, y)$ *for all* $x, y \in \mathbf{L}$, *and such that* $V(\cdot)$ *is unitary and strictly stable on this Hilbert space,* **H**. *Moreover,* (K, ϕ, Γ, v) *is unitarily equivalent to the free fermion field over* **H**, *apart from the restriction of* Γ to $V(\cdot)$.

Conversely, if the latter conditions are fulfilled for some complex Hilbert space structure **H** *on* $(\mathbf{L}, S)$, *then there exists a quantization of the indicated type, which is unique within unitary equivalence.*

PROOF. Let ϕ be extended to the real Clifford algebra **C** over $(\mathbf{L}, S)$ as an algebraic homomorphism, likewise denoted by ϕ, and let $E(u)$ denote $\langle\phi(u)v, v\rangle$. To begin with it will be shown that the "n-point function" $E(x_1x_2 \cdots x_n)$ is uniquely determined by **L**, S, and V under the hypotheses of the theorem. Consider first $E(x)$, $x \in \mathbf{L}$. This is a continuous real linear functional of x by virtue of the Clifford relations, which imply that $\|\phi(x)v\| = 2^{1/2}\|x\|$. Hence there exist unique vectors y and y' in **L** such that $E(x) = S(x,y) + iS(x, y')$. On the other hand, $E(V(t)x) = E(x)$, which by the uniqueness of y and y' implies that $V(-t)y = y$ and $V(-t)y' = y'$ for all t. But by hypothesis, V leaves no nonzero vector fixed, so that $y = y' = 0$, showing that $E(x) = 0$ for all $x \in \mathbf{L}$.

Now let x and y be arbitrary in **L**, let $x_t = V(t)x$, and apply E to both sides of the relation $x_ty + yx_t = S(x_t, y)$, obtaining

$$\mathrm{Re}(E(x_ty)) = \mathrm{Re}(E(yx_t)) = \tfrac{1}{2}S(x_t, y).$$

Now $E(x,y)$ is as a function of t the boundary value function on the real axis of a bounded analytic function in the upper half-plane. Observe next the

LEMMA 3.6.1. *For arbitrary $x \in \mathbf{L}$ and $u \in \mathbf{K}$, $\langle \Gamma(t)\phi(x)v, u\rangle$ is the boundary value for t on the real axis of a bounded analytic function $f(z)$, $z = t + is$, in the upper half-plane, and $f(is) \to 0$ as $s \to \infty$.*

PROOF. The nonnegativity of the selfadjoint generator H of $\Gamma(t)$ implies the existence of the indicated bounded analytic extension by the spectral theorem. If P is the projection of $\mathbf{K}$ onto the null space of H, then

$$\lim_{s\to\infty} f(is) = \lim_{s\to\infty} \langle e^{-sH}\phi(x)v, u\rangle = \langle P\phi(x)v, u\rangle$$

This is a continuous linear functional of x—which is moreover invariant under $V(\cdot)$, by the hypothesized intertwining relations—and must vanish by the same argument as in the proof that $E(x) = 0$. □

PROOF OF THEOREM, CONTINUED. Lemma 3.6.1 implies that the analytic function in the upper half-plane extending $E(x,y)$ tends to 0 as $s \to \infty$; since its real part on the real line is uniquely determined in terms of $(\mathbf{L}, S, V)$, it follows that the two-point function is uniquely determined in terms of $(\mathbf{L}, S, V)$.

The n-point functions for $n > 2$ may be determined by induction as in the proof of Theorem 2.1. Thus the linear functional E on $\mathbf{C}$ is determined by $(\mathbf{L}, S, V)$, and it follows by the canonical association of representations with positive linear functionals that $\mathbf{K}$, ϕ, and Γ are similarly determined. Now consider the real-linear map from $\mathbf{L}$ into $\mathbf{K}$, $T\colon x \mapsto 2^{1/2}\phi(x)v$, which is orthogonal by virtue of the anticommutation relations. The range $\mathbf{R}$ of T is accordingly a closed real-linear subspace of $\mathbf{K}$. Moreover $\mathbf{R}$ is orthogonal in $\mathbf{K}$ to the null space $\mathbf{N}$ of the generator H of $\Gamma(\cdot)$, as a consequence of Lemma 3.6.1. Now the restriction $\Gamma_+(t)$ of $\Gamma(t)$ to the orthocomplement $\mathbf{N}^\perp$ of $\mathbf{N}$ has a positive selfadjoint generator. But it follows from Scholium 3.1 that an invariant real subspace of such a one-parameter unitary group is necessarily complex-linear, and the intertwining relation $\Gamma(t)Tx = TV(t)x$ shows that $\mathbf{R}$ is indeed invariant under $\Gamma_+(\cdot)$. Transferring the consequent complex structure in $\mathbf{R}$ to $\mathbf{L}$ by the isometry T, the result is a complex Hilbert space structure on $\mathbf{L}$ such that $\mathrm{Re}(\langle x, y\rangle) = S(x, y)$. As T interwines the action of V and of Γ_+, $V(\cdot)$ becomes a unitary one-parameter group with positive generator. The unicity of this complex structure is then a consequence of general theory as earlier.

The converse follows similarly from Theorem 2.1. □

Problems

1. Let **L** denote the space of real C^∞ solutions having compact support in space at every time to the equation $\square\varphi + (c + V)\varphi = 0$ on $\mathbf{R} \times \mathbf{R}^n$, where c is a constant and V is a given bounded continuous function. Let $S(t)$ denote the one-parameter symplectic group in **L** defined by the equation. Let A denote the symplectic form defined in Section 3.5.

a) Show that if $c \geq 0$ and V is nonnegative there exists a unique stable unitarization for $S(\cdot)$. Compute the requisite complex structure.

b) Show that if $c < 0$ and $V = 0$ there exists no stable unitarization for $S(\cdot)$.

2. Develop the analog to Problem 1 in the context of the abstract equation $\varphi''(t) + A\varphi(t) = 0$, where A is a given selfadjoint operator in a real Hilbert space $\mathbf{H}'$, using Cauchy data in subspaces contained in the ranges of the spectral projections for A corresponding to bounded intervals.

3. Derive the ergodicity of the flow corresponding to Brownian motion, which may be defined as follows: let $x(t)$ be the stochastic process on $\mathbf{R}$ determined by the conditions that for any finite set of times $t_1 < t_2 < \cdots < t_n$, the $x(t_1), x(t_2), \ldots, x(t_n)$ have a joint Gaussian distribution such that $x(t_2) - x(t_1)$ has mean 0 and variance $t_2 - t_1$ and is stochastically independent of $x(t_4) - x(t_3)$. Let M denote the corresponding probability measure space, and let **K** denote the closure in $L_2(M)$ of the set of all polynomials in the differences $x(t) - x(t')$, as t and t' vary. The flow is given by the unique unitary transformation $U(s)$ on **K** that is multiplicative and carries $x(t) - x(t')$ into $x(t + s) - x(t' + s)$. (Hint: consider the isonormal process on $L_2(\mathbf{R})$, $f \rightarrow \int f(t)dx(t)$, and observe that no finite-dimensional subspace of $L_2(\mathbf{R})$ is invariant under temporal translation.)

3. Let $U(\cdot)$ be a continuous one-parameter unitary group on the Hilbert space **H** that leaves no nonzero vector fixed. Show that there exists a unique complex structure J on **H** that stably unitarizes $U(\cdot)$ as an orthogonal group in the following sense: let $S(x, y) = \mathrm{Re}(\langle x, y\rangle)$ and consider the orthogonal space $(\mathbf{H}^\#, S)$. Then J is required to be orthogonal; to commute with the $U(t)$; and the generator of $U(t)$ relative to the new inner product

$$\langle x, y\rangle' = S(x, y) - iS(Jx, y)$$

in the Hilbert space $\mathbf{H}'$ in which the action of i is changed to J, is required to be positive. Compute J explicitly.

Bibliographical Notes on Chapter 3

The measure-preserving action of the orthogonal subgroup of the unitary group on boson fields in the real wave representation was treated by Segal

(1957b). This derives from the concept of Koopman (1931) and generalizes Wiener's application of this to Brownian motion. The analog of the Koopman concept and the ergodicity question for fermion fields were also treated by Segal. The ergodicity of the flow associated with Brownian motion was treated by Anzai (1949), and its spectrum by Kakutani (1950). Unicity of positive-energy quantization was treated by Segal (1962) and generalized by Weinless (1969).

4

Absolute Continuity and Unitary Implementability

4.1. Introduction

A certain class of divergences in quantum field theory originates in the failure of the Stone–von Neumann uniqueness theorem. In general, there is no unitary equivalence between different irreducible canonical systems over the same space; or, put in terms of theoretical physical practice, the putative unitary equivalence in question is "divergent."

A simple example may be provided by the canonical transformation $p_j' = p_j + q_j$, $q_j' = q_j$ $(j = 1,\ldots,n)$, and just what happens to it when n becomes infinite. (We are working here modulo rigorous details, which follow from the Weyl relations; it is simpler conceptually to use the Heisenberg relations.) It is evident that the p_j' and q_j' are again a *canonical system*, i.e., satisfy the Heisenberg relations. The unitary equivalence is easily written down explicitly, as in all cases of linear canonical transformations:

$$(p', q') = U^{-1}(p, q)U, \text{ where } U = \exp(i(q_1^2 + q_2^2 + \cdots)/2);$$

the operator U is of course unique within a constant factor.

Now consider the case of an infinite number of dimensions, in which the p's and q's are those for a free boson field, relative to an orthonormal basis in the underlying single-particle Hilbert space. The primed operators still exist and satisfy the Weyl relations (modulo rigorous details of formation of closures, etc., readily supplied by Weyl-relation techniques similar to ones previously employed). However, the operator U makes no sense, the infinite sum $\Sigma_j q_j^2$ having no meaning, and the slight freedom in the definition of U, namely the ambiguous constant factor, is of no help in the matter. In a word, the "operator" U is divergent; this does not prevent theoretical physicists from trying to use it as if it existed, but clarifies the emergence of "infinities" (such as the matrix elements of the nonexistent operator U) and presents a mathematical problem.

Does there exist, in the infinite-dimensional case, any unitary operator U on the boson field Hilbert space **K** such that

$$p_j' = U^{-1}p_jU, \qquad q_j' = U^{-1}q_jU?$$

We shall develop techniques for dealing with such questions, and see that the answer to this specific question is negative (as it usually is in the infinite-dimensional case, except in a few simple—though interesting—instances).

Having come up against a crucial apparent difficulty, it will be seen in Chapter 5 that a partial remedy, which at least restores uniqueness analogous to that of the Stone–von Neumann theorem, is the consideration of an appropriate C^*-algebra, on which the canonical transformation can be effected as an automorphism.

It turns out that the criteria for unitary implementability are closely related to those for absolute continuity of probability measures (or noncommutative analogs thereof) on infinite-dimensional linear spaces. At first glance this might seem odd, but on reflection one sees that, if two measures are mutually absolutely continuous, the operation of multiplication by the square root of the Radon-Nikodym derivative of one with respect to the other is a unitary map of one of the L_2 spaces into the other, which transforms the multiplication algebra of one into the other. It is also true that any algebraic isomorphism of one multiplication algebra onto the other is unitarily implementable in this fashion; i.e., the respective measures are mutually absolutely continuous, the unitary operator just described exists, and so on. In the case of a boson field the canonical q's generate the multiplication algebra of a relevant measure space (cf. the real wave representation), and these considerations thereby apply. With the usual ex post facto reinterpretation, the same comments apply to fermion fields.

The results obtained will permit suitable extensions of the harmonic and spin representations to the infinite-dimensional case. It turns out that general symplectic (resp. orthogonal) transformations on the single-particle space can be represented by automorphisms of C^*-algebras, but (in general) have no unitary implementation in the free field representation. There is only an invariant subgroup of elements that, roughly speaking, commute with i within a Hilbert-Schmidt operator, that can be so represented in extension of the finite-dimensional case.

4.2. Equivalence of distributions

We begin by recalling and extending some earlier functional integration considerations. One type of equivalence of distributions that has already been considered is that defined by mutual absolute continuity. If **L** is a given real topological vector space, then two distributions m and n are equivalent in this sense if and only if there is a $*$-isomorphism between the respective algebras of measurables that carries $m(f)$ into $n(f)$, for all f in $\mathbf{L}^*$. Alternatively, m and n are equivalent in this sense if there exists a unitary transformation U from $L_2(\mathbf{L}, m)$ onto $L_2(\mathbf{L}, n)$ such that the operator $M_{m(f)}$ of multiplication by $m(f)$ is carried by U into that of multiplication by $n(f)$ on the latter Hilbert space: $UM_{m(f)}U^{-1} = M_{n(f)}$, for all f in $\mathbf{L}^*$. When **L** is finite-dimensional, m and n are equivalent in this sense if and only if the corresponding countably additive probability measures in **L** have the same null sets.

The notion just described was called algebraic equivalence, to distinguish it from the following stronger notion. Two distributions m and n are *metrically equivalent*, denoted $m \approx n$, in case there exists an isomorphism between their respective algebras of measurables that not only carries $m(f)$ into $n(f)$ for all f, but is integral- or expectation-preserving. The distinction is similar to that between a *measurable* transformation and a *measure-preserving* transformation.

One criterion for metric equivalence involves the concept of a *multiplicative* unitary transformation from one L_2 space to another. This is a unitary transformation U with the property that if f, g, and fg are all in L_2, then $U(fg) = U(f)U(g)$.

THEOREM 4.1. *Let m and n be distributions on the real topological vector space* **L**. *Then $m \approx n$ if and only if there exists a multiplicative unitary mapping u from $L_2(\mathbf{L}, m)$ onto $\mathbf{L}_2(\mathbf{L}, n)$ such that $UM_{m(f)}U^{-1} = M_{n(f)}$ for all $f \in \mathbf{L}^*$.*

PROOF. We simplify the notation here by setting M_f for $M_{m(f)}$ and N_f for $M_{n(f)}$. Regarding the "if" part, observe that for arbitrary $f_1, \ldots, f_k \in \mathbf{L}^*$ and for any Borel set B in $\mathbf{R}^k$ the probability that $(m(f_1), \ldots, m(f_k)) \in B$ may be represented as $\langle c_B(M_{f_1}, \ldots, M_{f_k})1, 1\rangle$, where c_B is the characteristic function of B, and 1 is the functional identically 1 on **L**. A multiplicative unitary mapping U from one L_2 space over a probability measure space to another commutes with the formation of Borel functions, and maps 1 into 1, by standard commutative spectral theory. It follows that $\langle c_B(M_{f_1}, \ldots, M_{f_k})1, 1\rangle = \langle c_B(N_{f_1}, \ldots, N_{f_k})1, 1\rangle$, showing that $m \approx n$.

If, on the other hand, it is given that $m \approx$ n, then a unitary transformation U may be defined from $L_2(\mathbf{L}, m)$ onto $L_2(\mathbf{L}, n)$ by first defining U_0 on the dense domain of bounded Borel functions $b(m(f_1), \ldots, m(f_k))$ of finitely many of the $m(f)$ as the map carrying this functional into $b(n(f_1), \ldots, n(f_k))$. Then it is not difficult to verify that U_0 is isometric and multiplicative on this domain and has dense range, and so extends uniquely to a unitary transformation U from all of $L_2(\mathbf{L}, m)$ onto $L_2(\mathbf{L}, n)$ which is readily seen to remain multiplicative. □

COROLLARY 4.1.1. *Let m be a given distribution on a real topological vector space* $\mathbf{L}$, *and let G be the group of all nonsingular continuous linear transformations T on* $\mathbf{L}$ *such that $m \approx m_T$. Then there exists a unique unitary representation $U(\cdot)$ of G on $L_2(\mathbf{L}, m)$ such that*

$$U(T)^{-1}M_{m(f)}U(T) = M_{m(T^*f)}; \qquad f \in \mathbf{L}^*, T \in G.$$

PROOF. This is an immediate deduction from Theorem 4.1 for the special case in which $L_2(\mathbf{L}, m)$ and $L_2(\mathbf{L}, n)$ are identical. □

EXAMPLE 4.1. Let g be the isonormal distribution on the real Hilbert space $\mathbf{H}'$. Then for any orthogonal transformation T on $\mathbf{H}'$, $g_T \approx g$. Consequently there exists a unitary representation $U(\cdot)$ of $O(\mathbf{H}')$ on $L_2(\mathbf{H}', n)$ which carries g into g_T, in the sense that multiplication by $g(x)$ is carried into multiplication by $g(T^*x)$, for all $x \in \mathbf{H}'$. This unitary operator $U(T)$ is the same as the operator $\Gamma(T_c)$, where T_c denotes the complex-linear extension of T to the complexification $\mathbf{H}$ of $\mathbf{H}'$, acting on the Hilbert space $\mathbf{K}$ of the free boson field over $\mathbf{H}$ in the real wave representation.

More generally, let n be the normal distribution on the real Hilbert space $\mathbf{H}'$ of mean 0 and covariance operator A, where A is nonsingular. If T is any orthogonal transformation on $\mathbf{H}'$ that commutes with A, it is easily seen that $n_T \approx n$, the normal distribution being determined by its mean and covariance operator, which are not affected by the transformation T. Consequently, there exists a unitary representation Γ of the centralizer of A in $O(\mathbf{H}')$ on $L_2(\mathbf{H}', n)$ with the property that $\Gamma(T)^{-1}M_{n(x)}\Gamma(T) = M_{n(T^*x)}$, with the same notation as earlier.

4.3. Quasi-invariant distributions and Weyl systems

The construction in Chapter 1 of a Weyl system from a quasi-invariant distribution can be regarded as an appropriate parallel to the construction of the Schrödinger representation. Conversely, a quasi-invariant distribution may be derived from a Weyl system. This is described in the next theorem, the general idea of which is briefly as follows: let $(\mathbf{K}, W)$ be a cyclic Weyl system over $\mathbf{L}\oplus\mathbf{L}^*$, where $\mathbf{L}$ is a given real linear topological space. Setting $W|\mathbf{L}^* = V$, and denoting as $\partial V(x)$ the selfadjoint generator of the one-parameter group $\{V(tx): t \in \mathbf{R}\}$, then the mapping $\partial: x \mapsto \partial V(x)$ is essentially linear from $\mathbf{L}^*$ into a commutative set of selfadjoint operators in $\mathbf{K}$. Here *essentially linear* means linear with respect to *strong operations* in the (partial) algebra of unbounded operators, in which the usual operations are followed by the operation of forming the closure. These closures exist because of the commutativity of the range of ∂V. More specifically, it is known that the totality of normal (possibly unbounded) operators whose spectral projections are contained in a commutative W^*-algebra form a linear associative algebra over the complex number field, with respect to strong operations. By spectral theory, any such W^*-algebra is $*$-algebraically isomorphic to the bounded measurables on a measure space, and the indicated normal operators (which are affiliated with the given W^*-algebra) then correspond to possibly unbounded measurables. Thus, ∂ fails to specify a predistribution essentially only in that a specific measure, and not merely an absolute continuity class, needs to be specified, and that this measure should be normalized to a total of unity on the whole space. The latter normalization is always possible in a separable Hilbert space, and once a distribution is obtained this way, the Weyl relations show that is it quasi-invariant.

However, given a quasi-invariant distribution, the full specification of a corresponding Weyl system involves in addition a *multiplier* (or *cocycle*). If S is any continuous representation of a group G on a Hilbert space $\mathbf{K}$, a multiplier for S is defined as a function T from G to the bounded linear operators on $\mathbf{K}$ such that $S(\cdot)T(\cdot)$ is again a continuous representation of G on $\mathbf{K}$. Such a multiplier is said to be *relative to a given ring* $\mathbf{A}$ of operators on $\mathbf{K}$ in case its values lie in the commutor $\mathbf{A}'$. Two multipliers are said to be *equivalent relative to* $\mathbf{A}$ if they are unitarily equivalent via a unitary operator in $\mathbf{A}'$.

THEOREM 4.2. *Let m be a quasi-invariant distribution on a real linear space $\mathbf{L}$ (topologized algebraically). For arbitrary $x \in \mathbf{L}$, let m_x denote the distribution $m_x(\lambda) = m(\lambda) + \lambda(x)$, $\lambda \in \mathbf{L}^*$. For arbitrary $x \in \mathbf{L}$ and $\lambda \in \mathbf{L}^*$, let $U(x)$ and $V(\lambda)$ be the operators on $\mathbf{K} = L_2(\mathbf{L}, m)$ defined as follows:*

$$U(x)\theta(F) = (dm_x/dm)^{1/2}\,\theta(F_x)$$
$$V(\lambda)\theta(F) = e^{im(\lambda)}\,\theta(F).$$

Here $\theta(F)$ is the element of $L_2(\mathbf{L}, m)$ corresponding to the tame function F, and $F_x(y) = F(x + y)$. Then (U, V) form a Weyl pair over $(\mathbf{L}, \mathbf{L}^*)$, with the algebraic topology on $\mathbf{L}^*$.

Conversely, if (U, V) is any Weyl pair over $(\mathbf{L}, \mathbf{L}^*)$ such that $\{V(\lambda): \lambda \in \mathbf{L}^*\}$ has a cyclic vector, then there exists a quasi-invariant distribution m on $\mathbf{L}$ and a multiplier S for the component V_0 of the Weyl pair (U_0, V_0), obtained from m as in the preceding paragraph, such that (U, V) is unitarily equivalent to (U_0, SV_0).

PROOF. The first part of the theorem is just a restatement of Theorem 1.5. Conversely, suppose (U, V) is a Weyl pair over $(\mathbf{L}, \mathbf{L}^*)$, with representation space $\mathbf{K}$ and a cyclic unit vector v for V in $\mathbf{K}$. Let $\mathbf{A}$ denote the (abelian) W^*-algebra generated by the $V(\lambda)$. By spectral theory, the system $(\mathbf{K}, \mathbf{A}, v)$ is unitarily equivalent to $(L_2(M), \mathbf{M}, 1)$ for some probability measure space M with multiplication algebra $\mathbf{M}$. In particular, if $Q(\lambda)$ denotes the selfadjoint generator of the one-parameter unitary group $V(t\lambda)$, $t \in \mathbf{R}$, $Q(\lambda)$ may be represented as a random variable on M, and $Q(\cdot)$ then defines a distribution over $\mathbf{L}$. The Weyl relations show that

$$U(x)Q(\lambda)U(x)^{-1} = Q(x) - \lambda(x),$$

implying that $Q(\cdot)$ is quasi-invariant, and moreover that $U(x)$ induces the automorphism of $\mathbf{M}$ corresponding to the translation through x. This means that $U(x)$ must coincide with the operator similarly designated in the first part of the theorem, apart from an x-dependent factor that commutes with every element of $\mathbf{M}$, and is consequently a multiplier. □

This result is readily extended to the case of a given Weyl pair for which V is not necessarily cyclic, through the use of spectral multiplicity theory, applied to the ring generated by the $V(\lambda)$. The general system is then seen as a direct integral of parts of uniform multiplicity, each part being a tensor product of a system such as that just treated, with a Hilbert space of the appropriate multiplicity, on which the Weyl operators act trivially.

In the case when $\mathbf{L}$ is finite-dimensional, the Stone–von Neumann theorem shows that the multiplier may be eliminated, i.e., is inessential. This can be interpreted as related to the vanishing one-dimensional cohomology of a finite-dimensional $\mathbf{L}$, the cohomology class being defined as the quotient of the closed modulo the exact 1-forms, where however, a generalized notion of form must be employed.

To this end, an *exact 1-form* on $\mathbf{L}$ is defined as an equivalence class of measurable functionals on $\mathbf{L}$, relative to the given distribution m, where the equivalence $f \sim g$ means that $f - g$ is constant. A *closed 1-form* is defined as an assignment to each finite-dimensional subspace $\mathbf{F}$ of $\mathbf{L}$ of an equivalence class of measurable functionals on $\mathbf{L}$, where the equivalence $f \sim_{\mathbf{F}} g$ means that $f - g$ is based on $\{\lambda \in \mathbf{L}^*: \lambda(x) = 0 \text{ for all } x \in \mathbf{F}\}$—this assignment being required to have the consistency feature that if $\mathbf{F} \subset \mathbf{F}'$, $\mathbf{F}'$ also being finite-dimensional, then restriction to $\mathbf{F}$ of the equivalence class assigned to $\mathbf{F}'$ yields that assigned to $\mathbf{F}$.

These notions are formally identifiable with corresponding ones for differential 1-forms on finite-dimensional spaces whose coefficients are appropriately generalized functions. A representative example of a form which is closed but not exact on an infinite-dimensional real Hilbert space $\mathbf{H}$ is (symbolically) $\Sigma_{1 \leq k < \infty} x_k dx_k$, where the x_j are coordinates relative to an orthonormal basis. This form assigns to each finite-dimensional subspace $\mathbf{F}$ of $\mathbf{H}$ the measurable functional $\frac{1}{2} \Sigma_{1 \leq k \leq n} y_k^2$, where the y_k are coordinates relative to any orthonormal basis of $\mathbf{F}$. Thus in each finite number of dimensions this form restricts to an exact form, but there is no actual *measurable* functional corresponding to $\Sigma_{1 \le k < \infty} x_k^2$ on **H**, so the form is inexact although closed on all of **H** (cf. Segal 1959a).

4.4. Ergodicity and irreducibility of Weyl pairs

In case **L** is finite-dimensional, the Weyl pair (U, V) associated with any given quasi-invariant distribution on **L**, as in the first part of the preceding theorem, is irreducible. This is not the case for arbitrary quasi-invariant distributions, or, to put it another way, a quasi-invariant distribution need not be ergodic with respect to the group of translations on **L**. One useful case in which irreducibility may be concluded is described in Theorem 4.3. In this connection, a distribution on a real Hilbert space $\mathbf{H}'$ is said to be the direct sum of distributions m_j on submanifolds $\mathbf{H}_j$ of which $\mathbf{H}'$ is the direct sum in case $m|\mathbf{H}_j = m_j$ and the m_j are stochastically independent for different j.

THEOREM 4.3. *The Weyl pair determined by an L_2-continuous quasi-invariant distribution on a real Hilbert space is irreducible provided it is a direct sum of (stochastically independent) distributions on finite-dimensional submanifolds.*

PROOF. It will be convenient here to consider the distribution as selfadjoint-operator-valued, rather than random-variable-valued. This can be attained by replacing each random variable by the operation of multiplication by it, acting on L_2; expectation of random variables is then replaced by expectation of operators, in the state defined by the state vector 1.

Note now that the $U(x)$ generate the same W^*-algebra **R** as that determined by the $m(x)$ (i.e., generated by the spectral projections of the multiplications by the $m(x)$). This is maximal abelian, being the multiplication algebra of the underlying probability space. Hence any bounded linear operator T on $L_2(\mathbf{H}', m)$ that commutes with both the $U(x)$ and $V(y)$ is an element of **R** that commutes with all of the $V(y)$.

To conclude the proof of irreducibility, it therefore suffices to show that the only operators T in **R** such that $\alpha_y(T) = T$ for all y, where α_y is the automorphism **R**, $S \to V(y)^{-1}SV(y)$ ($S \in \mathbf{R}$), are scalar multiples of the identity. To this end, let $P_{\mathbf{M}}$ denote the operator of projecting $L_2(\mathbf{H}', m)$ onto the subspace $L_2(\mathbf{M}, m)$ of operators based on **M** (i.e., affiliated with the W^*-algebra determined by the $m(x)$ with $x \in \mathbf{M}$). Then $P_{\mathbf{M}} \to I$ strongly as **M** ranges over the directed system of finite direct sums of the given stochastically independent subspaces, since these subspaces span and m is continuous.

Noting the validity of the corollary when $\dim(\mathbf{H}) < \infty$, it follows that to conclude the proof, it suffices to show that $P_{\mathbf{M}}$ commutes with the action of α_y for $y \in \mathbf{M}$, i.e., $P_{\mathbf{M}}\alpha_y(S) = \alpha_y(P_{\mathbf{M}}S)$ for arbitrary $S \in \mathbf{R}$. To this end, note first that both sides of this putative equality, to be denoted (E), are in the subalgebra $\mathbf{R}_{\mathbf{M}}$ determined by $m|\mathbf{M}$. For the left side this follows from the observation that $P_{\mathbf{M}}S$ is identical to $E[S|\mathbf{R}_{\mathbf{M}}]$ for arbitrary $S \in \mathbf{R}$. For the right side of (E), this follows from the defining characteristic of $V(y)$, showing that for $y \in \mathbf{M}$, $V(y)^{-1}SV(y) \in \mathbf{R}_{\mathbf{M}}$ if $S \in \mathbf{R}_{\mathbf{M}}$.

Thus to establish (E) it suffices to show that both sides have the same conditional expectation with respect to $\mathbf{R}_{\mathbf{M}}$, which by the definition of conditional expectation reduces to showing that for arbitrary selfadjoint $X \in \mathbf{R}_{\mathbf{M}}$,

$$E[P_{\mathbf{M}}\alpha_y(S)X] = E[\alpha_y(P_{\mathbf{M}}(S))X].$$

The left side of this proposed equation is identical to $E[\alpha_y(S)X]$, which in turn equals $E[S\alpha_y^{-1}(X)D_y]$, where $D_y = dm_y/dm$. On the other hand, since $\alpha_y^{-1}(X)$ is in $\mathbf{R}_{\mathbf{M}}$,

$$E[\alpha_y(P_{\mathbf{M}}(S))X] = E[P_{\mathbf{M}}(S)\alpha_y^{-1}(X)D_y] = E[E\{P_{\mathbf{M}}(S)D_y|\mathbf{R}_{\mathbf{M}}\}\alpha_y^{-1}(X)].$$

Using the multiplicative property of conditional expectation when one factor is affiliated with the subalgebra, the last expression reduces to $E[P_{\mathbf{M}}(S)\alpha_y^{-1}(X)E\{D_y|\mathbf{R}_{\mathbf{M}}\}]$, which by the definition of conditional expectation equals $E[S\alpha_y^{-1}(X)E\{D_y|\mathbf{R}_{\mathbf{M}}\}]$. Hence the stated equality holds if (and only if) $E\{D_y|\mathbf{R}_{\mathbf{M}}\} = D_y$, which follows directly from the circumstances that $\mathbf{M}$ and its orthocomplement are stochastically independent and that $y \in \mathbf{M}$. □

EXAMPLE 4.2. The isonormal distribution on a real Hilbert space $\mathbf{H}'$ is the direct sum of the isonormal distributions on the one-dimensional subspaces spanned by any orthonormal basis. These are stochastically independent, and Theorem 4.3 yields an alternative proof of the irreducibility of the Weyl system of a free boson field.

4.5. Infinite products of Hilbert spaces

Many of the structures of quantum field theory have useful realizations in terms of a type of infinite product introduced by von Neumann (1938). Although there appears to be no appropriate definition of an infinite product of measure spaces, except in the case when all but a finite number are probability measure spaces, von Neumann gave an effective and invariant definition of an infinite product of Hilbert spaces. But of greater practical utility than this product, which he described as "complete," are those he described as "incomplete," which resemble somewhat infinite products of probability spaces. The incomplete product is a product not simply of Hilbert spaces, but of Hilbert spaces with a distinguished unit vector, which plays the role of the unit random variable in the case of a probability space, or physically of a vacuum or ground state vector.

DEFINITION. For each index λ in the index set Λ, let z_λ be a unit vector in the complex Hilbert space $\mathbf{K}_\lambda$. A system $(\mathbf{K}, z, \tau_\Delta)$ consisting of a complex Hilbert space $\mathbf{K}$; a unit vector z in $\mathbf{K}$; and, for each finite set Δ of indices, an isometric map τ_Δ of the finite tensor product $\otimes_{\lambda\in\Delta}\mathbf{K}_\lambda$ into $\mathbf{K}$, is called a *direct product* of the $(\mathbf{K}_\lambda, z_\lambda)$ $(\lambda \in \Lambda)$ in case the following conditions are satisfied:

1) if Ξ is any finite set of indices, if $\Delta \subset \Xi$, and if $x_\lambda = z_\lambda$ for $\lambda \notin \Delta$, then $\tau_\Xi \otimes_{\lambda\in\Xi} x_\lambda = \tau_\Delta \otimes_{\lambda\in\Delta} x_\lambda$ (thus, tensoring with the z_λ does not materially affect the result, much as tensoring with the function 1 on probability measure spaces does not affect the result);
2) $\tau_\Delta \otimes_{\lambda\in\Delta} z_\lambda = z$, for all Δ; and
3) the union of the ranges of the τ_Δ is dense in $\mathbf{K}$.

A pair $(\mathbf{H}, z)$ consisting of a complex Hilbert space $\mathbf{H}$ and a unit vector $z \in \mathbf{H}$ may be called a *grounded Hilbert space*, since z plays the role of a ground state vector.

SCHOLIUM 4.1. *The direct product of grounded Hilbert spaces exists and is unique within unitary equivalence.*

PROOF. A proof can be given that is formally analogous to proofs of the existence of infinite products of probability spaces, made more algebraic. Alternatively, a direct limit type of argument may be used. The details are omitted. □

THEOREM 4.4. *Let $(\mathbf{K}, z, \tau_\Delta)$ be the direct product of the $(\mathbf{K}_\lambda, z_\lambda)$, and let A_λ be an automorphism of the ring of all bounded linear operators on $\mathbf{K}_\lambda$. Then there exists an automorphism of the ring of all bounded linear operators on $\mathbf{K}$ that coincides on the range of τ_Δ with $\tau_\Delta A_\Delta \tau_\Delta^{-1}$, where A_Δ is the tensor product of the A_λ, $\lambda \in \Delta$, if and only if the product $\Pi_{\lambda\in\Lambda}|\langle A_\lambda(P_\lambda)z_\lambda, z_\lambda\rangle|$ is convergent. Here P_λ denotes the projection of $\mathbf{K}_\lambda$ onto the one-dimensional subspace spanned by z_λ.*

The last condition asserts essentially that A_λ does not alter P_λ very much, or, equivalently, that the unitary operator that induces A_λ changes z only slightly, apart from phase. The tensor product of automorphisms need not be introduced as an additional concept, but may be treated adequately for present purposes in terms of the tensor products of implementing unitary operators.

LEMMA 4.4.1. *If V_λ is a given unitary operator on $\mathbf{K}_\lambda$, then there exists a unitary operator V on $\mathbf{K}$ such that $V \otimes_\lambda x_\lambda = \otimes_\lambda V_\lambda x_\lambda$ for all convergent $\{x_\lambda\}$ with $x_\lambda \in \mathbf{K}_\lambda$ (in the sense that $\tau_\Delta \otimes_{\lambda \in \Delta} x_\lambda$ is convergent in $\mathbf{K}$ as $\Delta \to \Lambda$, in which case $\otimes_\lambda x_\lambda$ is defined as the limit) and $\|x_\lambda\| = 1$, if and only if the numerical product $\Pi_\lambda \langle V_\lambda z_\lambda, z_\lambda \rangle$ is convergent.*

PROOF OF LEMMA. Note first that a necessary and sufficient condition that $\tau_\Delta \otimes_{\lambda \in \Delta} x_\lambda$ converge in $\mathbf{K}$ is that the product $\Pi_\lambda \langle x_\lambda, z_\lambda \rangle$ be convergent. For if $\Delta \subset \Xi$, where Δ and Ξ are finite subsets of Λ, then

$$|\tau_\Delta \bigotimes_{\lambda \in \Delta} x_\lambda - \tau_\Xi \bigotimes_{\lambda \in \Xi} x_\lambda|^2 = 2 - 2 \operatorname{Re} \prod_{\lambda \in \Xi - \Delta} \langle x_\lambda, z_\lambda \rangle$$

Hence Re $\Pi_{\lambda \in \Xi - \Delta} \langle x_\lambda, z_\lambda \rangle$ must tend to unity as $\Delta, \Xi \to \Lambda$, but as $|\Pi_{\lambda \in \Xi - \Delta} \langle x_\lambda, z_\lambda \rangle| \leq 1$, this implies that Im $\Pi_{\lambda \in \Xi - \Delta} \langle x_\lambda, z_\lambda \rangle \to 0$, so that $\Pi_{\lambda \in \Xi - \Delta} \langle x_\lambda, z_\lambda \rangle \to 1$, i.e., the product is convergent.

Note next that if T_λ is a continuous linear operator on $\mathbf{K}_\lambda$ with $T_\lambda = I_\lambda$ for all but a finite number of λ, then there exists a unique continuous linear operator T on $\mathbf{K}$ such that

$$T \tau_\Delta \bigotimes_{\lambda \in \Delta} x_\lambda = \tau_\Delta \bigotimes_{\lambda \in \Delta} T_\lambda x_\lambda$$

for every finite subset Δ of Λ and arbitrary $x_\lambda \in \mathbf{K}_\lambda$. Now observe that if V_λ is a given unitary operator on $\mathbf{K}_\lambda$, there exists a unitary operator V on $\mathbf{K}$ such that $V \otimes_{\lambda \in \Lambda} x_\lambda = \otimes_{\lambda \in \Lambda} V_\lambda x_\lambda$ for all $\otimes_{\lambda \in \Lambda} x_\lambda$ with $x_\lambda = z_\lambda$ for all but a finite set of λ, if and only if the numerical product $\Pi_\lambda \langle V_\lambda z_\lambda, z_\lambda \rangle$ is convergent. The necessity is clear from the preceding paragraph. To show the sufficiency, note that $\otimes_{\lambda \in \Lambda} V_\lambda x_\lambda$ is convergent in case $\Pi_\lambda \langle V_\lambda x_\lambda, z_\lambda \rangle$ is convergent, and this product is identical to the product $\Pi_\lambda \langle V_\lambda z_\lambda, z_\lambda \rangle$ except for a finite number of factors. □

PROOF OF THEOREM. With the notation of the lemma, $\otimes_\lambda V_\lambda$ is the strong limit of the net $\otimes_{\lambda \in \Delta} V_\lambda$, where $\otimes_{\lambda \in \Delta} V_\lambda$ denotes $\otimes_\lambda V'_\lambda$ with $V'_\lambda = V_\lambda$ for $\lambda \in \Delta$, and $V'_\lambda = I_\lambda$ for $\lambda \notin \Delta$. Note also that it is readily proved that if $\otimes_\lambda S_\lambda$ and $\otimes_\lambda T_\lambda$ are convergent products of unitary operators, then $\otimes_\lambda (S_\lambda T_\lambda)$ and $\otimes_\lambda S_\lambda^*$ are convergent, and $(\otimes_\lambda S_\lambda)(\otimes_\lambda T_\lambda) = \otimes_\lambda S_\lambda T_\lambda$ and $(\otimes_\lambda S_\lambda)^* = \otimes_\lambda S_\lambda^*$.

Now assume the indicated automorphism A exists, and let U be a unitary operator on $\mathbf{K}$ that induces A. Then, for any convergent product $\otimes_\lambda V_\lambda$ of unitary operators, $U^*(\otimes_\lambda V_\lambda)U = \otimes_\lambda A_\lambda(V_\lambda)$. Setting $V_\lambda = P_\lambda + b(I - P_\lambda)$ where $|b| = 1$, then V_λ is unitary and $V_\lambda z_\lambda = z_\lambda$, so that $\otimes_\lambda V_\lambda$ is convergent, implying that $\otimes_\lambda A_\lambda(V_\lambda)$ is also convergent, which in turn implies the convergence of the numerical product $\Pi_\lambda \langle A_\lambda(V_\lambda) z_\lambda, z_\lambda \rangle$. The λth factor in the latter product is 1 +

$(b - 1)(1 - \langle A_\lambda(P_\lambda)z_\lambda, z_\lambda\rangle)$, and assuming $b \neq 1$, a product of the form $\Pi_\lambda(1 + (b - 1)c_\lambda)$ with $0 \leq c_\lambda \leq 1$ is easily seen to be convergent if and only if $\Sigma_\lambda c_\lambda < \infty$, implying that $\Sigma_\lambda(1 - \langle A_\lambda(P_\lambda)z_\lambda, z_\lambda\rangle)$ is convergent. This implies the convergence of the infinite product given in the conclusion of the theorem.

Conversely, let U_λ be a unitary operator on $\mathbf{K}_\lambda$ that induces the given automorphism A_λ and is such that $\langle U_\lambda z_\lambda, z_\lambda\rangle \geq 0$. Then the given product is termwise the square of the product $\Pi_\lambda \langle U_\lambda z_\lambda, z_\lambda\rangle$. It follows that $\otimes_\lambda U_\lambda$ exists, and this operator evidently induces an automorphism A with the stated property.

□

The methods of this chapter apply in principle to the question of the unitary implementability of nonlinear canonical transformations, as well as to linear ones, but naturally it is much more difficult to carry through effective applications in the former case. As an example of how these may proceed, however, consider the case of nonlinear transformations of the form

$$Q_k \mapsto F_k(Q_k)$$

where $Q_1, Q_2, \ldots$ and $P_1, P_2, \ldots$ are an indexed set of boson field operators, the F_k being smooth one-to-one transformations of $\mathbf{R}$. In the finite-dimensional case, there always exists a unitary implementation for a canonical transformation extending the given transformation; namely, the implementing unitary operator U carries $f(x_1, \ldots, x_n)$ into $f(F_1(x_1), \ldots, F_n(x_n))m(x_1, \ldots, x_n)$ where m is the usual multiplier, the square root of the Radon-Nikodym derivative of the transformation $(x_1, \ldots, x_n) \mapsto (F_1(x_1), \ldots, F_n(x_n))$. In the infinite-dimensional case, it is to be expected that there is unitary implementability if and only if the F_k are asymptotically close to the identity transformation $x \mapsto x$, but precisely how close? In the case when the (P_k, Q_k) arise from stochastically independent quasi-invariant distributions, the answer is as follows:

Corollary 4.4.1. *Let $P_1, P_2, \ldots$ and $Q_1, Q_2, \ldots$ be the Heisenberg system over a real vector space $\mathbf{H}$, relative to a basis of vectors $e_1, e_2, \ldots$ that are stochastically independent with respect to the quasi-invariant distribution m on $\mathbf{H}$ that determines the given Heisenberg system. Let F_1, $F_2, \ldots$ be a sequence of continuously differentiable real-valued functions on $\mathbf{R}$ with $F_k'(x) > 0$ for all x and k.*

Then in order that there exist a unitary transformation U such that $U^{-1}Q_kU = F_k(Q_k)$ for all k, it is necessary and sufficient that the following product be convergent:

$$\prod_k \int_{-\infty}^{\infty} [p_k(t)p_k(F_k(t))F_k'(t)]^{1/2}\, dt,$$

where p_k is the probability density of the distribution of the operator $m(e_k)$. In *the event that such a unitary operator exists, it may be chosen so that $U^{-1}P_kU$ is affiliated with the ring of operators determined by P_k and Q_k ($k = 1, 2, \ldots$).*

PROOF. If there exists a unitary operator U such that $U^{-1}Q_kU = F_k(Q_k)$ for all k, let m' be the distribution on **H** such that $M_{m'(x)} = U^{-1}M_{m(x)}U$. Then by Scholium 1.5 m' is absolutely continuous with respect to m, so by Corollary 1.6.2 the product

$$\prod_k E_m(dm'_k/dm_k)^{1/2}$$

must be convergent. (Note that the basis vectors e_k in **H**, with which the P_k and Q_k are associated, are stochastically independent with respect to m' as well as m.) A simple computation shows that this is termwise equal to the given product.

Now suppose the given product is convergent. Let $\mathbf{K}_k = L_2(\mathbf{H}_k, m_k)$, where $\mathbf{H}_k$ is the one-dimensional subspace spanned by e_k, and m_k denotes the restriction of m to $\mathbf{H}_k$. Then $\mathbf{K}_k$ is unitarily equivalent to the space $L_2(\mathbf{R}, p_k)$ of all square-integrable complex-valued functions on $\mathbf{R}$, relative to the probability density p_k giving the distribution of $m(e_k)$, in such a way that Q_k corresponds to multiplication by the function x. (Note that the distribution of $m(e_k)$ is mutually absolutely continuous with Lebesgue measure because of its quasi-invariance.) Let U_k denote the unitary operator on $L_2(\mathbf{R}, p_k)$,

$$g \mapsto g(F_k^{-1}(\cdot))[p_k(F_k^{-1}(\cdot))F'_k(F_k^{-1}(\cdot))p_k(\cdot)]^{1/2}$$

g being arbitrary in $L_2(\mathbf{R}, p_k)$, and let U'_k be the corresponding unitary operator on $\mathbf{K}_k$, via the unitary equivalence described (which is not unique, but let a fixed one be chosen for each k and adhered to thereafter). This induces an automorphism

$$A_k: X \to U'^{-1}_k X U'_k$$

on the ring of all bounded linear operators on $\mathbf{K}_k$. Now consider the absolute continuity condition with reference to the formulation of the grounded Hilbert space $(L_2(\mathbf{H}, m), 1)$ as the product of the $(\mathbf{K}_k, 1_k)$. It is easy to see that the kth factor is $\langle U'_k z_k, z_k\rangle$. Using the given form for U_k, and noting that the unitary equivalence taking U'_k into U_k carries I into the function identically 1 on $\mathbf{R}$, the kth factor is readily evaluated as identical with the kth factor in the given product.

Thus there exists an automorphism A extending all the A_k in the fashion covered by our earlier theorem. This automorphism has the form $X \mapsto U^{-1}XU$ for some unitary operator U, whose restriction to $L_2(\mathbf{H}_k, m_k)$ extends the transformation induced by W'_k. It is clear that this takes Q_k into Q'_k, and takes P_k into some selfadjoint operator on $L_2(\mathbf{H}_k, m_k)$; any such operator is affiliated with the ring determined by P_k and Q_k because the system determined by a quasi-invariant distribution in one dimension is irreducible. □

In Corollary 4.4.1, the Q'_k were specified, but the P'_k were not completely specified. The method employed can be adapted to deal with the case when the P'_k are specified as well, as in

COROLLARY 4.4.2. *Let $P_1, P_2, \ldots$ and $Q_1, Q_2, \ldots$ be as in the preceding corollary and let $P'_1, P'_2, \ldots$ and $Q'_1, Q'_2, \ldots$ be a second Heisenberg system over* **H**, *relative to the same basis and on the same representation space, and such that for each k, P'_k and Q'_k are affiliated with the ring $\mathbf{R}_k$ determined by the P_k and Q_k.*

Necessary and sufficient for the existence of a unitary transformation simultaneously transforming the P_k and Q_k into P'_k and Q'_k (respectively, for $k = 1, 2 \ldots$) is that the following conditions be satisfied:

1) *for each k, there exist a unitary transformation U_k transforming the P_k and Q_k simultaneously into P'_k and Q'_k (respectively), which unitary transformation is affiliated with $\mathbf{R}_k$; and*
2) *the product $\Pi_k \left| \int_{-\infty}^{\infty} p_k(t)\, u_k(t)\, dt \right|$ be convergent, where p_k is as in the preceding corollary, and u_k is the transform of the function identically* 1 *on* **R** *under the transform by a unitary equivalence of $L_2(\mathbf{H}_k, m_k)$ with $L_2(\mathbf{R}, p_k)$ that takes Q_k into multiplication by x and P_k into the selfadjoint generator of translation.*

PROOF. Since P_k and Q_k jointly act irreducibly on $L_2(\mathbf{H}_k, m_k)$, any automorphism of all bounded operators on this space is determined by its extended action on P_k and Q_k. Similarly $P_1, P_2, \ldots$ and $Q_1, Q_2, \ldots$ form an irreducible set on $L_2(\mathbf{H}, m)$, and so any automorphism of all bounded operators on this space is determined by its action on the P's and Q's. Thus there exists a unitary transformation making the indicated transformation between the two canonical systems if and only if there exists an automorphism A as in the theorem earlier, i.e., if and only if the product given there is convergent. □

4.6. Affine transforms of the isonormal distribution

The question of the absolute continuity of affine transforms of the isonormal distribution arises in a variety of guises, some of which will be indicated later. In its conceptually simplest form the basic result is

THEOREM 4.5. *Let T be an invertible bounded linear operator on the real Hilbert space* $\mathbf{H}'$. *The transform g^T of the isonormal distribution g on $\mathbf{H}'$ by T: $g^T(x) = g(T^{-1}x)$, is mutually absolutely continuous with g if and only if TT^* has the form $I + B$, where B is Hilbert-Schmidt.*

If, moreover, B is trace class, then the derivative dg^T/dg is $(\det A)^{-1/2} \exp(-\frac{1}{2}\langle(A^{-1} - I)x, x\rangle)$ *as a functional on* $\mathbf{H}'$, *where $A = TT^*$.*

PROOF. Note first that in the foregoing expression for the derivative when B is trace class, for succinctness the interpretation of the given function on $\mathbf{H}'$ as a measurable with respect to the isonormal distribution has not been detailed. To amplify, if C is any selfadjoint trace class operator on $\mathbf{H}'$, $\det(I + C)$ is definable as the (evidently convergent) infinite product $\Pi_j(1 + \lambda_j)$, where $\lambda_1, \lambda_2, \ldots$ are the eigenvalues of C. The expression $e^{\langle Cx, x\rangle}$ is definable as a measurable with respect to n by its representation in terms of the $x_j = \langle x, e_j\rangle$, where the e_j are the eigenvectors of C, as the infinite product $\Pi_j\, e^{\lambda_j x_j^2}$, which is convergent in $L_1(g)$, each partial product being interpretable as a tame function.

Now for the "if" part, choose a basis in which TT^* is diagonal, with eigenvalue λ_j for the eigenvector e_j. Then g^T is absolutely continuous with respect to g if and only if the following infinite product is convergent:

$$\prod_j \int_{-\infty}^{\infty} [(2\pi)\lambda_j c]^{-1/2} [e^{-x^2/2\lambda_j c - x^2/c}]^{1/2}\, dx.$$

The jth factor in this product is readily evaluated as $[4\lambda_j/(1 + \lambda_j)^2]^{-1/4}$. There is no difficulty in verifying that the product is convergent when $\lambda_j = 1 + \varepsilon_j$ with $\Sigma_j\, \varepsilon_j^2 < \infty$.

To treat the "only if" part, let $\mathbf{H}_0$ be the closed linear subspace of $\mathbf{H}'$ spanned by the eigenvectors of B, and let $\mathbf{H}_1$ be the orthocomplement of $\mathbf{H}_0$ in $\mathbf{H}'$. Then B leaves both $\mathbf{H}_0$ and $\mathbf{H}_1$ invariant, and if $g^T << g$, then $g^T|\mathbf{H}_j << g|\mathbf{H}_j$ $(j = 0, 1)$. It will first be shown that $\mathbf{H}_1$ consists only of the 0 vector.

For $g|\mathbf{H}_1$ is the isonormal distribution on $\mathbf{H}_1$, and $g^T|\mathbf{H}_1$ is simply the normal distribution of mean 0 and covariance operator $A = TT^*$, restricted to $\mathbf{H}_1$. Thus it suffices to show that if A is any selfadjoint nonnegative invertible linear operator in a real Hilbert space $\mathbf{H}_1$ then the normal distribution $N(0, A^{1/2})$ (the notation $N(x, C)$, with $x \in \mathbf{H}_1$ and C a bounded selfadjoint operator on $\mathbf{H}_1$, signifying the normal distribution of mean 0 and covariance operator C^2) is not absolutely continuous with respect to $N(0, I)$, provided the point spectrum of C is empty. Now $N(0, C)$ is invariant under the group G of all orthogonal transformations of the form $f(C)$, where f is any Borel function such that $|f(x)| = 1$ for all $x \in \mathbf{R}$. This group leaves no finite-dimensional subspace of $\mathbf{H}_1$

invariant, since otherwise C itself would leave such a subspace invariant, contradicting the assumption that C was free of point spectrum. On the other hand, if $N(0, A^{1/2}) << N(0, I)$, then the derivative $dN(0, A^{1/2})/dN(0, I)$ would be invariant under the subgroup of $O(\mathbf{H}')$ leaving both of these distributions invariant, since each distribution is itself invariant. This subgroup would then include the group G, but this acts ergodically on $L_2(\mathbf{H}', g)$ by Theorem 3.4. Accordingly the derivative must then be a constant, and hence 1, showing that $A = I$, and so contradicting the assumption that A had no point spectrum.

Thus $\mathbf{H}_0$ is all of $\mathbf{H}'$, so it may be assumed that A has eigenvalues λ_i with eigenvectors e_i spanning $\mathbf{H}'$. The convergence of the infinite product given in the preceding paragraph is then necessary for absolute continuity. It is immediate that this convergence implies that $\lambda_i \to 1$. Setting $\lambda_i = 1 + \varepsilon_i$, it is easily seen that the ith factor has the form $1 - \frac{1}{4}\varepsilon_i^2 + 0(\varepsilon_i^3)$, from which it follows that the product converges only if $\Sigma_i \varepsilon_i^2 < \infty$.

The given form of the derivative follows as earlier, as a limit in $L_1(\mathbf{H}', g)$ of the corresponding derivatives of restrictions to finite-dimensional subspaces.

□

In addition to the applications of the absolute continuity criterion for transformations on Hilbert space to Wiener space, and in the theory of normal stationary time series, it applies directly to quantum fields.

COROLLARY 4.5.1. *Let S be a (linear) symplectic transformation on the complex Hilbert space* $\mathbf{H}$. *Then there exists a unitary transformation V on* $\mathbf{K}$ *such that $W(Sz) = VW(z)V^{-1}$ if and only if $SS^* - I$ is Hilbert-Schmidt, where $*$ denotes the adjoint of S as a real linear transformation on the real Hilbert space* $\mathbf{H}^\#$.

PROOF. Consider first the special case in which there exists a real part $\mathbf{H}_\kappa$ of $\mathbf{H}$ such that S has the form $S(x + iy) = Tx + iT^{-1}y$, for arbitrary x and y in $\mathbf{H}_\kappa$, where T is a nonnegative selfadjoint operator on $\mathbf{H}_\kappa$. The condition that $W(Sz) = VW(z)V^{-1}$ is equivalent to the condition that $\partial W(Sz) = V\partial W(z)V^{-1}$, which in turn is equivalent, by linearity, to the condition that $P(Tx) = VP(x)V^{-1}$ and $Q(T^{-1}y) = VQ(y)V^{-1}$.

Now the mapping $y \mapsto Q(y)$ defines a distribution on $\mathbf{H}_\kappa$ relative to the expectation functional given by the vacuum vector v, and as seen earlier there exists a unitary operator V such that $Q(T^{-1}y) = VQ(y)V^{-1}$ if and only if the transformation T is absolutely continuous on $\mathbf{H}_\kappa$. By the preceding theorem, this is the case if and only if $TT^* - I$ is Hilbert-Schmidt. Thus this condition is necessary in the present case. If, on the other hand, it is satisfied, then let V denote the unitarized operation of transformation induced from T, acting on

$L_2(\mathbf{H}_\varkappa, g)$. Here g is the isonormal distribution in $\mathbf{H}_\varkappa$, and $L_2(\mathbf{H}_\varkappa, g)$ is identified with $\mathbf{K}$, via the real wave representation. It is straightforward to verify that V transforms $P(x)$ into $P(Tx)$ as well as $Q(x)$ into $Q(T^{-1}x)$.

The general case may be reduced to the case just treated by the

LEMMA 4.5.1. *For any symplectic transformation R on a complex Hilbert space, there exists a unitary operator U on* $\mathbf{H}$, *and* $\mathbf{H}_\varkappa$ *and T as in the preceding proof, such that $R = US$, where $S(x + iy) = Tx + iT^{-1}y$ for $x, y \in \mathbf{H}_\varkappa$.*

PROOF. Note that an invertible continuous linear transformation R on $\mathbf{H}^\#$ is symplectic if and only if $R^\# JR = J$, where J is multiplication by i regarded as a real linear transformation on $\mathbf{H}^\#$.

Let $R = US$ denote the polar decomposition of R as a bounded linear operator on $\mathbf{H}^\#$, U being orthogonal and S being nonnegative and selfadjoint. Then, as is easily seen, $R^\#$ is also symplectic, so that $R^\# R$ is symplectic, i.e., S^2 is symplectic. Hence $S^2JS^2 = J$, whence $JS^2J^{-1} = S^{-2}$. It follows that J carries the spectral subspace $\mathbf{S}_+$ on which $S^2 > \mathrm{I}$ into the spectral subspace $\mathbf{S}_-$ on which $S^2 < \mathrm{I}$, and leaves invariant the subspace $\mathbf{S}_0$ on which S^2 acts as the identity. Thus $\mathbf{S}_0$ is a complex-linear subspace of $\mathbf{H}$. Setting $\mathbf{S}_0'$ for an arbitrary real part of $\mathbf{S}_0$, and $\mathbf{H}_\varkappa = \mathbf{S}_+ \oplus \mathbf{S}_0'$, it follows that $\mathbf{H}_\varkappa$ is a real part of $\mathbf{H}$. Moreover, for x and y in $\mathbf{H}_\varkappa$, $S^2(x + Jy) = Mx + JM^{-1}y$, where M is a nonnegative invertible linear operator on $\mathbf{H}_\varkappa$. Setting $T = M^{1/2}$, then $S(x + Jy) = Tx + JT^{-1}y$, and S is symplectic. It follows that U is symplectic, but, being also orthogonal, it is unitary. □

An alternative criterion for unitary implementability whose analogue for fermion fields is also valid is given by

COROLLARY 4.5.2. *With the notation and hypothesis of Corollary 4.5.1, the canonical transformation $W(z) \mapsto W(Sz)$ is unitarily implementable if and only if the commutator of S with i is Hilbert-Schmidt on* $\mathbf{H}$ *as a real Hilbert space.*

PROOF. It suffices to consider the case in which S has the form given in Lemma 4.5.1. In real terms, S then has the form $\begin{pmatrix} T & 0 \\ 0 & T^{-1} \end{pmatrix}$ and the action of i on $\mathbf{H}$ is represented by the matrix $\begin{pmatrix} 0 & -I \\ I & 0 \end{pmatrix}$. The commutator $[i,S]$ has the corresponding form

$$\begin{pmatrix} 0 & T - T^{-1} \\ T - T^{-1} & 0 \end{pmatrix}.$$

Now suppose $SS^* - I$ is Hilbert-Schmidt. This is equivalent to $TT^* - I$ being Hilbert-Schmidt. But S and hence T are nonnegative and selfadjoint, on $\mathbf{H}$ and $\mathbf{H}_\kappa$ respectively, so $TT^* - I = T^2 - I$. Multiplication by a bounded operator preserves the Hilbert-Schmidt property, so $T - T^{-1}$, which equals $T^{-1}(T^2 - I)$, is Hilbert-Schmidt. It follows that $[i, S]$ is also Hilbert-Schmidt.

Conversely, suppose that $[i, S]$ is Hilbert-Schmidt, i.e., $T - T^{-1}$ is Hilbert-Schmidt. By the argument just made, this implies that $T^2 - I$ is Hilbert-Schmidt, and hence that $SS^* - I$ is Hilbert-Schmidt as well. □

Problems

1. Let $\mathbf{H}'$ be a real Hilbert space, let T be a bounded invertible linear operator on $\mathbf{H}'$, and let a be arbitrary in $\mathbf{H}'$. Show that the affine transformation $x \mapsto Tx + a$ is absolutely continuous with respect to the isonormal distribution if and only if the same is true of the homogeneous constituent, $x \mapsto Tx$.

2. Develop an explicit expression for the derivative dn_T/dn in the case in which T is a Hilbert-Schmidt selfadjoint operator in the real Hilbert space $\mathbf{H}'$ given in diagonal form. (Use the infinite direct sum formulation.)

3. Let g be the isonormal distribution of unit variance on the real $L_2[0, 1]$ Let $\mathbf{W}$ denote Wiener space, i.e., the space of real continuous functions on $[0, 1]$ that vanish at 0. Let T be the continuous linear map from $L_2[0, 1]$ to $\mathbf{W}$ given by $(Tf)(x) = \int_0^x f(s)ds$. Define the distribution m on $\mathbf{W}$ by $m(\lambda) = n(T^*\lambda)$, where $\lambda \in \mathbf{W}^*$ and T^* is the adjoint of T. Show that m is strict, i.e., m defines a countably addditive measure (known as *Wiener measure*) on $\mathbf{W}$. (Cf. Doob, 1965.)

4. Show that if k is any absoutely continuous element of $\mathbf{W}$, then the transformation $x(\cdot) \mapsto x(\cdot) + k(\cdot)$ on this space is absolutely continuous with respect to Wiener measure. (Hint: use Problem 3 to deduce this from the quasi-invariance of the isonormal distribution.)

5. Let $K(t, s)$ be a given real Borel function on $[0, 1] \times [0, 1]$ such that 1) $K(t, s)$ is absolutely continuous as a function of t for almost all fixed s; and 2) $\int_0^1 \int_0^1 |(\partial/\partial t)K(t, s)|^2\, ds\, dt < \infty$. Show that the transformation

$$x(t) \to x(t) + \int_0^1 K(t, s)x(s)\, ds$$

maps Wiener space absolutely continuously into itself. (Hint: use Problem 3 and consider the transformation in $L_2[0, 1]$

$$f(t) \to f(t) + \int_0^1 (\partial/\partial s)K(t, s)f(s)\, ds.$$

Show that the integral kernel here is Hilbert-Schmidt and apply Theorem 4.5 noting that

$$\int_0^1 [f(t) + \int_0^1 K_s(t,s)x(s)ds]\, dx(t) = \int_0^1 f(t)\partial_t[x(t) + \int_0^1 K(t, s)dx(s)]\, dt$$

for sufficiently smooth $f(\cdot)$ and $x(\cdot)$.)

4.7. Implementability of orthogonal transformations on the fermion field

In this section we extend the basic results of Section 4.6 to the fermionic case. We define the orthogonal group $O(\mathbf{H})$ on a complex Hilbert space $\mathbf{H}$ to consist of all invertible real-linear transformations on $\mathbf{H}$ that leave invariant the real symmetric form $\mathrm{Re}\langle\cdot,\cdot\rangle$. This is the analog in the fermionic case to the symplectic group in the bosonic case. A given transformation $S \in O(\mathbf{H})$ will be said to be *unitarily implementable* (more precisely, the automorphism of the Clifford algebra that it induces is unitarily implementable in the free representation) in case the following is true: letting $(\mathbf{K}, \phi, \Gamma, \nu)$ denote the free fermion field over $\mathbf{H}$, there exists a unitary operator U on $\mathbf{K}$ such that

$$\phi(Sz) = U^{-1}\phi(z)U \qquad (z \in \mathbf{H}).$$

The basic result is

THEOREM 4.6. *An orthogonal transformation S on a complex Hilbert space $\mathbf{H}$ is unitarily implementable if and only if $[i, S]$ is Hilbert-Schmidt on $\mathbf{H}$ as a real Hilbert space.*

PROOF. As in the bosonic case, the primary case is that in which the given transformation S has a special form relative to a conjugation on $\mathbf{H}$ and has pure point spectrum. Reduction to this case involves in particular

LEMMA 4.6.1. *Let S_j be given orthogonal transformations on the complex Hilbert space $\mathbf{H}_j$ $(j = 1, 2)$, and let $(\mathbf{K}_j, \phi_j, \Gamma_j, \nu)$ denote the corresponding free fermion fields. Let $S = S_1 \oplus S_2$, $\mathbf{H} = \mathbf{H}_1 \oplus \mathbf{H}_2$, and let $(\mathbf{K}, \phi, \Gamma, \nu)$ denote the free field over $\mathbf{H}$. Then S is unitarily implementable on $\mathbf{K}$ if and only if each S_j is unitarily implementable on $\mathbf{K}_j$.*

PROOF. According to Theorem 3.2, $\mathbf{K}$ is unitarily equivalent to $\mathbf{K}_1 \otimes \mathbf{K}_2$ in such a way that $\phi(z_1 \oplus z_2) = \phi(z_1) \otimes I_2 + \Omega_1 \otimes \phi_2(z_2)$, where $\Omega_1 = \Gamma(-I_{\mathbf{H}_1})$; and in addition, $v = v_1 \otimes v_2$. From this representation, the "if" part of the lemma is immediate. To treat the "only if" part, we extend the ϕ_j and ϕ from linear maps of the underlying spaces $\mathbf{H}_j$ and $\mathbf{H}$ into the respective algebras $\mathbf{B}_j$ and $\mathbf{B}$ of all bounded operators on the spaces $\mathbf{K}_j$ and $\mathbf{K}$, to isomorphisms of the corresponding C^*-Clifford algebras $\mathbf{C}(\mathbf{H}_j)$ and $\mathbf{C}(\mathbf{H})$ into the $\mathbf{B}_j$ and $\mathbf{B}$; the extensions will also be denoted as ϕ_j and ϕ. Now suppose that S is unitarily implementable by the unitary operator U on $\mathbf{K}$.

We denote as E_1 the state of $\mathbf{C}_1 \equiv \mathbf{C}(\mathbf{H}_1)$ given by the equation $E_1(A) = \langle \phi(A)v_1, v_1 \rangle$ $A \in \mathbf{C}_1$. If e_j denotes the identity in $\mathbf{C}_j \equiv \mathbf{C}(\mathbf{H}_j)$, then $\phi(A \otimes e_2) = \phi_1(A) \otimes \mathrm{I}_2$ (where I_j denotes the identity on $\mathbf{K}_j$). Accordingly, E_1 may also be expressed as $E_1(A) = E(A \otimes e_2)$, where E denotes the state of $\mathbf{C} \equiv \mathbf{C}(\mathbf{H})$ given by the equation $E(B) = \langle \phi(B)v, v \rangle$.

Let α denote the automorphism of $\mathbf{C}$ induced from the orthogonal transformation S on $\mathbf{H}$; i.e., α is the unique automorphism that carries z into Sz, $z \in \mathbf{H}$. Let E' denote the state of $\mathbf{C}$ into which E is transformed by α, i.e., $E'(B) = E(\alpha(B))$. By virtue of the unitary implementability of S via U, E' may be expressed in the form $E'(B) = \langle \phi(B)v', v' \rangle$, where $v' = U^{-1}v$. Since $\mathbf{K} = \mathbf{K}_1 \otimes \mathbf{K}_2$, v' can be decomposed as $v' = \Sigma_j x_j \otimes e_j$, where the e_j form an orthonormal basis for $\mathbf{K}_2$, and the x_j are in $\mathbf{K}_1$, but not necessarily orthogonal; however, $\|v'\|^2 = \Sigma_j \|x_j\|^2$. It follows that $E'(A) = \Sigma_j \langle \phi_1(A)x_j, x_j \rangle$.

On the other hand, E_1 is a pure state of $\mathbf{C}_1$, since $\mathbf{C}_1$ acts irreducibly on $\mathbf{K}_1$ via the isomorphism ϕ_1. The transform of E_1 by any automorphism of $\mathbf{C}_1$ must likewise be a pure state. Now α extends to $\mathbf{C}$ the automorphsim α_1 of $\mathbf{C}_1$ that is induced by S_1, since for $z \in \mathbf{H}_1$, $Sz = S_1 z$. Thus the restriction $E'|\mathbf{C}_1$ coincides with the transform of E_1 by α_1. Accordingly, $E'|\mathbf{C}_1$ must also be a pure state of $\mathbf{C}_1$. It follows from the irreducibility of $\mathbf{C}_1$ and the definition of a pure state that the x_k are proportional to a fixed unit vector x, implying that $E'(A) = \langle \phi_1(A)x, x \rangle$. It follows that the mapping $\phi_1(A)v_1 \to \phi(\alpha^{-1}(A))x$ $(A \in \mathbf{C}_1)$ is isometric, and that it has a unique extension to a unitary operator on all of $\mathbf{K}_1$ that implements S_1. By symmetry, the same applies to S_2. □

Resuming the proof of the theorem, we focus on the special case parallel to the basic one in the case of boson field. Specifically, we assume the given operator S is such that relative to some conjugation $\varkappa$ on $\mathbf{H}$, $S(x + iy) = Rx + iy$ for arbitrary real x, y in $\mathbf{H}$, where R is orthogonal on $\mathbf{H}_\varkappa$ to itself. By virtue of the lemma, it may also be assumed that the spectrum of R is either totally devoid of point spectrum or consists entirely of point spectrum. Taking first the case in which the point spectrum is vacuous, then by spectral theory the group G of all orthogonal transformations on $\mathbf{H}_\varkappa$ that commute with R leaves no finite-dimensional subspace of $\mathbf{H}_\varkappa$ invariant. The same is true of its complex extension G' consisting of all transformations T' on $\mathbf{H}$ of the form $T'(x + iy) = Tx + iTy$, for arbitrary $x, y \in \mathbf{H}_\kappa$. By Theorem 3.4, the $\Gamma(T')$ leave no nonscalar operator in the W^*-algebra $\mathbf{R}_Q$ generated by the fermion canonical Q's invariant under the automorphisms of $\mathbf{R}_Q$ obtained by conjugation with the $\Gamma(T')$. On the other hand, $\Gamma(R')Q(x)\Gamma(R')^{-1} = Q(Rx) = UQ(x)U^{-1}$, if U is a unitary operator on $\mathbf{K}$ that induces the canonical transformation $\phi(z) \to \phi(Sz)$.

In the real wave representation, the $Q(x)$ are represented as left multiplications on an algebra on which the $P(x)$ are corresponding right multiplications, apart from the factor $\Gamma(-I)$. Since S commutes with $-I$, U and $\Gamma(-I)$ must commute within a scalar factor a: $U^{-1}\,\Gamma(-I)U = a\,\Gamma(-I)$. Squaring this relation, it follows that $a = \pm 1$. If $a = 1$, conjugation by U leaves the $\Gamma(-I)P(x)$ pointwise invariant. But as right multiplications, these operators commute with the left multiplications represented by the $Q(x)$. Together, the algebras of left and right multiplications are irreducible on $\mathbf{K}$ and are commutors of each other. Accordingly, the commutativity of U with all right multiplications implies that it is in the algebra generated by the left multiplications. But this means that U is in $\mathbf{R}_Q$. It follows that U is a scalar, implying in turn that $R = \mathrm{I}$, a contradiction.

If, on the other hand, $a = -1$, then conjugation by U carries the $\Gamma(-I)P(x)$ into their negatives, from which it follows that U^2 leaves them invariant. Then, by the argument just made, $U = I$, whence $R^2 = I$, which contradicts the assumption that R is devoid of point spectrum.

Thus R has pure point spectrum. By virtue of Lemma 4.6.1, it is no essential loss of generality to assume that 1 is absent from the spectrum of R, and that if -1 is in the spectrum, the corresponding invariant submanifold is infinite-dimensional. Thus $\mathbf{H}_\kappa$ is the direct sum of two-dimensional R-invariant subspaces, in each of which the action of R may be represented by a matrix of the form $\begin{pmatrix} \cos\theta & \sin\theta \\ -\sin\theta & \cos\theta \end{pmatrix}$. The various subspaces are stochastically independent as regards vacuum expectation values of operators in $\mathbf{R}_Q$. The entire space $\mathbf{K}$ may be represented as in the boson case as the tensor product of the fermion fields over the two-dimensional complexifications of these real R-invariant subspaces. Relative to the jth of these finite-dimensional fields, R is unitarily implemented by the operator U_j, and according to the theory earlier developed in this chapter, R will be unitarily implemented on all of $\mathbf{K}$ if and only if the following infinite product is convergent: $\Pi\,\langle U_j v_j, v_j\rangle$, where v_j represents the vacuum in the jth factor-field.

To evaluate the terms of this infinite product, we may represent the factor-field in the real wave representation as follows. Let $\mathbf{C}$ denote the Clifford algebra over the two-dimensinal real vector space $\mathbf{S}$ with orthogonal basis vectors $e_\pm$, satisfying the relations $e_\pm^2 = 1, e_+e_- + e_-e_+ = 0$; let e_0 denote the identity in **C**. **C** forms a complex Hilbert space **K** with the inner product $\langle A, B\rangle = \mathrm{tr}(B^*A)$. The canonical Q's are represented by left multiplications $L(x)$ by vectors x in **S**, and we set $Q_\pm = 2^{-1/2}L(e_\pm)$. The vacuum vector is represented by $2^{-1/2}e_0$. The orthogonal transformation indicated is implemented by the unitary operator U_j given by the equation $U_jw = u_j^{-1}wu_j$ $(w \in \mathbf{C})$, where $u_j = \cos(\frac{1}{2}\theta_j) + e_+e_- \sin(\frac{1}{2}\theta_j)$. It follows that the infinite product is convergent if and only if $\Sigma\theta_j^2$ is convergent, which is equivalent to $R - I$ being Hilbert-Schmidt. On the other hand

$$[R', i] = \left[\begin{pmatrix} R & 0 \\ 0 & I \end{pmatrix}, \begin{pmatrix} 0 & -I \\ I & 0 \end{pmatrix}\right] = 2\begin{pmatrix} 0 & I - R \\ I - R & 0 \end{pmatrix},$$

so this is equivalent to the criterion given in the theorem.

Having established the conclusion of the theorem for the case in which the given orthogonal tranformation S is of the form $S(x + iy) = Rx + iy$ for some R on a real part of **H**, it now suffices for the proof of the theorem to show that every orthogonal transformation T on **H** has the property that the orthogonal transformation $T\oplus T$ on $\mathbf{H}\oplus\mathbf{H}$ is of the form US', where U is unitary and S' has the special form indicated. This is the substance of

LEMMA 4.6.2. *For any orthogonal transformation V on the complex Hilbert space* **H**, *there exists a conjugation $\varkappa$ on* $\mathbf{H}\oplus\mathbf{H}$ *and a unitary operator U on* $\mathbf{H}\oplus\mathbf{H}$ *such that $V\oplus V = US$, where S is of the form $S(x + iy) = Rx + iy$ if x and y are real relative to $\varkappa$, R being orthogonal on the real subspace* $(\mathbf{H}\oplus\mathbf{H})_\varkappa$.

PROOF. Let **G** denote the complex Hilbert space whose underlying real linear structure is $\mathbf{H}^\#\oplus\mathbf{H}^\#$, whose complex structure J takes the form $J(x\oplus y) = -y\oplus x$ $(x, y \in \mathbf{H})$, and whose complex inner product, denoted by $\langle\langle\cdot,\cdot\rangle\rangle$, is given in terms of the usual direct sum inner product, which will be denoted as $\langle\cdot,\cdot\rangle_2$, by the equation $\langle\langle z, z'\rangle\rangle = \langle z, z'\rangle_2 - i\langle Jz, z'\rangle_2$.

Note that **G** and $\mathbf{H}\oplus\mathbf{H}$ have the same orthogonal structure, but different complex structures. We denote the usual complex structure $i\oplus i$ as j, and note that both j and $V\oplus V$ are unitary on **G**. Setting $T = V\oplus V$ and $W = j^{-1}T^{-1}jT$, then $j^{-1}Wj = W^{-1}$.

Applying the spectral theorem to the unitary operator W, **G** is the direct sum of spectral manifolds $G(1)\oplus G(-1)\oplus G(M_0)\oplus G(N_0)$, where $M_0 = \exp[i(0, \pi)]$ and $N_0 = \exp[(\pi, 2\pi)]$, corresponding to the indicated possible subsets of the spectral range of a unitary operator. From the relation $j^{-1}Wj = W^{-1}$ it follows that j leaves $G(1)$ and $G(-1)$ invariant, and exchanges $G(M_0)$ and $G(N_0)$. Taking these subspaces in sequence, $G(1)$ consists of all vectors of the form $z = x\oplus y$, where x and y are in the subspace of all vectors $u \in \mathbf{H}$ such that $iVu =$

Viu. This means that $G(1)$ is either of even or infinite dimension. A similar argument shows that the same is true of $G(-1)$. It follows that there exist j-conjugations $\varkappa(\pm 1)$ on $G(\pm 1)$ (i.e., conjugations relative to the complex structure j, regarding $G(\pm 1)$ as real spaces) with corresponding real invariant manifolds of eigenvalues ± 1, which we denote as $\mathbf{M}_1$ and $\mathbf{N}_1$ (such that $G(1) = \mathbf{M}_1 \oplus \mathbf{N}_1$), and $\mathbf{M}_{-1}$, $\mathbf{N}_{-1}$ (such that $G(-1) = \mathbf{M}_{-1} \oplus \mathbf{N}_{-1}$). Now set $\mathbf{M}_0 = G(M_0)$ and $\mathbf{N}_0 = G(N_0)$, and set j_0 and W_0 for the restrictions to $\mathbf{M}_0 \oplus \mathbf{N}_0$, which we denote as $\mathbf{L}_0$, of j and W. It follows from the unitarity of j on $\mathbf{H} \oplus \mathbf{H}$ and the fact that $j_2 = -I$ that relative to this decomposition of $\mathbf{L}_0$, j_0 has the form

$$j_0 = \begin{pmatrix} 0 & k_0 \\ k_0^{-1} & 0 \end{pmatrix},$$

where k_0 is an orthogonal map from $\mathbf{N}_0$ onto $\mathbf{M}_0$. Similarly, the relations $j^{-1}Wj = W^{-1}$ and $JW = WJ$ imply that W_0 has a matrix of the form

$$\begin{pmatrix} R_0 & 0 \\ 0 & k_0^{-1}R_0^{-1}k_0 \end{pmatrix},$$

where R_0 is an orthogonal map from $\mathbf{M}_0$ onto $\mathbf{M}_0$. Setting $k = k_1 \oplus k_0 \oplus k_{-1}$, where $k_{\pm 1}$ is orthogonal from $\mathbf{M}_{\pm 1}$ onto $\mathbf{M}_{\pm 1}$, then $j = j_1 \oplus j_0 \oplus j_{-1}$ where

$$j_{\pm 1} = \begin{pmatrix} 0 & k_{\pm 1} \\ k_{\pm 1}^{-1} & 0 \end{pmatrix}.$$

Setting $R = R_1 \oplus R_0 \oplus R_{-1}$ where $R_{\pm 1} = \pm I_{\mathbf{M}_{\pm 1}}$, the following representation results:

$$j = \begin{pmatrix} 0 & k \\ -k^{-1} & 0 \end{pmatrix},$$

where k is orthogonal from $\mathbf{N} = \mathbf{N}_1 \oplus \mathbf{N}_0 \oplus \mathbf{N}_{-1}$ to $\mathbf{M} = \mathbf{M}_1 \oplus \mathbf{M}_0 \oplus \mathbf{M}_{-1}$, and

$w = \begin{pmatrix} R & 0 \\ 0 & k^{-1}R^{-1}k \end{pmatrix}$. Here R is orthogonal from $\mathbf{M}$ onto $\mathbf{M}$.

Now set $U = T\begin{pmatrix} R^{-1} & 0 \\ 0 & I \end{pmatrix}$, where the matrix is relative to the decomposition of $\mathbf{H} \oplus \mathbf{H}$ as $\mathbf{M} \oplus \mathbf{N}$. Then

$$j^{-1}U^{-1}jU = \begin{pmatrix} 0 & -k \\ k^{-1} & 0 \end{pmatrix} \begin{pmatrix} R & 0 \\ 0 & I \end{pmatrix} T^{-1}jT \begin{pmatrix} R^{-1} & 0 \\ 0 & I \end{pmatrix},$$

which on replacing $T^{-1}jT$ by jW and using the matrix representations for j and W in the preceding paragraph, is seen to be the identity on $\mathbf{H}\oplus\mathbf{H}$. Thus U is complex linear with respect to j and orthogonal on $\mathbf{H}\oplus\mathbf{H}$. On the other hand, $T = U\begin{pmatrix} R & 0 \\ 0 & I \end{pmatrix}$, so that with $\varkappa$ represented by the matrix $\begin{pmatrix} I & 0 \\ 0 & -I \end{pmatrix}$, the conclusion of Lemma 4.6.2 is established, and the proof of the theorem is complete. □

Problems

1. Let $\mathbf{H}$ be a finite-dimensional Hilbert space, and let $(\mathbf{K}, W, \Gamma, \nu)$ denote the free boson field over $\mathbf{H}$. The *harmonic representation* of $Sp(\mathbf{H})$ is defined as follows: for arbitrary S in $Sp(\mathbf{H})$ there exists a unitary operator U on $\mathbf{K}$ such that $UW(z)U^{-1} = W(Sz)$ for all $z \in \mathbf{H}$; show that U is unique within a phase, or constant of absolute value 1. The mapping $S \rightarrow U$, to be denoted as Γ in extension of earlier notation, is a *projective* representation $Sp(\mathbf{H})$, i.e., $\Gamma(S)\Gamma(S') = c(S, S')\Gamma(SS')$, where $c(S, S')$ is a constant of absolute value 1.

a) Derive the condition on the function $c(S, S')$ that is implied by the projective representation character of Γ. (A function $c(S, S')$ satisfying this condition is called a *multipler* or *cocycle*, in extension of the terminology given in Section 4.3.)

b) Determine how $c(S, S')$ is transformed if $\Gamma(S)$ is changed by multiplication with a scalar of absolute value 1, $\Gamma(S) \rightarrow b(S)\Gamma(S)$.

c) Show that there is no choice for b that will reduce the factor $c(S, S')$ identically to unity, i.e., make Γ into a strict representation. (The cocycle is then said to be *essential*, cf. Shale, 1962.)

2. Let $\mathbf{H}$ be a Hilbert space of arbitrary dimension, and let $Sp_2(\mathbf{H})$ denote the subgroup of $Sp(\mathbf{H})$ consisting of operators whose commutator with i is Hilbert-Schmidt. Let the harmonic representation Γ be defined as in Problem 1. Show that the cocycle is inessential on the subgroup of all symplectics that commute with a given conjugation.

3. Let $\mathbf{H}$ be a Hilbert space. The *spin* representation of the subgroup of $O(\mathbf{H})$ having Hilbert-Schmidt commutator with i, denoted $O_2(\mathbf{H})$, is defined as in Problems 1 and 2 with the substitution of the fermion for the boson field. Develop the analogs of Problems 1 and 2, and show that the cocycle is essential on the full group $O_2(\mathbf{H})$ (cf. Araki, 1988).

4. Let $\mathbf{H}$ be a Hilbert space and let S be a symplectic transformation on $\mathbf{H}$. S is said to be *unitarily quantizable* in case there exists $(\mathbf{K}, W, T, \nu)$ where $\mathbf{K}$ is a complex Hilbert space; W is a Weyl system over $\mathbf{H}$ on $\mathbf{K}$; T is a unitary

transformation on **K** such that $TW(z)T^{-1} = W(Sz)$ for all $z \in$ H; and v is a cyclic vector for the $W(z)$ and T such that $Tv = v$. Show that S is unitarily quantizable if and only if it is conjugate to a unitary transformation in $Sp(\mathbf{H})$. (Cf. Segal, 1981.)

Bibliographical Notes on Chapter 4

The theory of infinite direct products of Hilbert spaces is due to von Neumann (1938). Application of the theory to the absolute continuity question for real transformations on Hilbert space was made by Segal (1958b), resulting in the Hilbert-Schmidt criterion for unitary implementability. The latter question for complex transformations was reduced to the real case for boson fields by Shale (1962) and for fermion fields by Shale and Stinespring (1965).

5

C^*-Algebraic Quantization

5.1. Introduction

In Chapters 1 and 2, free fields have been established in a form dependent on an underlying complex Hilbert space. There may or may not be a given such structure in the underlying so-called single particle or "classical field" space **L**. Moreover, as seen in the preceding chapter, there are many unitarily inequivalent quantizations over a given space **L**, when **L** is infinite-dimensional. This creates a plethora of technical problems that are not clearly germane to the underlying physical ideas, but to which the precise mathematical situation is sensitive.

Such problems do not intervene when **L** is finite-dimensional, as a consequence of the Stone–von Neumann theorem in the case of a symplectic structure, and of the structure of Clifford algebras in the case of an orthogonal structure, given in **L**. Fortunately it is possible in the infinite-dimensional case to cut down significantly on the technical problems and to bring the physical ideas to the fore by the use of representation-independent formalisms.

Taking for specificity the use of an infinite-dimensional symplectic vector space $(\mathbf{L}, A)$, it might be argued heuristically that the field observables F that depend only on a finite number of modes z—say, functions of $W(z_1)$, $W(z_2)$,..., $W(z_n)$—should be independent of the representation (or particular Weyl system), essentially as in the finite-dimensional case. Moreover, a uniform limit of such observables has a straightforward physical interpretation as a derived observable that can be directly approximated arbitrarily closely by the F, so these appear as natural observables also. But approximation in weaker senses—weak or strong operator topologies, etc.—are less clearly interpretable physically, and have a technical cast. On the mathematical side, it turns out that the class of uniform limits of the observables F is essentially independent of the particular Weyl system, but the class of weaker types of limits are quite materially dependent on the particular Weyl system.

More specifically, in the case of the free boson field over a given complex Hilbert space **H**, if one interprets "function" of a given set of bounded linear operators to mean "element of the W^*-algebra generated by the given operators," which is natural from a mathematical position, then this is representation-independent when only a finite number of $W(z)$ are involved, by virtue of the Stone–von Neumann theorem. But the W^*-algebra generated by all the $W(z)$ is the algebra **B**(**K**) of all bounded linear operators on **K**, all of whose *-automorphisms are unitarily induced, so there is no prospect of representing canonical transformations in general by such automorphisms. If, however, one takes the uniform closure of the totality of these finite-generated algebras, this concrete C^*-algebra **A** on **K** is not equal to **B**(**K**), and there is consequently the possibility, which turns out to be realized, that linear canonical transformations can be appropriately represented by automorphisms of **A**.

Among other physical applications, this leads to a general quantization procedure that is applicable to tachyons, which are represented by unstable fields lacking a vacuum. More broadly, an arbitrary linear canonical transformation gives rise to a corresponding well-defined transformation on the states of the physical system, though rarely on the state vectors in a given Hilbert space. To put it another way, linear canonical transformations are in general divergent in the Schrödinger picture, as Chapter 4 makes clear, but are convergently representable in the C^*-algebraic Heisenberg picture.

5.2. Weyl algebras over a linear symplectic space

Let $(\mathbf{L}, A)$ denote a given symplectic vector space. For any finite-dimensional linear subspace **M** of **L** such that $A|\mathbf{M}$ is nondegenerate, let $\mathbf{A}(\mathbf{M}, W)$ denote the W^*-algebra generated by the $W(z)$ with $z \in \mathbf{M}$, for the Weyl system $(\mathbf{K}, W)$ over $(\mathbf{L}, A)$. Let $\mathbf{A}_0(W)$ denote the union of the $\mathbf{A}(\mathbf{M}, W)$ as **M** varies, and let $\mathbf{A}(W)$ denote the uniform closure of $\mathbf{A}_0(W)$.

THEOREM 5.1. $\mathbf{A}(W)$ *is $*$-algebraically independent of W in the sense that if* $(\mathbf{K}', W')$ *is any other Weyl system over* $(\mathbf{L}, A)$, *then there is a unique $*$-algebraic isomorphism from* $\mathbf{A}(W)$ *onto* $\mathbf{A}(W')$ *that carries $W(z)$ into $W'(z)$.*

PROOF. Let **F** denote the totality of finite-dimensional subspaces **M** of **L** such that $A|\mathbf{M}$ is nondegenerate. If $(\mathbf{K}, W)$ and $(\mathbf{K}', W')$ are any two Weyl systems over $(\mathbf{L}, A)$, then for any $\mathbf{M} \in \mathbf{F}$ there is a unique algebraic $*$-isomorphism of $\mathbf{A}(\mathbf{M}, W)$ onto $\mathbf{A}(\mathbf{M}, W')$ that carries W(z) into $W'(z)$, for all $z \in \mathbf{M}$. For $W|\mathbf{M}$ and $W'|\mathbf{M}$ differ from the irreducible Weyl system over $(\mathbf{M}, A|\mathbf{M})$ only in multiplicity, by the Stone-von Neumann theorem, within unitary equivalence. The W^*-algebras generated by $W|\mathbf{M}$ and $W'|\mathbf{M}$ are hence $*$-algebraically equivalent to the algebra $\mathbf{B}(\mathbf{K}(\mathbf{M}))$ of all bounded linear operators on the state vector space $\mathbf{K}(\mathbf{M})$ for any copy of the irreducible Weyl system over $(\mathbf{M}, A|\mathbf{M})$. Accordingly there exists a $*$-algebraic isomorphism $\alpha(\mathbf{M})$ from $\mathbf{A}(\mathbf{M}, W)$ onto $\mathbf{A}(\mathbf{M}, W')$ that carries $W(z)$ into $W'(z)$, and by the triviality of the center of $\mathbf{B}(\mathbf{K}(\mathbf{M}))$, this isomorphism is unique.

On the other hand, if $\mathbf{M}$ is contained in the subspace $\mathbf{N} \in \mathbf{F}$, then the restriction of $\alpha(\mathbf{N})$ to $\mathbf{A}(\mathbf{M})$ will have the properties defining $\alpha(\mathbf{M})$, and so coincide with it. It follows that there exists a unique isomorphism α of $\mathbf{A}_0(W)$ onto $\mathbf{A}_0(W')$ that extends all of the $\alpha(\mathbf{M})$, $\mathbf{M} \in \mathbf{F}$. On each $\mathbf{A}(\mathbf{M}, W)$, α preserves the operator norm, being a C^*-isomorphism, and so extends uniquely to an isomorphism $\bar{\alpha}$ from $\mathbf{A}(W)$ onto $\mathbf{A}(W')$. This shows the existence of the isomorphism claimed in the theorem, and its uniqueness follows as above from the triviality of the centers of the $\mathbf{B}(\mathbf{K}(\mathbf{M}))$. □

DEFINITION. The equivalence class of all the $\mathbf{A}(W)$, under $*$-algebraic isomorphisms exchanging Weyl operators corresponding to the same vector in $\mathbf{L}$, is called the *mode-finite* Weyl algebra over $(\mathbf{L}, A)$, and denoted $\mathbf{W}(\mathbf{L}, A)$. Note that by its construction $\mathbf{W}(\mathbf{L}, A)$ has a unique norm giving it the structure of a C^*-algebra.

COROLLARY 5.1.1. *If $T \in Sp(\mathbf{L}, A)$ [the group of all continuous invertible linear symplectic transformations on $(\mathbf{L}, A)$], there exists a unique automorphism $\gamma(T)$ of $\mathbf{W}(\mathbf{L}, A)$ that carries $W(z)$ into $W(T^{-1}z)$ (in any concrete representation); and $\gamma(ST) = \gamma(S)\gamma(T)$ for $S, T \in Sp(\mathbf{L}, A)$.*

PROOF. Essentially immediate, and omitted. □

Given any net of proper subspaces $\mathbf{N}$ of $\mathbf{L}$, on each of which the restriction of A is nondegenerate and whose totality spans $\mathbf{L}$, a similar construction of a C^*-algebra is possible. For example, if $\mathbf{L}$ is the totality of C^∞ solutions of the wave equation on Minkowski space, having compact support in space at every fixed time, with the natural Poincaré-invariant symplectic structure, one may take as $\mathbf{M}$ the subspace of solutions whose Cauchy data at a fixed time t_0 are supported by the compact set $K \subset \mathbf{R}^n$. The net of all such K relative to inclusion then defines an analogous C^*-algebra, the *space-finite* Weyl algebra. This and similar algebras have been used in some connections, but are fundamentally different from the mode-finte Weyl algebra in *not* being representation-independent.

On the other hand, the mode-finite Weyl algebra is by no means the only algebra of its general type that does enjoy representation-independence. For example, let $\mathbf{W}_0$ denote the algebra of all finite linear combinations of the

$W(z)$; we leave it as an exercise to show that $\mathbf{W}_0$ is $*$-algebraically representation-independent in the same sense as the mode-finite Weyl algebra, and that the same is true of the C^*-closure of $\mathbf{W}_0$.

Which of these algebras is "correct" from a physical standpoint, or most correct, if any? The answer to this derives from the understanding that the basic goal of physical theory is to determine the evolution of the *states* of a physical system, rather than of the "observables," which are, in major part, of a conceptual rather than truly measurable character. In part, however, the observables serve to "label" the states, by their expectation values in the states, according to a standard practice in quantum physics. All of the representation-independent algebras indicated above, enveloping in some sense the $W(z)$, may be used to describe the evolution of the regular states, which are treated in the next section and represent all states for which there seems to be any "practical" possibility of their being empirically meaningful.

For example, in elementary quantum mechanics on the line, with the canonical Heisenberg variables p and q, $\cos p + (1 + q^2)^{-1}$ corresponds to a bounded linear hermitian operator, for which no known "Gedanken experiment" will actually directly determine the spectrum, and so represents an observable in a purely conceptual sense. Similarly, not all mathematical states, in the sense of positive linear functionals on the algebra of all bounded linear operators, generated by the spectral projections of p and q, are truly observable. For example, there are mathematical states that vanish on all $f(p)$ and $f(q)$ for all continuous functions f of compact support, implying that with probability 1, p and q will have infinite values in such states, which are consequently not empirically accessible.

Example 5.1. The representation-independence of the (mode-finite) Weyl algebra implies that in the case of the free boson field over an infinite-dimensional Hilbert space $\mathbf{H}$, the concrete representative for the Weyl algebra—i.e., the algebra of operators on the free field Hilbert space $\mathbf{K}$ to which it corresponds in this representation—must be a proper subalgebra of the algebra of all bounded linear operators $\mathbf{B}(\mathbf{K})$. For if it were equal to $\mathbf{B}(\mathbf{K})$, all of its $*$-automorphisms would be unitarily implementable, and it has been seen that this is not the case. This raises the question of what sorts of operators are in $\mathbf{B}(\mathbf{K})$ but not in the image on $\mathbf{K}$ of the Weyl algebra.

A simple example indicating an answer to this question is provided by the bounded functions of the total number of particles. (From a physical standpoint, the total number is observed only for particles whose energy is bounded below by a positive constant; in the general case, as of photons in conventional theory, the total number is not a true observable, and states with many photons of extremely low energy may present an "infrared divergence.") To show this, let b denote an arbitrary bounded Borel function for which $b(0) \neq b(1)$, and let N denote the total number of particles operator in the free boson field; it will be shown that $b(N)$ is not in the image of the Weyl algebra in $\mathbf{B}(\mathbf{K})$. Let R be in the W^*-algebra generated by the $W(z)$, with z in the finite-dimensional subspace $\mathbf{M}$ of the complex Hilbert space $\mathbf{H}$, and $e_1, e_2, \cdots$ be an orthonormal sequence in $\mathbf{H}$ including a basis for $\mathbf{M}$. Consider the sequence $\langle \mathrm{R}e_n', e_n' \rangle$, where e_n' is the vector in the one-particle subspace of $\mathbf{K}$ that is canonically unitarily equivalent to $\mathbf{H}$, corresponding to the vector e_n in $\mathbf{H}$. Since $\mathbf{H} = \mathbf{M} \oplus \mathbf{M}^\perp$, where $\mathbf{M}^\perp$ denotes the orthocomplement of $\mathbf{M}$, $\mathbf{K}(\mathbf{H})$ is canonically $\mathbf{K}(\mathbf{M}) \otimes \mathbf{K}(\mathbf{M}^\perp)$, from which it follows that $\langle Re_n', e_n' \rangle \to \langle Rv, v \rangle$ for $R = W(z)$ with $z \in \mathbf{M}$, and thence for all R, by approximation. Since $\mathbf{M}$ is arbitrary among finite-dimensional complex-linear subspaces of $\mathbf{H}$, and since every real-linear finite-dimensional subspace is contained in a complex-linear finite-dimensional subspace, it follows by an approximation argument that $\langle Re_n', e_n' \rangle \to \langle Rv, v \rangle$ for all $R \in \mathbf{W}$.

To show that $b(N)$ is not in $\mathbf{W}$, it therefore suffices to show that $\langle b(N)e_n', e_n' \rangle$ does not converge to $\langle b(N)v, v \rangle$. Now $b(N)v = b(0)v$, so that $\langle b(N)v, v \rangle = b(0)$. On the other hand, $b(N)e_n' = b(1)e_n'$, so that $\langle b(N)e_n', e_n' \rangle = b(1)$. It follows that if $b(0) \neq b(1)$, then $b(N)$ cannot be in $\mathbf{W}$. More generally, if $b(k) \neq b(k+1)$, then a similar argument with v replaced by the symmetric tensor product of $e_1, e_2, \ldots, e_k$ shows that $b(N)$ is not in $\mathbf{W}$, and it follows that the only bounded function of N that is in $\mathbf{W}$ is constant.

Example 5.2. C^*-algebraic quantization of a tachyonic structure can be illustrated by consideration of the equation $\Box\varphi - m^2\varphi = 0$, where $m > 0$, in Minkowski space $\mathbf{M}_0$. Let $\mathbf{L}$ denote the space of all C^∞ solutions of this equation having compact support in space at each time, and let the form A on $\mathbf{L}$ be defined as

$$A(\varphi_1, \varphi_2) = \langle \partial_t\varphi_1(t_0, \cdot), \varphi_2(t_0, \cdot) \rangle - \langle \varphi_1(t_0, \cdot), \partial_t\varphi_2(t_0, \cdot) \rangle,$$

where t_0 is arbitrary and $\langle \cdot, \cdot \rangle$ denotes the usual inner product in $L_2(\mathbf{R}^3)$. Let $\mathbf{P}$ denote the restricted Poincaré group, consisting of all transformations on $\mathbf{M}_0$ that are products of translations $x_j \mapsto x_j + a_j$ ($j = 0, 1, 2, 3$) with Lorentz transformations L (connected to the identity), where the linear transformation L leaves invariant the quadratic form $X \cdot X = x_0^2 - x_1^2 - x_2^2 - x_3^2$. $\mathbf{P}$ acts on $\mathbf{L}$ in accordance with the representation V, where for arbitrary $g \in \mathbf{P}$, $V(g)$ sends $\varphi(X)$ into $\varphi(g^{-1}X)$. Moreover, this action leaves invariant the form A. Thus $V(\cdot)$ is a representation of $\mathbf{P}$ as a group of automorphisms on the linear symplectic space $(\mathbf{L}, A)$, and continuity is enjoyed relative to the earlier defined topology.

There is, however, no stable quantization, even for the temporal evolution subgroup. To show this it suffices to show that the action of temporal evolution on $(\mathbf{L}, A)$ is not stably unitarizable, which is Problem 2 of Section 6.6.

Problems

1. Show that for any symplectic vector space $(\mathbf{L}, A)$ the Weyl algebra $\mathbf{W}(\mathbf{L}, A)$ has trivial center.

2. Show that for any symplectic vector space $(\mathbf{L}, A)$ the algebra $\mathbf{W}_0(\mathbf{L}, A)$ generated by finite linear combinations of the $W(z)$ as above is nonseparable. (Hint: show that if f is a nonzero element of $\mathbf{L}$ the spectrum of $W(f)$ is the unit circle. This implies that $\|W(tf) - W(t'f)\| = \|W((t - t')f) - I\| = 2$ for all $t \neq t'$.)

5.3. Regular states of the general boson field

The concept of the general boson field over a given symplectic vector space complements that of the free boson field over a given complex Hilbert space. Specifically, if $(\mathbf{L}, A)$ is a given symplectic vector space, the *general boson field over* $(\mathbf{L}, A)$ is defined as the pair (W, γ), where W is the mapping from $\mathbf{L}$ to the Weyl algebra $\mathbf{W}(\mathbf{L}, A)$, and γ is the representation of $Sp(\mathbf{L}, A)$ by automorphisms of $\mathbf{W}(\mathbf{L}, A)$ given by Corollary 5.1.1. There is no distinguished quantized field Hilbert space $\mathbf{K}$, and no distinguished state identifiable with the free vacuum. The Weyl system mapping W takes a more abstract form, and the unitary representation Γ of the full unitary group on the single-particle space is replaced by the automorphic representation γ of the symplectic group.

DEFINITION. The *generating function* of a state E of the Weyl algebra $\mathbf{W}(\mathbf{L}, A)$ over a linear symplectic space $(\mathbf{L}, A)$ is the function μ on $\mathbf{L}$ given by the equation $\mu(z) = E(W(z))$, $z \in \mathbf{L}$. A *regular state* of $\mathbf{W}(\mathbf{L}, A)$ is a state E whose restriction to each Weyl subalgebra over a finite-dimensional subspace $\mathbf{M}$ of $\mathbf{L}$ on which the restriction of A is nondegenerate (for short, nondegenerate subspace) has a trace-class density operator: i.e., there exists a trace-class element $D_{\mathbf{M}}$ relative to $A(\mathbf{M}, W)$ such that for $X \in A(\mathbf{M}, W)$, $E(X) = tr(XD_{\mathbf{M}})$. Here $A(\mathbf{M}, W)$ is the unique $*$-algebra equivalence class defined by all Weyl systems over $(\mathbf{L}, A)$, and so is devoid of any concrete representation Hilbert space; and "trace class" is defined correspondingly algebraically. In particular, $D_{\mathbf{M}}$ will be represented by an operator of finite trace in the conventional sense if $A(\mathbf{M}, W)$ is represented irreducibly, but may have infinite conventional trace if $A(\mathbf{M}, W)$ is represented as acting with infinite multiplicity. The following theorem treats the basic case in which $\mathbf{L}$ is topologized algebraically.

THEOREM 5.2. *A regular state is uniquely determined by its generating function* μ, *which is characterized by the properties that i)* $\mu(0) = 1$; *ii)* μ *is continuous on* $\mathbf{L}$; *and iii)*

$$\sum_{j,k} \alpha_j\bar{\alpha}_k \, \mu(z_j - z_k) \, e^{\iota A(z_j,z_k)/2} \geq 0$$

for arbitrary $\alpha_1, \ldots, \alpha_n$ *in* $\mathbf{C}$ *and* $z_1, \ldots, z_n$ *in* $\mathbf{L}$. *If* $T \in Sp(\mathbf{L}, A)$, *the automorphism* $\gamma(T)$ *of* $\mathbf{W}$ *carries (by its dual action) a state of generating function* $\mu(z)$ *into one of generating fuction* $\mu(T^{-1}z)$.

Proof. Suppose that μ is the generating function of a regular state; condition i) is evident. The trace has the property that for an arbitrary trace-class operator A, $\mathrm{tr}(AX)$ is strongly continuous in the operator X restricted to the unit ball in the algebra of all bounded operators. Thus for any finite-dimensional nondegenerate subspace $\mathbf{M}$ of $\mathbf{L}$, $tr(W(z)D_{\mathbf{M}})$ is a continuous function of z relative to $\mathbf{M}$. Condition iii) follows from the fact that for arbitrary $\alpha_1, \ldots, \alpha_n \in \mathbf{C}$ and $z_1, \ldots, z_n \in \mathbf{L}$, $(\Sigma_j\alpha_j W(z_j))^*(\Sigma_j\alpha_j W(z_j))$ is a nonnegative operator, together with the Weyl relations. Thus conditions i)–iii) are necessary.

To establish the converse, suppose that μ is given function satisfying i)–iii), and set $\mathbf{K}_0$ for the set of all complex-valued functions on $\mathbf{L}$ that vanish except at a finite set of points. For f and g in $\mathbf{K}_0$, let $\langle f, g\rangle = \Sigma_{z,z'\in\mathbf{L}} f(z)\bar{g}(z')\mu(z - z')e^{\iota A(z,z'))/2}$; then $\langle\cdot,\cdot\rangle$ is a nonnegative hermitian form on $\mathbf{K}_0$. Let $\mathbf{K}_1$ be the quotient of $\mathbf{K}_0$ modulo the subspace $\mathbf{N}$ of vectors of zero norm, where $\|f\|^2 = \langle f, f\rangle$, and let $\mathbf{K}$ be the completion of $\mathbf{K}_1$ with respect to the inner product deriving from that in $\mathbf{K}_0$. A Weyl system may be defined on $\mathbf{K}$ as follows: let $U_0(z)$ denote the operator on $\mathbf{K}_0$ carrying $f(u)$ to $e^{-\iota A(u,z)/2} f(u - z)$. It is straightforward to check that $U_0(z)$ is an isometry on $\mathbf{K}_0$ and that the relations

$$U_0(z)U_0(z') = e^{\iota A(z,z')/2} \, U_0(z + z')$$

are satisfied for arbitrary z and z' in $\mathbf{L}$. It follows that the induced operators $U_1(z)$ on $\mathbf{K}_1$: $U_1(z)(f + \mathbf{N}) = U_0(z)f + \mathbf{N}$, are iosmetric on $\mathbf{K}_1$ and satisfy the Weyl relations. Accordingly, $U_1(z)$ extends uniquely to a unitary operator $W(z)$ on $\mathbf{K}$, and $(\mathbf{K}, W)$ satisfies the Weyl relations over $(\mathbf{L}, A)$. To show that it forms a Weyl system, it is only necessary to establish continuity of $W(\cdot)$. By virtue of the unitarity of the $W(z)$ and the density of $\mathbf{K}_1$ in $\mathbf{K}$, it suffices to show that $\langle W(z)f_1, g_1\rangle$ is continuous as a function of z relative to any finite-dimensional subspace of $\mathbf{L}$, where f_1 and g_1 are the residue classes modulo $\mathbf{N}$ corresponding to f and g in $\mathbf{K}_0$. Explicit computation of $\langle W(z)f_1, g_1\rangle$ in terms of μ shows that this follows directly from the continuity of μ relative to finite-dimensional subspaces of $\mathbf{L}$. If h denotes the function on $\mathbf{L}$ with $h(0) = 1$ and $h(u) = 0$ if $u \neq 0$, then for $h_1 = h + \mathbf{N}$, $\langle W(z)h_1, h_1\rangle = \mu(z)$. It follows that the state E defined by the equation $E(A) = \langle Ah_1, h_1\rangle$, for arbitrary A in the Weyl algebra as represented on $\mathbf{K}$ via U, is regular and has generating function μ.

To show that μ determines the state E uniquely, let **M** be an arbitrary finite-dimensional nondegenerate subspace of **L**. If $\mathbf{R}_0$ is a C^*-algebra of bounded linear operators in Hilbert space that is strongly dense in the W^*-algebra **R**, then the unit ball in $\mathbf{R}_0$ is strongly dense in the unit ball of **R** (Dixmier, 1981). Taking as $\mathbf{R}_0$ the algebra of all finite linear combinations of the $W(z)$, it follows from the continuity property of the trace mentioned above that if D and D' are finite-trace density matrices of states on the algebra $\mathbf{R} = \mathbf{A}(\mathbf{M})$, then $tr(DX) = tr(D'X)$ for all X in the unit ball of **R**, provided $tr(DW(z)) = tr(D'W(z))$ for all $z \in \mathbf{M}$, i.e., if $\mu = \mu'$. By virtue of the Stone–von Neumann theorem, every regular state on **A(M)** has a trace-class density operator in the sense indicated above, which is purely algebraic relative to **A(M)**. This effectively suppresses the multiplicity of the Weyl system when not irreducible, completing the proof. □

5.4. The Clifford C^*-algebra

In the fermionic case the situation is parallel to that in the bosonic case just treated. Let $(\mathbf{L}, S)$ denote a given real orthogonal space with positive definite symmetric form S, such that **L** is either infinite-dimensional or even-dimensional. Let $(\mathbf{K}, \phi)$ be a Clifford system over $(\mathbf{L}, S)$. For any finite-dimensional subspace **M** of **L**, let $\mathbf{C}(\mathbf{M}, \phi)$ denote the W^*-algebra (finite-dimensional in fact) generated by the $\phi(z)$ with $z \in \mathbf{M}$. Let $\mathbf{C}_0(\phi)$ denote the union of the $\mathbf{C}(\mathbf{M}, \phi)$ as **M** varies, and let $\mathbf{C}(\phi)$ denote the uniform closure of $\mathbf{C}_0(\phi)$.

THEOREM 5.3. $\mathbf{C}(\phi)$ *is* $*$*-algebraically independent of* $(\mathbf{K}, \phi)$ *in the sense that if* $(\mathbf{K}', \phi')$ *is any other Clifford system over* $(\mathbf{L}, S)$ *then there is a unique* $*$*-algebraic isomorphism from* $\mathbf{C}(\phi)$ *to* $\mathbf{C}(\phi')$ *that carries* $\phi(z)$ *into* $\phi'(z)$, *for all* $z \in \mathbf{L}$.

PROOF. This is similar to the proof of Theorem 5.1 except that in place of the Stone–von Neumann theorem, the $*$-algebraic unicity of the Clifford algebra over an even-dimensional vector space relative to a given positive symmetric form is used. □

COROLLARY 5.3.1. *If* $T \in O(\mathbf{L}, S)$ [*the group of all orthogonal transformations on* $(\mathbf{L}, S)$], *there exists a unique automorphism* $\gamma(T)$ *of* $\mathbf{C}(\mathbf{L}, S)$ *that carries* $\phi(z)$ *into* $\phi(T^{-1}z)$ *(in any concrete representation); and* $\gamma(ST) = \gamma(S)\gamma(T)$ *for* $S, T \in O(\mathbf{L}, S)$.

PROOF. Again omitted. □

DEFINITION. The *mode-finite* Clifford algebra over $(\mathbf{L}, S)$ is the $*$-algebraic equivalence class of the isomorphic C^*-algebras $\mathbf{C}(\phi)$.

5.5. Lexicon: The distribution of occupation numbers

From the standpoint of an experimentalist, the occupation numbers $\partial\Gamma(P)$, where P is the projection on a subspace of the complex Hilbert space $\mathbf{H}$, are probably the most crucial observables. Interactions as observed are typically interpreted as transformations from states described by the occupation numbers for free incoming particles to the same for outgoing particles. On the other hand, in a theory in which the single-particle or classical field space $\mathbf{L}$ has no given complex Hilbert space structure, but only a symplectic or orthogonal one, the requisite projection P is not defined, or at best not uniquely defined. In order for the concept of particle to make sense, it appears necessary for an appropriate complex structure to be defined or obtained in $\mathbf{L}$. Typically this structure is derived from temporal invariance and stability constraints, as developed earlier.

The occupation numbers $\partial\Gamma(P)$ in the free field over a complex Hilbert space, on which P is an orthogonal projection, automatically have nonnegative integral proper values, as required for the physical interpretation. Moreover, they are coherent with the statistical interpretation of the occupation numbers in relation to the expected observables in a given state; that is, the sum of the products of the occupation numbers with the corresponding values of the physical observables, etc. Formally this means, as noted earlier, that if A is a selfadjoint operator in $\mathbf{H}$ with the spectral resolution $A \sim \lambda dE_\lambda$, then $\partial\Gamma(A) \sim \int \lambda\, \partial\Gamma(E_\lambda) = \int \lambda dN(\lambda)$, where $N(\lambda)$ is the number of particles for which A has a value $\leq \lambda$.

An occupation number $\partial\Gamma(P)$ is affiliated with the Weyl algebra if P is the projection onto a finite-dimensional subspace, in the sense that the bounded functions of $\partial\Gamma(P)$ are in $\mathbf{W}$; in particular, $e^{it\partial\Gamma(P)} = \Gamma(e^{itP})$ is in $\mathbf{W}$ for arbitrary real t. This implies that the joint distribution of such occupation numbers is well defined in any regular state—even if this state is "nonnormalizable" in the sense that there is no associated trace-class density matrix on $\mathbf{K}$, the free field state vector space. More specifically, if $e_1, e_2, \cdots$ is an orthogonal basis for $\mathbf{H}$, the joint distribution of the $n_j = \partial\Gamma(P_j)$, where P_j is the projection on the one-dimensional space spanned by e_j $(j = 1, 2, \ldots, k)$, has $E[e^{i(t_1 n_1 + t_2 n_2 + \cdots + t_k n_k)}] = f(t_1, t_2, \ldots, t_k)$, where E is the state in question. This assures that there exists a joint distribution of all the n_j, in any regular state, by Theorem 1.4, or by Kolmogorov's theorem, providing a particle interpretation for an arbitrary such state. In particular, $\Sigma_{1 \leq j < \infty} n_j$ is well defined as a random variable, which may however be infinite on a set of positive measure, or indeed on the entire underlying probability space.

A general symplectic T on **H** will induce an automorphism $\gamma(T)$ of **W**, as seen earlier, whose dual action on states will carry any given regular state into another such state. In general, the vacuum of the free field will be carried into a state that cannot be represented by vector in **K**, although it necessarily remains pure.

EXAMPLE 5.3. In particular, any unitary operator U on **H** induces a transformation on the regular states of the free boson or fermion field over **H**. The free vacuum is invariant under the induced actions of all of $U(\mathbf{H})$. It is the unique such state that is normalizable in the free field representation.

Examples of states of the Weyl algebra that are invariant under (the induced actions of) arbitrary unitaries on **H** are those with generating function $\mu_k(z) = \exp(-\frac{1}{4}k\|z\|^2)$, where $k \geq 1$. The average of such with respect to a probability distribution for k on $[1, \infty)$ is also automatically the generating function of a $U(\mathbf{H})$-invariant state, and it is the most general such regular state. A corresponding result is valid for the Clifford algebra. The joint distribution of the occupation numbers in the state of generating function μ_k can be computed explicitly, and it developes that the expected number of particles is infinite unless $k = 1$. In principle there may be other states invariant under a nontrivial unitary representation of the Poincaré group, or another group having a noncompact simple factor, but none have been rigorously constructed. This problem connects with constructive quantum field theory, in which the invariance properties of the putative physical vacuum are in question.

Problems

1. Let **H** be a complex Hilbert space and let **W** denote the Weyl algebra over **H**. A state E of **W** is said to be unitarily invariant in case $E(A) = E(\gamma(U)A)$ for all $A \in W$ and $U \in U(\mathbf{H})$.

a) Show that $F(z) = e^{-\frac{1}{4}k\|z\|^2}$ is the generating function of a regular state of **W** if and only if $k \geq 1$, and that this state is unitarily invariant.

b) Show that the most general unitarily invariant regular state can be expressed in terms of an integral of the $F(z)$ with respect to a probability measure in k-space (cf. Segal, 1962).

2. Determine the distribution of occupation numbers in the unitarily invariant states described in Problem 1a, and show that the total number of particles is a finite random variable only when $k = 1$.

3. Show that every regular state of the C^*-algebra **W** having finite particle number with probability one has a trace class density matrix in the free field representation (cf. Chaiken, 1967).

Bibliographical Notes on Chapter 5

Algebraic-quantum phenomenology was originated by von Neumann (1936) in essentially spatial terms. A purely algebraic variant of the von Neumann formalism was proposed by Segal (1947a), and integrated (1947b) with the algebraic characterization of C^*-algebras due to Gelfand and Neumark (1943). This was applied to the C^*-algebraic treatment of quantum fields by Segal (1959, 1961, 1963). The unicity of the Schrödinger representation was treated by Stone (1930) and shown in definitive form by von Neumann (1931).

6

Quantization of Linear Differential Equations

6.1. Introduction

Quantum field theory originated in extremely intuitive and heuristic work, which in part appears to some physicists and mathematicians as almost fortuitously successful. Briefly, Dirac was inspired by the success of Heisenberg's postulate $pq - qp = -i\hbar$ for treating systems of a finite number of degrees of freedom, and sought to extend it to the electromagnetic field, which has an infinite number of degrees of freedom. At a fixed time, the field values and their first time derivatives resemble analytically the p and q that describe the kinematics of a nonrelativistic particle on the line, apart from the infinite-dimensionality, which requires an infinite set of p's and q's, such as have been treated earlier. The commutation relations are determined in a natural way and are formally the same as those of Heisenberg, despite the considerable difference between the essentially geometrical variables in the original Heisenberg relation (position and momentum) and the function-space variables in quantum field theory (such as the components of the electromagnetic field).

Dirac's proposed theory was initially highly successful; its first-order implications correlated very well with empirical and earlier theoretical work (of Einstein, notably) on processes in which particles are emitted or absorbed. In particular, it explained why such processes occur, whereas the Heisenberg-Schrödinger theory described some of the key processes without explaining their origin. The theory of "quantized fields" was polished, extended, and rounded out by Dirac himself, Heisenberg and Pauli, Bohr and Rosenfeld, and many others in the several years following Dirac's original work in the middle 1920s. From an intuitive and formal mathematical standpoint it was extremely attractive and interesting, embodying as it did a rather complete, far-reaching, and plausible theory of the relations between particles and fields in a few compact and empirically motivated assumptions. There was, however, a certain fundamental vagueness in the formulation, which involved nonlinear functions of a quantized field. In modern terms these fields are recognized as operator-valued distributions, of which nonlinear functions are not in general well defined. This vagueness remains unresolved to this day, and represents probably the central foundational issue of quantum field theory. In any event, quantum field theory fell into some disrepute because of the persistent concomitant infinities in its higher-order implications.

The theory was basically too nebulous to survive mathematical attempts to isolate, clarify, or remove all such infinities. For many years physicists argued, as some continue to do, that the infinities were symptomatic of physical rather than mathematical deficiencies; that an essentially different theory, to be suggested perhaps by further accelerator experiments, was required. But no really concrete proposals for an effective alternative have come from the extensive experiments conducted over many decades. Purely mathematical work during this period has, however, indicated that reformulations and developments in line with contemporary mathematical ideas may well be the answer. And thereby hangs our tale.

6.2. The Schrödinger equation

From an abstract position, there is no problem in quantizing a linear differential equation that can be put in the form $\partial u/\partial t = iAu$, where u is a function from $\mathbf{R}$ to a complex Hilbert space $\mathbf{H}$, in which A is a given selfadjoint operator. Taking for specificity the case of boson quantization, the free boson field $(\mathbf{K}, W, \Gamma, \nu)$ earlier treated specializes by applying Γ to the "single-particle dynamics" e^{itA} to define the dynamics $\Gamma(e^{itA})$ of the corresponding quantized field. The number of particles in various states at later times, given the initial state, may then be computed according to conventional quantum phenomenology. But concretely, the interpretation of fields as existing in an ambient geometric space appears as a virtual conceptual necessity, even though the quantized field themselves are not directly observed. More significantly from a practical standpoint, the spatial dependence of fields is needed in order to form *local* interactions, which indirectly assures consistency with the basic physical constraint of causality.

The simplest case is that of the Schrödinger equation $\partial\psi/\partial t = iA\psi$, and its quantization serves to illustrate the basic ideas. In the *classical* (i.e., not quantized, or sometimes, not *second-quantized*, the "first" quantization describing a single particle) equation, $\psi(t, x)$ is a complex-valued function on space-time such that $\psi(t,\cdot) \in L^2(\mathbf{R}^3) = \mathbf{H}$, and A is an essentially given selfadjoint operator in $\mathbf{H}$. Heuristically, the process of quantization is the replacement of ψ by an operator-valued function $\boldsymbol{\psi}$ that satisfies the "same" differential equation (formally) but whose values do not mutually commute. Without some specification of the commutators, quantization is obviously indeterminate, but the simplest nontrivial assumption is that at any fixed time the commutators are scalars. Invariance and/or classical dynamical considerations (originating in Lie's theory of contact transformations) indicates the specific heuristic commutation relations

$$[\psi^*(t,x), \psi(t,y)] = \delta(x - y), \qquad [\psi(t,x), \psi(t,y)] = 0.$$

The intervention of the Dirac delta distribution here indicates that $\psi(t, x)$ must be a distribution or other generalized function, and an analysis from the standpoint of physical measurement (originated by Kramers and developed by Bohr and Rosenfeld) confirms the idea that only the averaged fields $\int\psi(t,x)f(x)dx$, where f is a reasonably regular (e.g., C_0^∞) function on space, have physical meaning as observables accessible by a conceptual measuring process. This idea is readily implemented mathematically. Indeed, generalized operators may be defined that represent the values of the fields $\psi(t,x)$ at individual points, but we begin with the simpler case that can be described in terms of densely defined operators in Hilbert space. Thus we deal with an operator-valued function $\Psi(t,f)$ that is interpreted physically as the average of the quantized field $\psi(t,x)$ at time t with respect to the given test function f. Symbolically, $\Psi(t,f) = \int\psi(t,x)f(x)dx$.

Naturally, regularity questions are involved in the treatment of differential equations. It is convenient to work, in part at least, with a domain that is invariant under the differential operators involved and related operators. In a given space-time, C^∞ regularity is normally sufficient. On the other hand, generality and succinctness are facilitated by an abstract treatment involving unbounded operators in Hilbert space. If A is a given partially defined linear operator in a topological linear space $\mathbf{L}$, the domain $\mathbf{D}_\infty(A)$ consisting of the intersection of the domains of A^n $(n = 0, 1, 2, \ldots)$ often intervenes in such a treatment.

Similar considerations apply to group representations rather than to individual operators. It will suffice here to consider the case of a given continuous one-parameter group V in a Banach space $\mathbf{B}$. $\mathbf{D}_\infty(V)$ is then definable as $\mathbf{D}_\infty(A)$, where A is the generator of V. When multi-parameter groups such as the Poincaré group in relativistic theory are involved, the relevant regularity considerations are largely reducible to the consideration of the time evolution subgroup, by virtue of a partially majorizing role of the energy. For this reason, little essential is lost by limiting consideration here to the case of a single operator or one-parameter group.

The next theorem gives a rigorous formulation of the quantization of the Schrödinger equation, in a slightly abstract form that enhances generality. For rigorous purposes it is convenient to begin with the consideration of hermitian fields $\Phi(t, f)$ rather than the nonnormal fields $\psi(t, x)$ discussed heuristically above.

In order to state properly the Schrödinger equation for the quantized field, which exists only as a generalized function, we first observe in formal terms how the classical equation is naturally reformulated to deal with generalized solutions. The equation $\partial_t\psi = iA\psi$, where $\psi = \psi(t, x)$ and A is an operator on functions of x, takes the following form on multiplication by an arbitrary C_0^∞ function followed by integration with respect to x:

$$\partial_t \int \psi(t, x) f(x) dx = i\int (A\psi)(t, x) f(x) dx = i\int \psi(t, x)(A^* f)(x) dx$$

Accordingly, if $\Psi(t, f)$ denotes $\int \psi(t, x) f(x) dx$, then the integrated counterpart to the Schrödinger equation takes the form

$$\partial_t \Psi(t, f) = i\,\Psi(t, A^* f) \tag{6.1}$$

for appropriately regular f. Equation 6.1 will be used to formulate the Schrödinger equation for the quantized field, first in terms of the hermitian components of the field.

THEOREM 6.1. *Let A be a given strictly positive selfadjoint operator in the complex Hilbert space* $\mathbf{H}$. *Let* $(\mathbf{K}, W, \Gamma, v)$ *denote the free boson field over* $\mathbf{H}$, *let $\phi(U)$ denote the selfadjoint generator of the unitary group $W(tu)$, and set* $\Phi(t, u) = \Gamma(e^{itA})\phi(u)\Gamma(e^{-itA})$. *Then $\Phi(t, u)$ satisfies the (quantized) Schrödinger equations in the following sense:*

i) the closure of $\varepsilon^{-1}[\Phi(t + \varepsilon, u) - \Phi(t, u)]$ *has the limit* $\Phi(t, iAu)$ *as* $\varepsilon \to 0$ *in the space of selfadjoint operators in* $\mathbf{K}$, *with its strong topology, if* $u \in \mathbf{D}(A)$; *and*

ii) for arbitrary $u \in \mathbf{D}_\infty(A)$ *and* $w \in \mathbf{D}_\infty(H)$, *where* $H = \partial\Gamma(A)$,

$$\partial_t \Phi(t, u) w = \Phi(t, iAu) w.$$

PROOF. Conclusion i) follows from Chapter 1, on noting that $\Phi(t + \varepsilon, u) = \Phi(t, e^{i\varepsilon A}u)$ and that the closure of $\varepsilon^{-1}[\Phi(t + \varepsilon, u) - \Phi(t, u)]$ is $\Phi(t, \varepsilon^{-1}(e^{i\varepsilon A} - I)u)$. For $\phi(x_m) \to \phi(x)$ in the strong operator topology as $x_m \to x$. Indeed this is equivalent to the convergence of the one-parameter unitary groups $W(tx_n)$ to the group $W(tx)$. This follows in turn from the strong continuity of $W(z)$ as a function of z.

Regarding conclusion ii), note first

LEMMA 6.1.1. *Given arbitrary* $x \in \mathbf{H}$ *and* $u \in \mathbf{D}(H^{1/2})$, *then* $u \in D(\phi(x))$ *and*

$$\|\phi(x)u\| \le c\,\|x\|\,\|(H + I)^{1/2}u\|$$

for some constant c *independent of x and u.*

PROOF. Write $\phi(x)u$ in the form $2^{-1/2}(C(x) + C(x)^*)u$, applicable to u in the common domain of $C(x)$ and $C(x)^*$. Then

$$\langle\phi(x)u,\ \phi(x)u\rangle = \tfrac{1}{2}[\langle C(x)u,\ C(x)u\rangle + \langle C(x)^*u,\ C(x)^*u\rangle + \langle C(x)^*u,\ C(x)u\rangle + \langle C(x)u,\ C(x)^*u\rangle].$$

Applying the Schwarz inequality, it follows that

$$\|\phi(x)u\|^2 \leq \|C(x)u\|^2 + \|C(x)^*u\|^2.$$

Let P denote the projection on the one-dimensional subspace spanned by x. It is easily seen that it is no essential loss of generality to assume that A is bounded below by I, so that $P \leq I \leq A$, implying that $\partial\Gamma(P) \leq N \leq H$, where $N = \partial\Gamma(I)$. It follows, noting that $\partial\Gamma(P)$ and N commute strongly, that $\mathbf{D}(N) \subset \mathbf{D}(\partial\Gamma(P))$. But it is readily verified, taking $\|x\| = 1$ (without loss of generality) that $C(x)C(x)^* = \partial\Gamma(P)$ and that $C(x)^*C(x) = \partial\Gamma(P) + I$. It follows in turn that $\|C(x)^*u\|^2 \leq \langle Nu, u\rangle$ and $\|C(x)u\|^2 \leq \langle (N + I)u, u\rangle$, whence

$$\|\phi(x)u\|^2 \leq \langle (2N + I)u,u\rangle \leq \|(2N + I)^{1/2}u\|^2.$$

By a simple approximation argument, this inequality remains valid for arbitrary $u \in \mathbf{D}(N^{1/2})$. But $2N + I \leq 2H + I$, so $(2N + I)^{1/2} \leq (2H + I)^{1/2}$, whence the inequality holds a fortiori when N is replaced by H. □

PROOF OF THEOREM, CONTINUED. The conclusion of the lemma may be expressed as $\|\phi(x)(H + I)^{-1/2}u\| \leq c\|u\|\,\|x\|$ for arbitrary $u \in \mathbf{K}$, i.e., $\phi(x)(H + I)^{-1/2}$ is a bounded operator. Now

$$\begin{aligned}\varepsilon^{-1}[\Phi(t + \varepsilon, u)w - \Phi(t, u)w] &= \phi(\varepsilon^{-1}(e^{i(t+\varepsilon)A} - e^{itA})u)w \\ &= [\phi(\varepsilon^{-1}(e^{i(t+\varepsilon)A} - e^{itA})u)\,(H + I)^{-1/2}]\,[(H + I)^{1/2}w].\end{aligned}$$

Applying the indicated boundedness, it follows that ii) is valid for arbitrary $w \in \mathbf{D}(H^{1/2})$, and in particular for $w \in \mathbf{D}_\infty(H)$. □

COROLLARY 6.1.1. *Let $\Psi(t,f)$ denote the closure of $2^{-1/2}(\Phi(t,f) - i\Phi(t, if))$. Then for arbitrary real s and t, the closures of*

$$[\Psi(s,f),\ \Psi(t,g)^*], \qquad [\Psi(s,f),\ \Psi(t,g)], \qquad [\Psi(s,f)^*,\ \Psi(t,g)^*]$$

exist and are respectively $\langle e^{i(s-t)A}f, g\rangle$, 0, and 0.

PROOF. The case when $s \neq t$ is reducible to the case $s = t$ by the observation that

$$\Phi(t,f) = \Gamma(e^{i(t-s)A})\Phi(s,f)\Gamma(e^{-i(t-s)A}) = \Phi(s, e^{i(t-s)A}f).$$

The case $s = t$ is similarly reducible to the case $t = 0$. The corollary then follows from the relations between creation operators and the hermitian fields developed in Chapter 1. □

6.3. Quantization of second-order equations

In general, quantization of second-order equations can be effected by reduction to the first-order case just treated. An important and typical case is that of an equation of the form $\Box\varphi + V(x)\varphi = 0$. Here $V(x)$ is a given function on physical space, e.g., $\mathbf{R}^n$, while φ is a real function on space-time, i.e., $\mathbf{M}_0 = \mathbf{R} \times \mathbf{R}^n$, and $\Box = \partial_t^2 - \Delta$. Assuming that V is nonnegative, bounded, and measurable, and introducing the selfadjoint operator $B = (V - \Delta)^{1/2}$, this equation may be expressed in the abstract form $\partial_t^2 u + B^2 u = 0$, where $u = u(t)$ is a function from $\mathbf{R}$ to a function on $\mathbf{R}^n$. This in turn becomes the first-order equation $\partial_t y = Ay$, where $y = y(t)$ has values in the space of pairs (f, g), each of f and g being a function over physical space, and A is represented by the matrix

$$\begin{pmatrix} 0 & I \\ -B^2 & 0 \end{pmatrix}.$$

The main difference from the Schrödinger equation is the more explicit role played by $\partial_t \varphi$, which is usually dealt with by a treatment in terms of "canonically conjugate" fields, i.e., analogues to p and q. The basic result may be stated as

THEOREM 6.2. *Let B be a given strictly positive selfadjoint operator in a real Hilbert space $\mathbf{H}'$. Let $\mathbf{H}$ denote the complex Hilbert space consisting of all pairs (x, y) with x and y in $\mathbf{H}'$, with the following action of i:*

$$i: (x, y) \mapsto (B^{-1}y, -Bx),$$

and the inner product

$$\langle (x, y), (x', y') \rangle = \langle Cx, Cx' \rangle + \langle C^{-1}y, C^{-1}y' \rangle + i(\langle x', y \rangle - \langle x, y' \rangle)$$

where $C = B^{1/2}$. Let A denote the selfadjoint generator in $\mathbf{H}$ of the one-parameter unitary group represented by the matrix

$$U(t) = \begin{pmatrix} \cos(tB) & B^{-1}\sin(tB) \\ -B\sin(tB) & \cos(tB) \end{pmatrix}.$$

Let $(\mathbf{K}, W, \Gamma, \nu)$ denote the free boson field over $\mathbf{H}$, and let $\phi(z)$ denote the associated field operator. Then setting

$$\Phi(t,x) = \Gamma(U(t))\phi(B^{-1}x,0)\Gamma(U(t))^{-1},$$
$$\Pi(t,x) = \Gamma(U(t))\phi(0,Bx)\Gamma(U(t))^{-1}$$

for arbitrary $x \in \mathbf{H}'$, *the following equations are satisfied:*

$$\partial_t\Phi(t,x)w = \Pi(t,x)w; \qquad \partial_t^2\Phi(t,x)w + \Phi(t,B^2x)w = 0$$

for arbitrary $x \in \mathbf{D}_\infty(B)$ *and* $w \in \mathbf{D}_\infty(\partial\Gamma(A))$.

Conversely, suppose that $\Phi(t,x)$ *and* $\Pi(t,x)$ *are maps from* $\mathbf{R} \times \mathbf{H}'$ *to the selfadjoint operators on the complex Hilbert space* $\mathbf{K}$, *in which is given the one-parameter unitary group* $\Gamma(\cdot)$ *with nonnegative generator* H, *such that* $Hv = 0$ *for a given vector* v *in* $\mathbf{K}$ *that is cyclic for the totality of all operators of the form* $e^{i\Phi(t_0,x)}$ *or* $e^{i\Pi(t_0,y)}$ $(x, y \in \mathbf{H}'$; t_0 *fixed), and that the Weyl relations at time* $t = 0$ *are satisfied together with the quantized differential equation in its integrated form:* $e^{itH}W(z)e^{-itH} = W(U(t)z)$ $(t \in \mathbf{R}, z \in \mathbf{H})$. *Then this structure* $(\mathbf{K}, \Phi, \Pi, \Gamma, v)$ *is unitarily equivalent to that just described.*

PROOF. This is essentially a special case of Theorem 1.11, in which the single-particle Hamiltonian is A. Details are straightforward and are omitted. □

In terms of *space-time averages*, rather than *space averages at fixed times*, Theorem 6.2 may be formulated as

COROLLARY 6.2.1. *Suppose* B *takes the form* $(V - \Delta)^{1/2}$ *in* $L_2(\mathbf{R}^n)$, *where* V *is a strictly positive measurable function. Then there exists a unique map* Φ *from* $C_0^\infty(\mathbf{M}_0)$ *to selfadjoint operators in* $\mathbf{K}$ *such that*

- *i) for any function* $f(t,x)$ *in* $C_0^\infty(\mathbf{M}_0)$, $\Phi(f)$ *is the closure of* $\int\Phi(t, f(t,\cdot))dt$, *defined as a weak integral relative to vectors in* $\mathbf{D}_\infty(H)$;
- *ii)* $\Phi((\Box + V)f) = 0$ *for arbitrary* f *in* $C_0^\infty(\mathbf{M}_0)$; *and*
- *iii) for arbitrary* w *and* w' *in* $\mathbf{D}_\infty(H)$, $\langle\Phi(f)w, w'\rangle$ *is as a function of* f, *a distribution on* $\mathbf{M}_0$.

Moreover, if G *is the subgroup of the Poincaré group that leaves* V *fixed, with corresponding action* U *on* $\mathbf{H}$, *then* Φ *intertwines with* $\Gamma(U(G))$: *for arbitrary* $g \in G$, $\Gamma(U(g))\,\Phi(f)\Gamma(U(g))^{-1} = \Phi(f_g)$, *where* $f_g(X) = f(g^{-1}(X))$.

PROOF. This follows, e.g., by the representation of functions in $C_0^\infty(\mathbf{M}_0)$ as limits of linear combinations of products of functions on time and space separately; the details are left as an exercise. □

LEXICON. In theoretical physical terms, the last theorem describes in essentially relativistic terms the quantization of the Klein-Gordon equation, $\Box\varphi +$

$m^2\varphi = 0$, and similar equations including those involving a given external potential V. The heuristic quantized field $\varphi(t, x)$ is to be thought of as the symbolic kernel for Φ, i.e., $\Phi(f) = \int\varphi(X)f(X)dx$. Later it will be seen that an appropriate interpretation of $\varphi(X)$ is available as a generalized operator, but such pseudo-operators cannot generally be multiplied, and for many purposes it is more effective to work directly with true operators.

Practical (perturbative) quantum field computations in heuristic theory often dispense with rigor and facilitate computations by working with the symbolic kernels $[\varphi(x), \varphi(y)] = iD(x - y)$ and $\langle\varphi(x)\varphi(y)v, v\rangle = G(x - y)$ of the fundamental antisymmetric and symmetric forms, $\mathrm{Im}\langle\cdot,\cdot\rangle$ and $\mathrm{Re}\langle\cdot,\cdot\rangle$, and related kernels. These are definable as distributions satisfying the underlying classical equation with appropriate initial conditions and inhomogeneous term. For constant V they are readily expressible via Fourier analysis in momentum space, or explicitly as Bessel functions in physical space. Thus, $D(x)$ is the unique distribution satisfying the equation $(\Box + V)D = 0$, with the Cauchy data: $D(0, x) = 0$, $\partial_t D(0, x) = \delta(x)$, while $G(x)$ is the unique distributional solution of the same equation with the Cauchy data: $G(0, x) = (B^{-1}\delta)(x)$, $\partial_t D(0, x) = 0$. These and similar singular functions primarily involve aspects of classical hyperbolic equations and space-time geometry, and will not be elaborated here.

6.4. Finite propagation velocity

An important aspect of causality, and one that distinguishes solution manifolds of wave equations from general section spaces, is the notion of finite propagation velocity. Einstein enunciated the principle that no observable physical effect can be propagated more rapidly than light. This has a variety of analytic interpretations, one of which is that the fundamental partial differential equations of physical theory must be hyperbolic. Such equations have well known domain of dependence properties, which roughly state that the value of the solution of the Cauchy problem at a given point of space and time t depends only on the Cauchy data in a region whose size grows with t, rather than on the Cauchy data throughout all of space. The rate of increase of the domain of dependence is substantially the propagation velocity, and Einstein's law, as well as intuitive physical perceptions, is quite effectively modeled in this way.

The finite propagation velocity property is a highly distinctive one, and characteristic of hyperbolic equations. For example, if B is a positive selfadjoint operator in $L_2(\mathbf{R}^n)$ that commutes with the action of the euclidean group on this space (and so may be represented by multiplication by an arbitrary positive rotation-invariant measurable function on the Fourier transform), the differential equation $(\partial_t^2 + B^2)\varphi = 0$ has the finite propagation velocity property only if B^2 differs from $-\Delta$ by a constant (Berman, 1974). In this section we show that simple types of quantized equations also exhibit finite propagation velocity, if dependence is defined in the manner appropriate to noncommuting operators.

In order to treat succinctly fairly general cases, we make the

DEFINITION. A *graduation* on a Hilbert space **H** is a function q from **H** to the interval $[0, \infty]$ such that

i) $\{x \in \mathbf{H}; q(x) < \infty\}$ is dense in **H**;
ii) $q(ax + y) \leq \max\{q(x), q(y)\}$ for any real a and $x, y \in \mathbf{H}$; and
iii) if $x_n \to x$ in **H**, then $q(x) \leq \liminf_{n\to\infty} q(x_n)$.

A given one-parameter group $S(\cdot)$ on **H** is said to have *speed c* in case $q(S(t)x) \leq c|t| + q(x)$.

EXAMPLE 6.1. Let **H** denote the Hilbert space of real normalizable solutions φ of the wave equation on $\mathbf{M}_0$, relative to the complex Hilbert structure given in Section 6.3. Let $q(\varphi)$ denote the diameter of the support of the Cauchy data for φ at time $t = 0$, and let $U(t)$ denote the one-parameter unitary group of time evolution. Then the foregoing conditions are satisfied. The same is true if more generally a bounded nonnegative potential $V(x)$ is incorporated into the equation. These results follow from the classical theory of hyperbolic second-order differential equations.

THEOREM 6.3. *Let* **H** *be a graduated Hilbert space, and let $U(\cdot)$ be a continuous one-parameter group on* **H** *of speed c. Let $(\mathbf{K}, W, \Gamma, \nu)$ denote the free boson field over* **H**. *Then $W(U(t)z)$ is in the C*-algebra of operators generated by the $W(z')$ with $q(z') \leq q(z) + c|t|$, for arbitrary $z \in \mathbf{H}$ and $t \in \mathbf{R}$.*

PROOF. $U(t)z$ is itself such a z', so the conclusion is immediate. □

COROLLARY 6.3.1. *For the Klein-Gordon field on $\mathbf{M}_0$ with external potential V, where V is a given nonnegative, bounded measurable function on space $\mathbf{R}^3$, and for any neighborhood N of a given point $x \in \mathbf{R}^3$, the W*-algebra $R(t, N)$ generated by the spectral projections of the $\Phi(t, u)$ and $\Pi(t, u)$ with u in C^∞ and vanishing outside N is contained in $R(0, N + B(t))$, where $B(t)$ denotes the ball of radius $|t|$ centered at the origin.*

PROOF. This is the special case of the theorem in which **H** is the solution manifold of the Klein-Gordon equation described above, and $U(t)$ is the temporal evolution group defined by the equation. □

Thus, the field $\varphi(t, x)$ is "virtually" in the W^*-algebra of operators generated by the $\varphi(0, y)$ and the $\partial_t\varphi(0, y)$ with $|y - x| \leq |t|$.

COROLLARY 6.3.2. *Theorem 6.3 remains valid in the case of a free fermion field, provided* W *is replaced by the corresponding fermion field* ϕ.

PROOF. The same argument applies. □

6.5. Quantization of the Dirac equation

As far as the algebraic framework is concerned, the Klein-Gordon equation could be quantized equally in fermionic terms, but this would violate the Einstein principle, as interpreted in terms of the commutativity of the field operators in space at equal times. There is, however, no violation of Einstein causality in the fermionic quantization of the Dirac equation, in the standard physical interpretation according to which the Dirac field itself is not directly observable, but only certain bilinear expression in this field, called "currents," for which appropriate commutation at different points of space at the same time follow from the fermionic anticommutation (Clifford) relations. The Dirac field is the simplest prototype for the modeling of fermionic particles and its quantization exemplifies the general case.

The theory of classical spinor fields is basically a matter of space-time geometry, rather than of algebraic quantization, and so will be described only briefly here. We take the elementary position according to which a general spinor field is a function on space-time $\mathbf{M}_0$ with values in a finite-dimensional "spin" space. Under a transformation g in the Poincaré group $\mathbf{P}$, such a function $\psi(x)$ transforms not only directly, displacing x into $g^{-1}(x)$, but the vector in spin space, $\psi(x)$, is transformed according to a given representation R of the universal (in fact, two-fold) cover $\tilde{\mathbf{P}}$ of $\mathbf{P}$. In the particle models of standard relativistic physics, R is trivial on space-time translations, which form an invariant subgroup of the Poincaré group, and so reduces effectively to a representation of the universal (two-fold) cover $\tilde{\mathbf{L}}$ of the homogeneous Lorentz group $\mathbf{L}$, which may be identified, in the case $\dim M_0 = 4$, to which we now restrict consideration, with the group $SL(2, \mathbf{C})$.

The basic and prototypical case is that of normal Dirac spinors, for which the spin space is four-dimensional, and the action of $\mathbf{L}$ can be described simply, in infinitesimal form, as follows: let $Q(X, Y)$ denote the symmetric form on $\mathbf{M}_0$, $X{\cdot}Y = x_0y_0 - x_1y_1 - x_2y_2 - x_3y_3$, and let γ denote the Clifford map over $\mathbf{M}_0$, i.e., the essentially unique linear map from $\mathbf{M}_0$ (as a vector space) to 4×4 matrices, such that $\gamma(X)^2 = Q(X, X)\, I$, any two such maps being conjugates by a fixed invertible matrix. Let the e_j ($j = 0, 1, 2, 3$) be the vectors in the original four-space with all components 0 except 1 in the jth position. The Lorentz group can be defined as that of homogeneous linear transformations leaving invariant Q, and its action on $\mathbf{M}_0$ is accordingly generated by the vector fields $L_{jk} = \varepsilon_j x_j \partial_k - \varepsilon_k x_k \partial_j$, where $\varepsilon_j = Q(e_j, e_j)$. The spin representation R of $\mathbf{L}$ can be defined as that for which the corresponding infinitesimal representation carries L_{jk} into $\frac{1}{2}\gamma_j\gamma_k$. The Dirac equation takes the form $\mathbf{D}\psi + im\psi = 0$ ($m > 0$), where $\mathbf{D}$ denotes the operator in the space of functions over $\mathbf{M}_0$ with values in the four-dimensional spin space, $\mathbf{D} = \sum \varepsilon_j \gamma_j \partial_j$. The null space of $\mathbf{D} + im$ is invariant under the following representation U of $\tilde{\mathbf{P}}$:

$$U(g)\colon \psi(X) \mapsto R(g)\psi(g^{-1}(X)).$$

Just as the C^∞ solutions of the Klein-Gordon equation having compact support in space at each time form a $\tilde{\mathbf{P}}$-invariant pre-Hilbert space in a unique way, the same is true for the Dirac equation, with some slight differences. In the latter case, the so-called "wave function" ψ is complex-valued; and the completion of the solution manifold is not irreducible under $\tilde{\mathbf{P}}$, but decomposes into four irreducible components, each of which is transformed in a unitary way by the action U of the Poincaré group. The spectrum of the self-adjoint generator of temporal evolution is positive for one pair of the components and negative for the other pair; each pair is interchanged by space reversal, which is not contained in the connected group $\tilde{\mathbf{P}}$. The physical interpretation of the apparent negative energy of one pair of components, which was advanced by Dirac and remains in heuristic use, is that this pair essentially represents antiparticles to the particles represented by the positive components. More specifically the antiparticles are represented by "holes" in a "sea" of negative energy particles. This has not appeared satisfactory to some physicists. We will see that the problem arises essentially because of the use of an inappropriate complex structure in the solution manifold of the Dirac equation. To begin with, we treat one of the components, namely the positive-frequency solutions of the indicated equation, having finite Hilbert norm; conventionally, this is the "particle" component.

Consider then the space $\mathbf{H}_m^+$ of all solutions ψ of the equation $(\mathbf{D} + im)\psi = 0$ having positive frequency. To specify this space precisely, we note that on multiplication by γ_0 the equation attains the form

$$i\partial_t\psi = \alpha_1\partial_1\psi + \alpha_2\partial_2\psi + \alpha_3\partial_3\psi + \alpha_4 m\psi,$$

where the α_j are hermitian matrices on spin space, for an appropriate representation to the γ's.

It follows that if $\langle\langle\cdot,\cdot\rangle\rangle$ denotes the usual positive definite inner product in $\mathbf{C}^4$, then $\langle\varphi, \psi\rangle = \int\langle\langle\varphi(t,x), \psi(t,x)\rangle\rangle d^3x$ is independent of t for any two solutions φ and ψ, and it can be shown to be $\tilde{\mathbf{P}}$-invariant. Accordingly, $\mathbf{H}_m$ is defined as the Hilbert space of all solutions ψ for which $\|\psi\|^2 = \langle\psi,\psi\rangle$ is finite. The subspaces $\mathbf{H}_m^{\pm}$ of positive/negative frequency are the corresponding eigenspaces of the selfadjoint generator of temporal evolution.

In these terms the quantized (positive-frequency) Dirac field may be formulated as follows: let $(\mathbf{K}, C, \Gamma, v)$ denote the free fermion field over the Hilbert space just defined, C denoting the complex-Clifford map

$$C(x)C(y)^* + C(y)^*C(x) = \langle x, y\rangle, \qquad C(x)C(y) + C(y)C(x) = 0$$

for $x, y \in \mathbf{H}_m$. Let $\Psi(t, u) = \Gamma(U(t))C(u)\Gamma(U(t))^{-1}$ where $t \in \tilde{\mathbf{P}}$ denotes temporal evolution (translation) by time t. The symbolic function $\psi(t, x)$ is the kernel of $\Psi(t, u)$, which has the symbolic form

$$\Psi(t, u) = \int \langle\langle\psi(t, x), u(x)\rangle\rangle\, d^3x.$$

The Dirac field just established satisfies all but one of the basic physical desiderata: (1) the Dirac equation, as adapted to operator-valued (generalized) functions; (2) the Clifford (or canonical anticommutation) relations; (3) positivity of the energy, as a selfadjoint operator generating the temporal evolution of the field; and (4) the existence of an essentially unique vacuum vector, i.e., the lowest eigenvector for the energy, which moreover is a cyclic vector for the field operators. It can be shown that these features alone uniquely determine the field, within unitary equivalence (cf. below), and that, moreover, the field is Lorentz-invariant, in essentially the same sense as the scalar field treated above. The one feature it lacks is invariance under "particle-antiparticle conjugation." To achieve this it is necessary to treat the full Dirac equation, including its negative frequency component. The treatment of this component involves an additional feature that is now simple and transparent, but was historically obscure and problematic; and to some extent, even recent physical literature reflects this obscurity.

Dirac's treatment was in part interpretive rather than analytical: as noted, he argued that the negative-frequency vectors in the full solution manifold for this equation represented "holes in a sea of negative electrons." A treatment that was operational rather than intuitive took some decades to evolve. A representative view, as expressed by G. Källén (1964), was that "a completely consistent formulation of [the interpretation of] the negative-frequency solutions of the Dirac equation can not be obtained without use of second quantization." It is universally accepted that with suitably modified quantization of the full Dirac equation, the energy operator is nonnegative. In the physical literature the modification consists in the replacement of the free-field representation of the Clifford relations by one that is inequivalent on the negative frequency subspace, together with consequent alterations of the basic formalism. This is arguably somewhat opportunistic, but the result agrees, as regards its practical implications, with what is obtained from modification of the initial complex structure on the basis of invariance and stability constraints, which is used here. This latter general approach is in keeping with other indications that it is not so much an equation that requires quantization, as a group representation in a real orthogonal or symplectic space, which arises as an invariant subspace of the section space of a homogeneous vector bundle over space-time, or physically speaking, a specified class of fields. The equation itself serves, from this point of view, as an alternative description of the action of time translation in the group representation.

Thus the single-particle *frequency* (i.e., dual variable to the time) must be distinguished from the *energy* of the particle as a state of the quantized field. This may be interpreted mathematically as a change in the complex structure of the single-particle Hilbert space **H**, according to which the original complex structure i must be replaced by the modified complex structure $i' = i\text{sgn}(k_0)$, where k_0 is the dual variable to the time coordinate x_0 (i.e., the frequency). Here $\text{sgn}(k_0)$ is defined by the operational calculus for selfadjoint operators (or by Fourier transformation). The operator i' is equally Lorentz-invariant, of square equal to -1, and orthogonal relative to the symmetric form $\text{Re}(\langle\cdot,\cdot\rangle)$; but temporal evolution on the negative frequency subspace has a positive generator, relative to the Hilbert space $\mathbf{H}'$ obtained by using i' in place of i as complex structure, and retaining the cited symmetric form.

The modified Hilbert space $\mathbf{H}'$ is correspondingly decomposable as the direct sum $\mathbf{H}^+ \oplus \mathbf{H}^-$, where $\mathbf{H}^+$ and $\mathbf{H}^-$ are the positive and negative frequency components in the original space **H**. This decomposition is Lorentz-invariant, and the representations of the Poincaré group $\tilde{\mathbf{P}}$ on $\mathbf{H}^+$ and $\mathbf{H}^-$ are abstractly identical, i.e., unitarily equivalent. Relative to the complex structure i', the two subspaces cannot be distinguished Poincaré-wise (e.g., by conventional "quantum numbers," which are represented by the actions of elements of the enveloping algebra of the Lie algebra of $\tilde{\mathbf{P}}$). They may be distinguished only by a two-valued "label," which is physically interpreted as *charge*, and given the values ± 1 in suitable units. The quantization of $\mathbf{H}'$ then results in a physically appropriate quantization for the full Dirac equation, including positivity of the energy and the other physical desiderata earlier cited, as well as a simple form of charge conjugation.

The unicity theorem of Chapter 3 shows that the result of the foregoing procedure is the unique positive-energy quantization of the full Dirac equation. The complex structure i' is unique, and any positive-energy quantization must be unitarily equivalent to that given above, by virtue of this theorem. However, since modification of an initial complex structure in accordance with physical constraints of invariance and stability is an idea that is not yet commonly encountered in the physics literature, the full "fermion-antifermion" quantized field is usually obtained in another manner. The device used for the quantization of the negative-frequency component of **H** has an appearance of opportunism, but is effectively (i.e., as regards "practical" implications) equivalent to that obtained by modification of the complex structure in **H**. This is implied by the unicity theorem as a consequence of its positive energy and other features characterizing free fermion fields.

More specifically, the basic device used commonly in the physical literature consists in the interchange of creation and annihilation operators for the antiparticle, and a corresponding redefinition of the field energy, particle numbers, etc., which in this literature are defined in terms of the creation and annihilation operators. Consider, to begin with, the problem of the positive-energy fermionic quantization of a classical field whose single-particle temporal evolution generator is negative. It is clear that on the free fermion field as defined earlier, the total energy of the free fermion field will be negative. To overcome this, one may interchange the creation and annihilation operators, which can be regarded as a mathematical implementation of Dirac's essential concept. This in itself would not change the representation Γ of the unitary group on the single-particle space, as formulated here; but in much of the literature, such operators as the energy, the number operators, and other generators of the Poincaré group are given explicit (if heuristic) expressions in terms of the creation and annihilation opertors, rather than defined by group invariance considerations, as here. In consequence, the effective Γ (more precisely, the restriction of Γ to $\tilde{\mathbf{P}}$ and to the "phase transformations" whose generators define the particle number operators, which are the aspects of Γ required for practical purposes) is altered in such a way that the field energy and particle numbers become positive operators.

This device was earlier viewed with some suspicion, in part because of the apparent lack of unitary implementability of the canonical transformation involved in the interchange of creation and annihilation operators. That this apparent lack is real may be confirmed by application of Theorem 4.6. Relative to an appropriate basis $e_1, e_2, \ldots$ for the single-particle space, the interchange of creation and annihilation operators is the implementation on the Clifford algebra of the orthogonal, nonunitary, transformation: $e_j \to e_j$, $ie_j \to -ie_j$. It is immediate that this is unitarily implementable only in the case of a finite-dimensional Hilbert space. The literature includes many attempts to obtain effective limits of the implementing unitary transformations in the n-dimensional case, as n becomes infinite, which limits, of course, do not exist. A simple, rigorous formulation of the stable quantization of a negative-frequency fermion field is as follows.

We first define the *anti-free-field* over a complex Hilbert space **H** with a given conjugation $\varkappa$. Let $(\mathbf{K}, C, \Gamma, v)$ denote the free field over **H**. Let $\mathbf{K}' = \mathbf{K}$, $C'(x) = C(x)^*$, $\Gamma'(U) = \Gamma(\varkappa U \varkappa)$, and $v' = v$. The anti-free-field over **H** is the system $(\mathbf{K}', C', \Gamma', v')$. Now let U be a continuous, unitary representation of the Poincaré group $\tilde{\mathbf{P}}$ on the complex Hilbert space **H**, and suppose the generator of temporal evolution is negative. Then the anti-free-field over **H** with respect to the unique conjugation $\varkappa$, such that $\varkappa U\varkappa$ has a positive generator for temporal evolution, is such that the selfadjoint generator of $\Gamma'(V(.))$, where $V(t)$ is the one-parameter group representing temporal evolution, is nonnegative.

It is clear that this construction yields one that is identical (within unitary equivalence) to that obtained by changing the complex structure i on the negative-energy Hilbert space to the complex structure $-i$. The main result, that this procedure is intrinsically a natural and essential one, however it may be formulated analytically, may be clarified by the following considerations: the canonical representation Γ of the unitary group $U(\mathbf{H})$ of a Hilbert space $\mathbf{H}$ is naturally extendable to a representation Γ^e of the group $U^e(\mathbf{H})$ ("e" for "extension") of all real-linear operators on $\mathbf{H}$ that are either unitary or antiunitary (i.e., anticommute with i). Thus, in the particle representation, $\Gamma^e(V)$ for any $V \in U^e(\mathbf{H})$ may be defined as the direct sum of all symmetrized or skew-symmetrized powers of V (depending on whether the quantization is bosonic or fermionic). Alternatively, it may be characterized by the properties that $\Gamma^e(V)$ is the unique unitary or antiunitary operator T on $\mathbf{K}$ such that $TC(z)T^{-1} = C(Vz)$ and $Tv = v$, taking for specificity the fermionic case.

Consider now the fermionic quantization of an arbitrary one-parameter unitary group of transformations on a Hilbert space. The application to wave equations naturally involves the formulation of the solution manifold as a Hilbert space on which the temporal evolution defined by the equation is represented unitarily.

SCHOLIUM 6.1. *Let $V(t)$, $t \in \mathbf{R}$, be a one-parameter continuous unitary group on a Hilbert space $\mathbf{H}$. Then there exists a system $\Sigma = (\mathbf{K}, \phi, \Gamma, v)$ such that $\mathbf{K}$ is a complex Hilbert space; ϕ is a real linear map from $\mathbf{H}$ to bounded hermitian operators on $\mathbf{K}$ such that $[\phi(x), \phi(y)]_+ = \mathrm{Re}\langle x, y\rangle$; $\Gamma(\cdot)$ is a one-parameter unitary group on $\mathbf{K}$ such that $\Gamma(t)\phi(z)\Gamma(t)^{-1} = \phi(V(t)z)$ ($t \in \mathbf{R}$, $z \in \mathbf{H}$), and whose selfadjoint generator is nonnegative; and v is a unit vector in $\mathbf{K}$ that is invariant under all $\Gamma(t)$ and cyclic for the $\phi(x)$, $x \in \mathbf{H}$. Σ is unique within unitary equivalence if and only if $V(\cdot)$ has no nontrivial fixed vector.*

Concretely, Σ may be represented as follows. Let A denote the selfadjoint generator of $V(\cdot)$, and let $\mathbf{H} = \mathbf{H}_+ \oplus \mathbf{H}_-$, where A is positive on $\mathbf{H}_+$ and nonpositive on $\mathbf{H}_-$, each of $\mathbf{H}_+$ and $\mathbf{H}_-$ being invariant under $V(\cdot)$. Let the free fermion fields over the $\mathbf{H}_+$ and $\mathbf{H}_-$ be denoted as $(\mathbf{K}_\pm, C_\pm, \Gamma_\pm, v_\pm)$. Then $\mathbf{K} = \mathbf{K}_+ \otimes \mathbf{K}_-$; $C(z_+ + z_-) = C(z_+)\otimes I + \Omega \otimes C^*(\varkappa z_-)$, $\Gamma(V(t)) = \Gamma(V_+(t))\otimes\Gamma_-(\varkappa V_-(t)\varkappa)$, where $V(t) = V_+(t)\oplus V_-(t)$, relative to the decomposition $\mathbf{H} = \mathbf{H}_- \oplus \mathbf{H}_-$; and $v = v_+ \otimes v_-$.

PROOF. This essentially recapitulates elements of the prior discussion, and details are left as an exercise.

Problems

1. Show that the representation U of $\tilde{\mathbf{P}}$ on the Hilbert space $\mathbf{H}_m$ described above is in fact strongly continuous, unitary, and splits into two irreducible components that are interchanged by space reversal. (Hint: use the Fourier transform.)

2. Develop analogues to Corollaries 6.2.1 and 6.3.1 for the Dirac equation.

3. Let $\varkappa$ be a conjugation on the given Hilbert space $\mathbf{H}$, and suppose that $\mathbf{H}$ has the form $\mathbf{H}_+ \oplus \mathbf{H}_-$, where $\varkappa$ carries $\mathbf{H}_+$ onto $\mathbf{H}_-$. Formulate an appropriate analog to the free fermion field for the free fermion-antifermion field over ($\mathbf{H}$, $\varkappa$), and show its essential unicity by adaption of the positive-energy constraint as applied in the proof of Theorem 3.6.

4. Let $\mathbf{S}$ denote all C_0^∞-spinor-valued functions on space $\mathbf{R}^3$, and let $\Psi(t,f)$ denote a function on $\mathbf{R} \otimes \mathbf{S}$ to bounded linear operators on a Hilbert space $\mathbf{K}$ that satisfies the following conditions:

   i) $(\partial/\partial)\, \Psi(t,f) = \Psi(t, (\mathbf{D} + im)^* f)$ ($\mathbf{D}$ = Dirac operator);
   ii) $[\Psi(t,f), \Psi(t,g)^*]_+ = \langle f, g\rangle$, $[\Psi(t,f), \Psi(t,g)]_+ = 0$;
   iii) there exists a nonnegative selfadjoint operator H in $\mathbf{K}$ such that $\Psi(t,f) = e^{itH}\Psi(0,f)e^{-itH}$ (for arbitrary t and f); and
   iv) there exists a vector v in $\mathbf{K}$ that is cyclic for the $\Psi(0,f)$ and such that if $Hy = 0$, then $y = cv$ for some scalar c.

Show that any other such function $\Psi'(t,f)$ is unitarily equivalent to $\Psi(t,f)$.

5. Develop the C^*-algebraic quantization of the Dirac equation with the inclusion of an arbitrary bounded measurable potential V, which is added to the mass m. Show that if V is bounded by m, then a positive-energy Hilbert space quantization exists and is unique within unitary equivalence.

6. Prove the following variant of Scholium 6.1: if V is replaced by a continuous unitary representation of a Lie group G, which includes a given one-parameter group designated as temporal evolution, and if the respective positive- and negative-frequency components of the spectral decomposition of the temporal evolution group are invariant under $V(G)$, then the conclusion of Scholium 6.1 remains valid for the full group G, with the extension of Γ to a representation of all of G, and the corresponding adaptation of the concrete representation given for Σ. (In particular, the positive-energy quantization of the Dirac equation is invariant under $\tilde{\mathbf{P}}$.)

6.6. Quantization of global spaces of wave functions

A useful way to regard the space of solutions of a classical wave equation, e.g., the Klein-Gordon equation $\Box f + m^2 f = 0$, is as an infinitesimal eigenspace in a spectral decomposition of the differential operator in question—here, $\Box$. Thus the space $\mathbf{L} = L_2(\mathbf{M}_0, \mathbf{R})$ of all real square-intergrable functions on $\mathbf{M}_0$ is acted on naturally by the Poincaré group $\mathbf{P}$ by $f(X) \mapsto f(g^{-1}(X))$, $g \in \mathbf{P}$, and by the $\mathbf{P}$-invariant operator $\Box$. But suitable global eigenspaces may also be quantized, and not merely the infinitesimal ones, which can be regarded as arising in the "direct integral" decomposition of $\mathbf{L}$ under the actions either of $\mathbf{P}$ or of the wave operator. The infinitesimal case treated above can be regarded as a limiting case—as a mass packet becomes of vanishing width, in physical terms. There are, however, some limitations; m^2 must be nonnegative, though the spectrum of $\Box$ extends into positive as well as negative values. The positive spectrum corresponds to negative values for m^2, or "imaginary mass," representing purely hypothetical fields described as tachyonic in the physical literature. Tachyonic subspaces require separate treatment, in part for the reason indicated in the following Scholium. (Concerning tachyonic equations, see also Problem 2 of this section).

SCHOLIUM 6.2. *Let* $\mathbf{M}$ *denote the closed linear subspace of* $\mathbf{L}_2(\mathbf{M}_0, \mathbf{R})$ *consisting of the eigenspace of* $\Box$ *in spectral range from* λ *to* λ'. *If* λ *and* λ' *are both negative, there exists a continuous* $\mathbf{P}$*-invariant antisymmetric form on* $\mathbf{M}$, *the direct integral of the forms earlier indicated for the Klein-Gordon equation. If* λ *and* λ' *are both positive, there exists no nonzero continuous antisymmetric form on* $\mathbf{M}$ *that is* $\mathbf{P}$*-invariant.*

PROOF. Taking Fourier transforms, the action of translations in $\mathbf{M}_0$ on $\mathbf{L}$ become multiplication operators by complex exponentials, on the subspace of all hermitian elements F of $\mathbf{L}_2(\mathbf{M}_0^*, \mathbf{C})$; that is, those F with $\bar{F}(K) = F(-K)$, $K \in \mathbf{M}_0^*$. These multiplication operators generate all multiplications by bounded measurable hermitian functions on subspaces of the indicated types, and any operator commuting with translations must itself be represented by a multiplication by a bounded hermitian function F, by the maximal abelian character of the totality of such. A continuous bilinear form will be representable by an operator in conjunction with the underlying inner product, and the form will be invariant if and only if the operator commutes with the action of $\mathbf{P}$, and antisymmetric if the corresponding function F satisfies $F(K) = -F(-K)$. When λ and λ' are both negative, the operation of multiplication by sgn k_0 (where $k_0(K)$ is the first component of K, i.e., the dual variable to x_0) is Lorentz-invariant and provides the operator for the form whose existence is claimed in the theorem. When they are both positive, the question is that of the existence of a bounded hermitian function defined on the region between hyperboloids corresponding to λ and λ' in $\mathbf{M}_0^*$, which is Lorentz-invariant and antisymmetric under reflection in the origin. However, in this region, K and

$-K$ are on the same Lorentz orbit, so there exists no nonzero antisymmetric Lorentz-invariant function in the region. □

The examples of single-particle space treated in this chapter have all been spaces of *functions*, although vector-valued on occasion. This is a noninvariant and special feature that derives from the simplicity of the representations involved, and of the space-time $\mathbf{M}_0$. More invariantly and generally, single-particle spaces are formulated as spaces of sections of a homogeneous vector bundle over the space-time under consideration, e.g., differential forms, spinor fields, etc.

It is beyond the scope of the present work to enter into the treatment of these matters, but we note especially that "the quantized field at a point" is not an invariant notion. More properly, the quantized field should be defined as an essentially linear map from the dual S^* of the section space S of the section space S of the vector bundle in question to the operators on the quantized field vector space $\mathbf{K}$. Formally, this map Ξ is of the form $\Xi(\zeta) = \langle\varphi, \zeta\rangle$, where ζ is arbitrary in S^* and φ is the quantized field. When the bundle is "parallelizable," S^* is representable as a space of smooth functions from the space-time manifold to a finite dimensional vector space, and φ can be represented as a generalized function. But this representation depends in an essential way on the parallelization, quite apart from the unitary equivalence of the structures treated earlier. In addition there are important cases, that of Maxwell's equations, for example, in which a *quotient* space of solutions, rather than a subspace, is involved. The selection of a representative in the residue class (or *choice of gauge*) then becomes an issue in the definition of the field at a point, which is involved in the formulation of a local interaction (which, however, is ultimately independent of the choice of gauge). Despite these significant complications, the underlying algebraic structures are primarily those of the free or general quantized fields, correlated with the given wave equation by variants of the procedures described in this chapter.

Problems

1. a) Let $\mathbf{L}$ denote $C_0^\infty(\mathbf{M}_0)$. Suppose that T is a $\mathbf{P}$-invariant linear map from $\mathbf{L}$ into the manifold of normalizable solutions of the Klein-Gordon equation, $\Box\varphi + m^2\varphi = 0$ in $\mathbf{M}_0$. Let (K, W, Γ, ν) denote the free boson field over this manifold, structured as a complex Hilbert space $\mathbf{H}$ as above, and set $\Xi(f) = \phi(Tf)$. Show that Ξ is a strongly linear and $\mathbf{P}$-invariant map from $\mathbf{L}$ into self-adjoint operators on $\mathbf{K}$. That is, $\Xi(af + h)$ is the closure of $a\,\Xi(f) + \Xi(h)$, for arbitrary real a and f, h in $\mathbf{L}$; and for any $g \in \mathbf{P}$, $\Xi(f_g) = \Gamma(U(g))\,\Xi(f)\,\Gamma(U(g))^{-1}$, where U denotes the action of $\mathbf{P}$ on $\mathbf{H}$, and $f_g(X) = f(g^{-1}(X))$.

b) Let D denote the distribution solution of the equation $\Box D + m^2 D = 0$, having the Cauchy data at time zero, $D(0, x) = 0$ and $\partial_t D(0, x) = \delta(x)$. Show that the map $T: f \mapsto \int D(X - Y) f(Y) d^4Y$ satisfies the conditions of a). (The distribution D has the properties: $D(-x) = -D(x)$, and D is invariant under the action of the Lorentz group.)

c) Conclude that the corresponding Ξ is continuous from $\mathbf{L}$ to the selfadjoint operators on $\mathbf{K}$, in the strong operator topology (for unbounded operators) on the latter and the standard topology in $\mathbf{L}$. Show that this definition of Ξ agrees with that given in Corollary 6.3.1 (with $V = 0$).

d) Show that there exists a map φ from $\mathbf{M}_0$ to sesquilinear forms on $\mathbf{D}_\infty(H)$, where H is the energy operator in $\mathbf{K}$, such that for arbitrary $u, u' \in \mathbf{D}_\infty(H)$ and $f \in \mathbf{L}$

$$\langle \Xi(f)u, u' \rangle = \int \varphi(X)(u, u') f(X)\, d^4X.$$

Establish a topology in the space of such forms in which the map φ is continuous and unique, and conclude its $\mathbf{P}$-invariance

$$\varphi(g^{-1}(X)) = \Gamma(U(g))\varphi(X)\Gamma(U(g))^{-1},$$

where unitary transformation of forms is defined to extend that on operators. Show also that $\langle \varphi(X)u, u' \rangle$ satisfies the classical Klein-Gordon equation, as a function of X.

2. Let $\mathbf{L}$ denote the class of all C^∞ real solutions in $\mathbf{M}_0$ of the tachyonic equation $(\Box - m^2)\varphi = 0$, $m > 0$, having compact spatial support at each time in the usual C_0^∞ topology (which is $\mathbf{P}$-invariant). For any two vectors φ and ψ in $\mathbf{L}$, let $A(\varphi, \psi) = \int (\psi \partial_t \varphi - \varphi \partial_t \psi) d^3x$ (integration being over space at a fixed time). Show that $(\mathbf{L}, A)$ is a symplectic vector space that is invariant under the action U of the Poincaré group: $U(g)$ carries $\varphi(X)$ into $\varphi(g^{-1}(X))$, but that the action of temporal evolution in $(\mathbf{L}, A)$ is not unitarizable, so that there is no quantization (of the type applicable to real mass, inclusive of a vacuum, etc.). (Hint: use Fourier transformation.)

3. Develop the fermionic quantization of the Klein-Gordon equation in analogy with its bosonic quantization, and show that Einstein causality is then violated.

4. Extend Theorem 6.2 to the case when A and B are not necessarily bounded by a positive multiple of I (but remain positive). (Hint: work in the completion of $\mathbf{D}(C)$ when C is a positive selfadjoint operator that is not bounded away from 0, with respect to the inner product $\langle x, y \rangle_C = \langle Cx, Cy \rangle$.) Apply the result to the quantization of the wave equation, and extend Corollary 6.3.1 to this case.

5. The Maxwell equations apply to differential 1-forms, or "potentials," A on $\mathbf{M}_0$. They assert that $\delta F = 0$, where F is the "field" dA. Although only the latter is construed as directly observable, it is important to work with 1-

forms rather than 2-forms since the fundamental interaction with charged particles is local only in terms of 1-forms; but the underlying single-particle Hilbert space is then a quotient space. Show that the Poincaré group acts unitarily in this space, when formulated as follows in terms of Fourier transforms $A(K)$ of the potentials:

Let **C** denote the class of all complex-valued measurable functions A defined on the cone $C = \{K \in \mathbf{M}_0^*: K\cdot K = 0\}$ that satisfy the hermiticity condition $A_j(-K) = \overline{A_j(K)}$ (corresponding to $\hat{A}(X)$ real) and the normalization $K\cdot A(K) = 0$ (i.e., A is in the "Lorentz gauge"), almost everywhere on C. Let $\mathbf{L}_0$ denote the set of all $A \in \mathbf{C}$ for which $\langle A, A\rangle = \int_C A(K)\cdot\bar{A}(K)\,dK$ is finite, where dK denotes the (unique) Lorentz-invariant element of volume on C, i.e., $|k_0|^{-1}dk_1dk_2dk_3$. Show that $\langle A, A\rangle$ is nonnegative for arbitrary A satisfying the normalization condition. Let $\mathbf{L}$ denote the quotient of $\mathbf{L}_0$ by the subspace of A for which $\langle A, A\rangle = 0$, and for any two residue classes $[A]$ and $[A']$, define $S([A],[A'])$ as $\langle A, A'\rangle$. Let J denote the complex structure on $\mathbf{L}$ that is the quotient of the structure J' that carries each A into its product with $i\,\mathrm{sgn}(k_0)$, and let **H** then be the complex Hilbert space obtained by completion of the inner product space for which the real part of the inner product is given by S, the complex structure by J, and the imaginary part of the inner product between x and y by $-S(Jx, y)$.

6. a) Treat the analog of Problem 1 for the wave equation on $\mathbf{R} \times S^1$, and develop explicit expansions for the quantized field into corresponding complex exponentials $e^{ij(t\pm\theta)}$, with coefficients that are operators of designated commutation relations. (Careful about **P**-invariance!)

b) Do the same for the fermionic quantization of the Dirac equation on $\mathbf{R} \times S^1$, $\gamma_0\partial_t\psi + \gamma_1\partial_\theta\psi = 0$, where $\gamma_0^2 = 1 = -\gamma_1^2$ and $\gamma_0\gamma_1 + \gamma_1\gamma_0 = 0$.

7. Let **H** denote the Hilbert space of normalizable solutions to the Klein-Gordon equation in Minkowski space $\mathbf{M}_0$, and let $(\mathbf{K}, W, \Gamma, v)$ denote the free boson field over **H**. Let V denote an arbitrary neighborhood of the origin in space ($\mathbf{R}^n$), and let $\mathbf{H}_V$ denote the (real-linear) subspace of **H** consisting of vectors whose Cauchy data at time t are supported in V. Show that v is a cyclic vector for the set of operators of the form $W(z)$ $z \in \mathbf{H}_V$. (This result may be interpreted as exhibiting the intrinsically global character of the vacuum, in contrast to the local character of the commutation relations. It is a variant of the Reeh-Schlieder theorem [1961], which was developed in the context of axiomatic quantum field theory. [Cf. Segal and Goodman, 1965.].)

8. In the context of Problem 7, show that even if V and V' have disjoint support, the field operators $W(f\oplus 0)$ and $W(f'\oplus 0)$, where $\mathrm{supp} f \subset V$ and $\mathrm{supp}\, f' \subset V'$, are not necessarily stochastically independent with respect to vacuum expectation values.

9. With the notation of Problem 1, show that if a particle is defined to be localized in the region R at time t if its Cauchy data $f\oplus g$ are such that Cf and

$C^{-1}g$ are supported in V, where $C = (m^2I - \Delta)^{1/4}$, then regions of space having disjoint closures have stochastically independent fields localized in them.

10. In the context of Problem 9, show that there are no selfadjoint operators X_j that are local in terms of the Cauchy data at a given time that transform as do the Minkowski coordinates x_j under eulidean transforms ($j = 1, 2, 3$). Show, however, that the nonlocal operators $C^{-1}M_jC$, where M_j denotes multiplication by x_j, have this property. (The latter operators are essentially those considered by Newton and Wigner [1949]. There is an extensive literature on position coordinates for relativistic wave equations, but locality, selfadjointness, and stochastic independence in disjoint regions cannot be simultaneously attained.)

Bibliographical Notes on Chapter 6

Quantization of general second-order equations in the sharp-time format was considered by Segal (1964a, 1969, 1970c). Finite propagation velocity issues for the quantization of such equations was treated by Segal (1967, 1970d). C^*-algebraic quantization of the Dirac equation coupled to an external field was studied by Bongaarts (1970). The role of complex structures in stable quantization was developed by Segal (1958a, 1964, 1969a, 1989) and extended by Weinless (1969).

7

Renormalized Products of Quantum Fields

7.1. The algebra of additive renormalization

Both multiplicative and additive renormalizations are used in practical quantum field theory, but only the latter is clearly involved at a foundational level. Additive renormalization in its practical form has sometimes been called "Subtraction Physics." This refers to the isolation of apparently meaningless or "infinite" terms in an additive symbolic expression, followed by their removal as rationalized by such considerations as locality and invariance. In this chapter we formulate the basic theory necessary to give mathematical viability to the program of quantization of nonlinear wave equations and in particular avoid infinities.

More specifically, we are concerned here to give appropriate mathematical meaning to the nonlinear putatively local functions of quantum fields that occur in symbolic quantized nonlinear wave equations. For example, the symbolic expression $\varphi(X)^p$, where φ is a given quantized field, and X is a given point of space-time, is such a putative function. However, as a distribution in space-time, even after integration against a function of space or space-time, $\varphi(X)^p$ has no a priori formulation. Indeed, there appears to be no useful definition of products even of classical distributions of a level of singularity comparable to that of the simplest quantized fields. Underlying this is the lack of any direct expression for the integrals $\int \varphi(X)^p f(X)dX$, where f is a smooth test function, in terms of the integrals $\int \varphi(X) f(X)dX$ that define the distribution φ.

It is remarkable that nonetheless, precisely because of the quantization, there is an effective theory of powers and other products of quantized fields. This involves the use of a certain "renormalization map." Intuitively, this map subtracts away infinite divergences in a local and otherwise physically reasonable way. It will promote logical clarity to develop the initial aspects of the subject in a purely algebraic setting.

DEFINITION. The *infinitesimal Weyl algebra* **E** (or **E**(**L**, A)) over a real symplectic vector space (**L**, A) is the associative algebra over **C** generated by **L** together with a unit e, with the relations

$$zz' - z'z = -iA(z, z')e$$

for arbitrary z, z' in **L**. The subalgebra generated by e and the elements z of a subspace **N** of **L** is denoted as **E**(**N**). The *degree* deg(u) of an element u of **E** is the least integer r such that u is in the linear span of $z_1 z_2 \cdots z_r$, the z_j being arbitrary in **L**; if $u = \alpha e$ with $\alpha \in$ **C** we define $\deg(u) = 0$ if $\alpha \neq 0$ and $\deg(u) = -\infty$ if $\alpha = 0$.

EXAMPLE 7.1. Let **L** denote the space of pairs of real numbers (a, b), with $A((a, b), (a', b')) = a'b - ab'$, and let F denote the linear map from **L** to the algebra **A** of all linear differential operators on **R**:

$$(a, b) \rightarrow ax - ib(d/dx).$$

This map is an isomorphism of **L** into **A** with the usual definition of the bracket in **A**. Accordingly, F can be extended to a homomorphism, which we also denote as F, from **E** into **A**. It is not difficult to verify that F is then an isomorphism, and that F(**E**) = **A**. Thus the infinitesimal Weyl algebra over a two-dimensional real symplectic vector space is simply representable as the algebra of all ordinary linear differential operators with polynomial coefficients.

More generally, if **L** is a $2n$-dimensional real symplectic vector space, **E** is isomorphic to the algebra of all partial differential linear operators on $\mathbf{R}^n$ with polynomial coefficients. This follows from the two-dimensional case by taking A to be in canonical form.

Still more generally, suppose **L** has the form **L** = **M**⊕**M***, where **M** is a given real locally convex topological vector space and **M*** is the space of all continuous linear functionals on **M**. Let A denote the form

$$A(x \oplus f, x \oplus f') = f(x') - f'(x).$$

Then the algebra **E** over the linear symplectic space (**L**, A) may conveniently be represented as follows: let **Q** denote the associative algebra over **C** freely generated by **M*** together with the function identically 1 on **M**; an element of **Q** will be called a *polynomial* on **M**. For x in **M**, let $\varrho(x)$ denote the derivation on **Q** which is uniquely determined by the requirement that it carry an arbitrary element f of **M*** into $if(x)$. For f in **M***, let $\varrho(f)$ denote the operation on **Q** of multiplication by f. The $\varrho(f)$ are then mutually commutative, and $[\varrho(x), \varrho(f)] = if(x)e$, where e denotes the identity operator on **Q**. From this it follows that

ϱ may be extended by linearity to a mapping defined on all of **L**, likewise denoted ϱ, and satisfying the relations

$$[\varrho(z), \varrho(z')] = -iA(z, z')e$$

for arbitrary z, z' in **L**. By general algebra, ϱ then extends uniquely to a representation ϱ of the enveloping algebra **E** as linear transformations of **Q**.

When **L** is finite-dimensional, $(\mathbf{L}, A)$ is always of the foregoing form, within isomorphism, as earlier noted, and **E** is isomorphic to the algebra of all linear differential operators having polynomial coefficients, acting on the polynomials **Q** on **M**.

DEFINITION. An *isotropic* subspace **N** of **L** is one such that $A(z, z') = 0$ for all $z, z' \in \mathbf{N}$. **N** is said to be *Lagrangian* if it is maximal isotropic.

SCHOLIUM 7.1. *The center of the infinitesimal Weyl algebra* **E** *of a given symplectic space* $(\mathbf{L}, A)$ *consists only of scalar multiplies of the identity. There are no nontrivial two-sided ideals in* **E**. *If* **L** *is finite-dimensional and* **N** *is Lagrangian in* **L**, *then* **E(N)** *is maximal abelian in* **E**.

PROOF. Recall first that every finite-dimensional subspace of **L** is contained in a finite-dimensional nondegenerate subspace of **L** (i.e., one such that the restriction of A to it is nondegenerate). An arbitrary element u of **E** is evidently in **E(M)** for some finite-dimensional subspace **M** of **L**, and hence in **E(M)** for some nondegenerate finite-dimensional subspace **M** of **L**. If u is central in **E**, it is central in **E(M)**, and hence it suffices to show that **E(M)** has trivial center. This follows without difficulty from the representation given in Example 7.1.

If u is a nonzero element of a proper ideal in **E**, u is contained in a proper ideal of **E(M)** for some finite-dimensional nondegenerate subspace **M**; hence, the simplicity of **E** follows from the simplicity of **E(M)**. This follows in turn from the observation that by the formation of successive commutators of any given nonzero linear differential operator with polynomial coefficients on **M** with multiplication by linear functionals, or differentiations, a nonzero element of $\mathbf{C}e$ is eventually attained.

It is immediate that if **N** is an isotropic subspace of **L**, then **E(N)** is abelian. Now, suppose that **N** is Lagrangian in **L**, and that **L** is finite-dimensional. If $\mathbf{N}'$ is any complement to **N** in **L**, it follows from the representation $\mathbf{L} = \mathbf{N} \oplus \mathbf{N}'$ that, up to isomorphism, $(\mathbf{L}, A)$ is the symplectic vector space built from a finite-dimensional space **M**, and $\mathbf{N} = \mathbf{M}^*$. The maximal abelian character of **E(N)** is then equivalent to the maximal abelian character of the algebra of all (multiplications by) polynomials, in the algebra of all linear differential operators with polynomial coefficients on a finite-dimensional vector space. This is easily seen. □

DEFINITION. A linear mapping T from a symplectic vector space $\mathbf{L}$ to a vector space $\mathbf{L}'$ is called *tame* if there exists a finite-dimensional nondegenerate subspace $\mathbf{M}$ of $\mathbf{L}$ such that $Tz = TP_{\mathbf{M}}z$ for all $z \in \mathbf{L}$, where $P_{\mathbf{M}}$ is the linear mapping from $\mathbf{L}$ into $\mathbf{M}$ defined by the equation $A(P_{\mathbf{M}}z, z') = A(z, z')$ for all $z' \in \mathbf{M}$. In this case T is said to be *based* on $\mathbf{M}$.

DEFINITION. A *monomial* in $\mathbf{E}$ is an element of the form $z_1z_2\cdots z_n$ with $z_1, \ldots, z_n$ in $\mathbf{L}$. Let E be a given linear functional on $\mathbf{E}$. A *renormalization map* on $\mathbf{E}$, relative to E, is a function N from the monomials in $\mathbf{E}$ to $\mathbf{E}$ that is not identically zero and has the following properties, where the ^ means that the vector is deleted:

$$[N(z_1z_2\cdots z_n), z'] = \sum_{j=1}^{n} [z_j, z']\, N(z_1z_2\cdots \hat{z}_j \cdots z_n) \tag{7.1}$$

$$E(N(z_1z_2\cdots z_n)) = 0 \tag{7.2}$$

for arbitrary $z_1, \ldots, z_n$ and z' in $\mathbf{L}$. Heuristically, these conditions state that $N(z_1z_2\cdots z_n)$ satisfies commutation relations analogous to those of the ordinary product,

$$[z_1z_2\cdots z_n, z'] = \sum_{j=1}^{n} [z_j, z']\, z_1z_2\cdots \hat{z}_j \cdots z_n,$$

but is "renormalized" so as to have zero expectation value relative to E. We shall prove that a unique renormalization map exists relative to any linear functional E with $E(e) \neq 0$, and that the renormalized product has many of the properties of the usual product, although with some notable differences. For example, $N(z_1\cdots z_n)$ is linear in each argument and lies in the subalgebra of $\mathbf{E}$ generated by $z_1, \ldots, z_n$ and the identity.

THEOREM 7.1. *If E is a given linear functional on the infinitesimal Weyl algebra $\mathbf{E}$ over $(\mathbf{L}, A)$ such that $E(e) \neq 0$, then there exists a unique renormalization map on $\mathbf{E}$, relative to* E.

The following result can be regarded as a quantized version of the Poincaré lemma concerning the exactness of a differential form.

LEMMA 7.1.1. *Let* K *be tame linear map from* $\mathbf{L}$ *into* $\mathbf{E}$ *such that* $[K(z), z'] = [K(z'), z]$ *for all* $z, z' \in \mathbf{L}$. *Then there exists an element* $u \in \mathbf{E}$ *such that* $K(z) = [u, z]$ *for all* $z \in \mathbf{L}$.

PROOF. Observe first that it suffices to treat the case in which $\mathbf{L}$ is finite-dimensional. For if K is based on the finite-dimensional nondegenerate subspace **M**, then $K|\mathbf{M}$ satisfies the indicated conditions relative to $\mathbf{E}(\mathbf{M})$; hence, by the hypothesized resolution of the finite-dimensional case, there exists an element $u \in \mathbf{E}(\mathbf{M})$ such that $K(z) = [u, z]$ for all $z \in \mathbf{M}$. If now z is arbitrary in $\mathbf{L}$, $K(z) = K(P_{\mathbf{M}}z) = [u, P_{\mathbf{M}}z] = [u, z]$, since $[z', z] = [z', P_{\mathbf{M}}z]$ for all z' in **M**.

Taking **L** as finite-dimensional, it is no essential loss of generality to assume that $(\mathbf{L}, A)$ is the symplectic vector space built from **M**. Taking a basis $r_1, \ldots, r_n$ for **M** and a dual basis $q_1, \ldots, q_n$ for $\mathbf{M}^*$, it suffices to show that if $A_1, \ldots, A_n$ and $B_1, \ldots, B_n$ are given in **E**, where n is the dimension of **M**, then there exists an element u of **E** such that

$$[u, r_j] = B_j; \qquad [u, q_j] = A_j \; (j = 1, \ldots, n),$$

provided the following relations equivalent to the condition that $[K(z), z'] = [K(z'), z]$ for all z and z' are satisfied:

$$[q_j, A_k] = [q_k, A_j]; \qquad [r_j, B_k] = [r_k, B_j]; \qquad [q_j, B_k] = [r_k, A_j],$$

for all j and k. This element u will be obtained by successive reduction to the cases in which i) $A_1 = 0$ and, ii) $A_1 = A_2 = 0$, etc.

It is easily seen that there exist unique elements $a_0, a_1, \ldots$ (only finitely many nonzero) of the subalgebra of **E** generated by $q_1, \ldots, q_n$ and $r_2, \ldots, r_n$, such that

$$A_1 = \sum_{\ell=0}^{\infty} a_\ell r_1^\ell.$$

Now let

$$u_1 = \sum_{\ell=0}^{\infty} i a_\ell r_1^{\ell+1}/(\ell+1);$$

then $[u_1, q_1] = A_1$. Setting

$$K_1(z) = K(z) - [u_1, z],$$

K_1 satisfies the condition hypothesized for k, and the question of the existence of u is reduced to the case in which $A_1 = 0$.

Taking this to be the case, apply the same procedure to A_2. Expressing A_2 in the form $A_2 = \sum_{\ell=0}^{\infty} b_\ell r_2^\ell$, define

$$u_2 = \sum_{\ell=0}^{\infty} i b_\ell r_2^{\ell+1}/(\ell+1),$$

and set $K_2(z) = K_1(z) - [u_2, z]$. K_2 again satisfies the same condition as K, and has the property that $K_2(q_2) = 0$; in addition, $K_2(q_1) = 0$. For, on noting that $[r_2, q_1] = 0$, it follows that

$$[u_2, q_1] = \sum_{\ell=0}^{\infty} i[b_\ell, q_1]\, r_2^{\ell+1}/(\ell+1).$$

On the other hand, since, $[q_1, A_2] = [q_2, A_1]$ and A_1 is now zero, it follows that $[q_1, A_2] = 0$, which means that

$$\sum_{\ell=0}^{\infty} [b_\ell, q_1]\, r_2^\ell = 0.$$

The uniqueness of the expression of any element of **E** as a polynomial in *one* of the vectors $q_1, \ldots q_n, r_1, \ldots, r_n$, with coefficients on the left in the subalgebra generated by the remaining vectors, shows that $[b_\ell, q_1] = 0$ for each ℓ. This implies that $[u_2, q_1] = 0$, which in turn implies that $K_2(q_1) = 0$.

The question of the existence of u has now been reduced to the case in which $A_1 = A_2 = 0$. It is evident that by continuing in the same fashion, the question may be reduced to the case in which all $A_j = 0$ $(j = 1, \ldots, n)$. In this case, $[B_k, q_j] = [A_j, r_k] = 0$, showing that each B_k commutes with all the q_j. Recall now that, by the last part of the preceding scholium, the $q_1, \ldots, q_n$ generate a maximal abelian subalgebra of **E**. Thus, each B_k is a polynomial in the $q_1, \ldots, q_n$.

The problem is now equivalent to finding a polynomial u in $q_1, \ldots, q_n$ such that

$$\partial u/\partial q_j = iB_j,$$

the B_j being given polynomials satisfying the conditions

$$\partial B_j/\partial q_k = \partial B_k/\partial q_j.$$

The usual method of solving this problem by a line integral shows that a solution exists and is a polynomial. □

Proof of theorem. Let $\mathbf{M}_n$ denote the collection of all monomials in **E** of degree d such that $1 \le d \le n$. As the basis of an induction argument, assume that there exists a unique nonzero map N_n from $\mathbf{M}_n$ into **E** such that equations 7.1 and 7.2 hold for all elements of $\mathbf{M}_n$. When $n = 1$, it is easily verified that this is the case, N_1 having the form

$$N_1(z) = z - E(z)/E(e).$$

To show that N_{n+1} exists given N_n $(n \ge 1)$, let $z_1, \ldots, z_{n+1} \in \mathbf{L}$ be given, set $w = z_1 \cdots z_{n+1}$, and let z be arbitrary in **L**. Set

$$K(z) = \sum_{j=1}^{n+1} [z_j, z]\, N_n(z_1 \cdots \hat{z}_j \cdots z_{n+1});$$

then K is a tame linear map from $\mathbf{L}$ into $\mathbf{E}$. If z' is also arbitrary in $\mathbf{L}$, then by the induction hypothesis

$$[K(z), z'] = \sum_{j,k=1}^{n+1} [z_j, z] [z_k, z'] N_n(z_1 \cdots \hat{z}_j \cdots \hat{z}_k \cdots z_{n+1}).$$

The interchange of z and z' on the right has the same effect termwise as the interchange of the indices j and k, and so leaves invariant the right side as a whole; i.e., $[K(z), z'] = [K(z'), z]$. By Lemma 7.1.1, there exists an element $u \in \mathbf{E}$ such that $[u, z] = K(z)$ for all $z \in \mathbf{L}$. By Scholium 7.1, u is determined within an additive constant (i.e., a scalar multiple of the identity), which may be fixed by the requirement that $E(u) = 0$. Defining $N_{n+1}(w) = u$ and $N_{n+1}(w') = N_n(w')$ for all $w' \in \mathbf{M}_n$, it is evident that equations 7.1 and 7.2 are satisfied.

To complete the induction argument it suffices to show that N_{n+1} is nonzero and unique. Since by construction $N_{n+1}|\mathbf{M}_n = N_n$, N_{n+1} is nonzero. To prove uniqueness, let $F: \mathbf{M}_n \to \mathbf{E}$ be the difference of two maps satisfying (7.1) and (7.2). By equation 7.1 and the induction hypothesis, $[F(w), z'] = 0$ for all $w \in \mathbf{M}_n$, $z' \in \mathbf{L}$. This implies by Scholium 7.1 that there exists a complex number α such that $F(w) = \alpha e$. Equation 7.2 then implies that $\alpha = 0$, hence $F = 0$.

Defining $N(w)$ for any monomial $w \in \mathbf{E}$ as the common value of $N_n(w)$ for all $n \geq \deg w$, N is a renormalization map. The uniqueness argument given in the previous paragraph also shows that N is the unique renormalization map.

□

COROLLARY 7.1.1. *$N(u) - u$ has degree less than that of u, for all monomials u in* $\mathbf{E}$.

PROOF. This is true when $\deg u = 1$. If the conclusion of the corollary is assumed for u of degree $\leq n$, then for any u of degree $n + 1$, say $u = z_1 z_2 \cdots z_{n+1}$, and arbitrary $z' \in \mathbf{L}$,

$$[N(u) - u, z'] = \sum_{j=1}^{n+1} [z_j, z] (N(z_1 \cdots \hat{z}_j \cdots z_{n+1}) - z_1 \cdots \hat{z}_j \cdots z_{n+1}).$$

By the induction assumption, $N(z_1 \cdots \hat{z}_j \cdots z_{n+1}) - z_1 \cdots \hat{z}_j \cdots z_{n+1}$ has degree at most $n - 1$. Setting $w = N(u) - u$, this means that $[w, z']$ also has degree at most $n - 1$. Now w is determined within an additive constant by K, where $K(z') = [w, z']$, and the proof of Lemma 7.1.1 shows that if $\deg(K(z)) \leq n - 1$ for all $z \in \mathbf{L}$, then $\deg w \leq n$. The corollary now follows by induction.

□

COROLLARY 7.1.2. *$N(z_1 \cdots z_n)$ is a symmetric function of $z_1, \ldots, z_n$ (for any fixed $n = 1, 2, \ldots$).*

PROOF. This is evident for $n = 1$. Proceeding by an induction argument and assuming the conclusion valid for lesser values of n, it is sufficient to show that $[N(z_1 \cdots z_n), z']$ is a symmetric function of $z_1, \ldots, z_n$ since these commutators, together with the symmetrical requirement of vanishing expectation value, uniquely determine $N(z_1 \cdots z_n)$. The commutator in question equals

$$\sum_{j=1}^{n} [z_j, z'] \, N(z_1 z_2 \cdots \hat{z}_j \cdots z_n)$$

which under the permutation π goes over into

$$\sum_{j=1}^{n} [z_{\pi(j)}, z'] \, N(z_{\pi(1)} z_{\pi(2)} \cdots \hat{z}_{\pi(j)} \cdots z_{\pi(n)}).$$

Employing the induction hypothesis, $N(z_{\pi(1)} z_{\pi(2)} \cdots \hat{z}_{\pi(j)} \cdots z_{\pi(n)}) = N(z_1 z_2 \cdots \hat{z}_{\pi(j)} \cdots z_n)$, so the sum in question is equal to

$$\sum_{j=1}^{n} [z_{\pi(j)}, z'] \, N(z_1 z_2 \cdots \hat{z}_{\pi(j)} \cdots z_n)$$

which, on making the substitution $\pi^{-1}(j)$ for the summation variable, yields the requisite result. □

COROLLARY 7.1.3. *$N(z_1 \cdots z_n)$ is linear in each z_i separately.*

PROOF. This is clear when $n = 1$. Again using an induction argument, and assuming the conclusion for all lesser n, it suffices to show that

$$[N(z_1 \cdots z_n), z] = K(z)$$

is a linear function of z_1, for all $z_2, \ldots, z_n$ and z in $\mathbf{L}$. This follows from the observation that the mapping between elements w of $\mathbf{E}$ such that $E(w) = 0$ and functions $K\colon \mathbf{L} \to \mathbf{E}$ satisfying the condition that $[K(z), z'] = [K(z'), z]$ for all $z, z' \in \mathbf{L}$ is a linear isomorphism. Now

$$K(z) = \sum_{j=1}^{n} [z_j, z'] \, N(z_1 z_2 \cdots \hat{z}_j \cdots z_n),$$

which is linear as a function of z_1 by the induction hypothesis. □

COROLLARY 7.1.4. *Suppose that $\mathbf{M}$ is a subspace of $\mathbf{L}$ that is either nondegenerate or isotropic. Given arbitrary $z_1, \ldots, z_n \in \mathbf{M}$, $N(z_1 \cdots z_n)$ lies in $\mathbf{E}(\mathbf{M})$.*

This is in fact true for an arbitrary subspace $\mathbf{M}$ of $\mathbf{L}$, but we will only need these special cases.

PROOF. It is no essential loss of generality to assume that **M** is finite-dimensional; the corollary obviously holds in the case $n = 1$. Proceeding by induction, and so assuming its validity for all lesser n, it follows from the definition of the renormalization map that $[N(z_1 \cdots z_n), z]$ lies in **E(M)** for all $z \in \mathbf{L}$. From the construction in Lemma 7.1.1, it is easily seen that $N(z_1, \ldots, z_n)$ also lies in **E(M)** if **M** is nondegenerate or isotropic (the latter case corresponding to the case in which the A_j are zero). □

COROLLARY 7.1.5. *If* **N** *is an isotropic subspace of* **L**, *then the restriction of N to the monomials in* **E(N)** *extends uniquely to a linear mapping N' from* **E(N)** *into* **E** *such that $N'(e) = e$.*

PROOF. The uniqueness is evident, since e and the monomials in **E(N)** span **E(N)**. To show that the extension is possible, it suffices to show that if $w_1, \ldots, w_n$ are monomials in the elements of **N** such that $w_1 + \cdots + w_r = 0$, then

$$N(w_1) + \cdots + N(w_r) = 0 \tag{7.3}$$

Since there exists a finite-dimensional nondegenerate subspace $\mathbf{L}'$ of **L** on which all the w_j are based, and $\mathbf{L}' \cap \mathbf{N}$ is isotropic, it suffices to establish (7.3) for the case in which **L** is finite-dimensional. In this case **N** may be extended to a Lagrangian subspace of **L**, and it is no essential loss of generality to take **N** to be identical with this subspace. Nor is it an essential loss of generality to further suppose that $(\mathbf{L}, A)$ is built from a finite-dimensional space **M**, and that $\mathbf{N} = \mathbf{M}^*$. **E(N)** is then the polynomial algebra over **M**, and is evidently invariant under ad **L** (i.e., the operators $\mathrm{ad}(w)$: $u \to [w, u]$, for $w \in \mathbf{L}$).

It is evident that N satisfies equation 7.3 when the w_j are all of degree at most 1. Now suppose as an induction hypothesis that (7.3) has been shown in the case when the w_j are all of degree at most n. Then there exists a unique linear mapping N_n' from the vector space $\mathbf{P}_n$ of polynomials of degree $\leq n$ in **N** to **E** that coincides with N on monomials. Note that by property 7.1 of N, if w is any monomial in **E(N)** of degree $\leq n + 1$, so that $[w, z] \in \mathbf{P}_n$ for all $z \in \mathbf{L}$, then

$$[N(w), z] = N_n'([w, z]).$$

To show that equation 7.3 holds when the w_j are monomials of degree at most $n + 1$, it suffices to show that the commutator of the left side with every element z of **L** vanishes; for this implies that the left side is a scalar, which must vanish since it must be carried into 0 by E. By the linearity of N_n' and the equation above,

$$[N(w_1) + \cdots + N(w_r), z] = [N(w_1), z] + \cdots + [N(w_r), z]$$
$$= N_n'([w_1, z]) + \cdots + N_n'([w_r, z])$$
$$= N_n'([w_1, z] + \cdots + [w_r, z]) = N_n'([w_1 + \cdots + w_r, z]) = 0. \qquad \square$$

COROLLARY 7.1.6. *The mapping N' of the previous corollary is the unique linear map from* **E**(**N**) *to* **E** *such that* $[N'(u), z] = N'([u, z])$ *and* $E(N'(u)) = \lambda(u)$ *for all* $u \in$ **E**(**N**) *and* $z \in$ **L**, *where* λ *is the unique linear functional on* **E**(**N**) *such that* $\lambda(e) = 1$ *and* $\lambda(z_1 \cdots z_n) = 0$ *for arbitrary* $z_1, \ldots, z_n$ $(n \geq 1)$ *in* **N**.

PROOF. That N' has the indicated properties follows from the foregoing proof. If N'' is another mapping with the same properties which is different from N', let u be an element of least degree such that $N'(u) \neq N''(u)$; evidently, $\deg u \geq 1$. Noting that for arbitrary $z \in$ **L**,

$$\deg [u, z] < \deg u,$$

it follows that

$$[N'(u), z] = N'([u, z]) = N''([u, z]) = [N''(u), z]$$

for all $z \in$ **L**. Hence $N'(u)$ and $N''(u)$ differ only by a scalar multiple of the identity, but since both are carried into 0 by E, this multiple must vanish. The assumption that N'' is different from N' thus leads to a contradiction, i.e., N' is unique. □

COROLLARY 7.1.7. *The mapping N' of Corollary 7.1.5. maps* **E**(**N**) *into* **E**(**N**).

PROOF. This is a direct consequence of the linearity of N' and Corollary 7.1.4. □

EXAMPLE 7.2. Assuming for convenience that $E(e) = 1$, the renormalization map is readily computed in each given order:

$$N(z) = z - E(z)e,$$
$$N(zz') = zz' - E(z)z' - E(z')z - E(zz')e + 2E(z)E(z')e,$$

etc. The expressions are slightly simpler in the case of renormalized powers. Assuming that $E(e) = 1$ and $E(z) = 0$ for all $z \in$ **L**,

$$N(z) = z,$$
$$N(z^2) = z^2 - E(z^2),$$

$$N(z^3) = z^3 - 3E(z^2)z - E(z^3),$$
$$N(z^4) = z^4 - 6E(z^2)z^2 - 4E(z^3)z - E(z^4) + 6E(z^2)^2.$$

Problems

1. Assuming that $E(e) = 1$, show that

$$N(z^n) = \sum_{i=0}^{n} \sum_{j_1+2j_2+\quad+(n-i)j_{n-i}=(n-i)} C(j_1,\ldots,j_{n-s},n,i)\, E(z)^{j_1}E(z^2)^{j_2}\cdots E(z^{n-i})^{j_{n-i}}\, z^i,$$

where $C(j_1,\ldots,j_{n-i},n,i)$ is equal to

$$(-1)^{j_1+j_2+\quad+j_{n-i}}n!(j_1+j_2+\cdots+j_{n-i})!\,[j_1!j_2!\cdots j_{n-i}!\,(2!)^{j_2}\cdots((n-i)!)^{j_{n-i}}]^{-1}.$$

2. Prove that the algebra of all polynomials is maximal abelian in the algebra of all linear differential operators with polynomial coefficients on a finite-dimensional vector space.

3. Extend the renormalization map to the fermion case as follows: let F be a given nondegenerate bilinear form on the real infinite-dimensional vector space $\mathbf{L}$; let $\mathbf{C}$ denote the Clifford algebra over $(\mathbf{L}, F)$; and let Ω denote the automorphism of $\mathbf{C}$ that extends the map $z \rightarrow -z$ on $\mathbf{L}$. For any eigenvector u of Ω, let its parity p be defined by the equation $\Omega u = pu$, and set $\{u, z\} = uz - (-1)^p zu$ for arbitrary z in $\mathbf{L}$. Let the operation $\{u, z\}$ be extended from the elements of exact parity to all $u \in \mathbf{C}$ by linearity, and let E and E_0 be two given linear functionals on $\mathbf{C}$ such that $E(e) \neq 0$ and $E_0(e) \neq 0$. Show that there exists a unique map N on monomials u in $\mathbf{C}$ such that

$$\{N(u), z\} = N(\{u, z\}), \qquad E(N(u)) = E_0(u)$$

for all monomials u and $z \in \mathbf{L}$. (Hint: consider first the case in which $\mathbf{L}$ is of finite even dimension.)

4. In the situation of Problem 3, show that $\deg(N(u) - u) < \deg(u)$ if $\deg(u)$ is defined as in the infinitesimal Weyl algebra and $u \neq 0$.

7.2. Renormalized products of the free boson field

A Weyl system $(\mathbf{K}, W)$ over a linear symplectic space $(\mathbf{L}, A)$ induces a natural representation of the infinitesimal Weyl algebra $\mathbf{E}$ over $(\mathbf{L}, A)$ by unbounded operators in $\mathbf{K}$. This induces a correspondence between the unique adjunction operator in $\mathbf{E}$ such that e and all elements of $\mathbf{L}$ are selfadjoint, and the usual adjoint operator in $\mathbf{K}$, restricted to the domain $\mathbf{D}$ given in

SCHOLIUM 7.2. *Let* ($\mathbf{K}, W$) *be a Weyl system over the symplectic vector space* ($\mathbf{L}, A$), *and let the map* $z \mapsto \phi(z)$ *from* $\mathbf{L}$ *into selfadjoint operators on* $\mathbf{K}$ *be the corresponding infinitesimal Weyl system. Let* $\mathbf{D}$ *denote the space of vectors in* $\mathbf{K}$ *that lie in the domains of all products of the form* $\phi(z_1)\cdots\phi(z_n)$, $z_1, \ldots, z_n \in \mathbf{L}$. *Then the map* $z \mapsto \phi(z)$ *extends uniquely to a* $*$*-representation* w *of the infinitesimal Weyl algebra* $\mathbf{E}$ *of* $\mathbf{L}$ *as linear transformations of* $\mathbf{D}$.

PROOF. The proof follows directly from the observations that $\mathbf{E}$ is generated by $\mathbf{L}$ and that $[\phi(x),\phi(y)]u = -i A(x, y)u$, $u \in \mathbf{D}$. □

We now consider renormalized products of the free boson field. In this case the symplectic vector space ($\mathbf{L}, A$) arises from a complex Hilbert space $\mathbf{H}$, and the complex structure may be used to define a distinguished linear functional on $\mathbf{E}$, the "normal vacuum," which is closely related to the vacuum state of the free boson field. The renormalization map relative to this linear functional, sometimes called the "normal-ordered product," satisfies a number of useful identities.

The *infinitesimal Weyl algebra* $\mathbf{E}$ over $\mathbf{H}$ is defined to be the infinitesimal Weyl algebra over ($\mathbf{H}^\#, A$). Given $z \in \mathbf{H}$, we define the *creator* $c(z) \in \mathbf{E}$ as follows, where J is the complex structure in $\mathbf{H}$:

$$c(z) = 2^{-1/2}(z - iJz)$$

and the *annihilator* $c^*(z) \in \mathbf{E}$ by

$$c^*(z) = c(z)^* = 2^{-1/2}(z + iJz).$$

To avoid possible confusion, it should be recalled that complex numbers are involved in the infinitesimal Weyl algebra over a *real symplectic space* only as elements of the coefficient field.

DEFINITION. Let $\mathbf{E}$ be the infinitesimal Weyl algebra over the complex Hilbert space $\mathbf{H}$. A *normal vacuum* on $\mathbf{E}$ is a linear functional E on $\mathbf{E}$ such that $E(e) = 1$ and

$$E(c(x_1)\cdots c(x_n)c^*(y_1)\cdots c^*(y_m)) = 0$$

for all $x_1, \ldots, x_n$ and $y_1, \ldots, y_m$ in $\mathbf{H}$ (where $n + m > 0$).

SCHOLIUM 7.3. *Let* $\mathbf{E}$ *be the infinitesimal Weyl algebra over a complex Hilbert space* $\mathbf{H}$. *Then there exists a unique normal vacuum* E *on* $\mathbf{E}$. *Let* ($\mathbf{K}, W, \Gamma, v$) *denote the free boson field over* $\mathbf{H}$. *The normal vacuum* E *satisfies*

$$E(u) = \langle w(u)v, v\rangle$$

for all $u \in \mathbf{E}$, *where the representation* w *is given as in Scholium 7.2.*

Proof. Noting that as elements of **E**,

$$x = 2^{-1/2}(c(x) + c^*(x))$$

for all $x \in \mathbf{H}$, and

$$[c(x), c^*(y)] = -\langle x, y\rangle e$$

for all $x, y \in \mathbf{H}$, it follows that **E** is spanned by products of the form $c(x_1)\cdots c(x_n)c^*(y_1)\cdots c^*(y_m)$, so that a normal vacuum is unique if it exists. It thus suffices to show that $\langle w(u)v, v\rangle$ is well defined for $u \in \mathbf{E}$ and vanishes if

$$u = c(x_1)\cdots c(x_n)\, c^*(y_1)\cdots c^*(y_m).$$

It is clear from the particle representation that v is in the domain of all the products $\phi(z_1)\cdots\phi(z_n)$, where $z_1, \ldots, z_n \in \mathbf{H}$, so that $\langle w(u)v, v\rangle$ is well defined for all $u \in \mathbf{E}$. It is also clear from the particle representation that

$$w(c(x_1)\cdots c(x_n)c(y_1)^*\cdots c(y_m)^*)v = C(x_1)\cdots C(x_n)C^*(y_1)\cdots C^*(y_m)v.$$

This vanishes if $m > 0$, and taking adjoints, $\langle w(u)v, v\rangle = 0$ if $m = 0$ and $n > 0$. □

Scholium 7.4. *Let* **E** *be the infinitesimal Weyl algebra over a complex Hilbert space* **H**. *There is then a unique representation* γ *of* $U(\mathbf{H})$ *as* $*$*-algebra automorphisms of* **E** *such that* $\gamma(U)z = Uz$ *for all* $z \in \mathbf{H}$, $U \in U(\mathbf{H})$. *If* E *is the normal vacuum on* **E**, *then* $E(\gamma(U)u) = E(u)$ *for all* $u \in \mathbf{E}$, $U \in U(\mathbf{H})$.

Proof. The proof is straightforward and is omitted. □

The renormalization map for the normal vacuum coincides effectively with the "Wick product," which was introduced as a means of standardizing the "correspondence principle" in the case of polynomials in the canonical p's and q's. The correspondence principle was the basis for associating with a given classical Hamiltonian a corresponding quantum Hamiltonian. A lengthy process of development from the original "Ansatz" of Heisenberg indicated that considerations of symmetry and of the Lie-theoretic connection between the Poisson bracket in the classical case and the operator bracket in the quantum case were insufficient to fix uniquely a mathematical form for the heuristic correspondence principle. At the same time, this process suggested giving primacy to the creation and annihilation operators rather than the hermitian p's and q's (e.g., as a means of avoiding unwanted "zero-point" energies). Using a definition that was intrinsically limited to the normal vacuum (cf. below), Wick standardized a renormalization map along such lines and established simple explicit rules for the computation of expectation values of products of renormalized monomials. Practical quantum mechanics in the form of Feynman graph theory has been based in significant part on these rules.

With the identification of the Wick product as the renormalization map relative to the normal vacuum, which is made later, these rules are given in

THEOREM 7.2. *Let* **E** *be the infinitesimal Weyl algebra over a complex Hilbert space* **H**, *let* E *be the normal vacuum on* **E**. *Then the renormalization map* N *relative to* E *satisfies the following relations, where the* z_j $(j = 1, 2, \ldots, r + 1)$ *are arbitrary in* **H**:

a) $N(z_1 \cdots z_r z_{r+1}) = N(z_1 \cdots z_r) z_{r+1} - \Sigma_{1 \le j \le r} N(z_1 \cdots \hat{z}_j \cdots z_r) E(z_j z_{r+1})$

b) $z_1 \cdots z_r = N(z_1 \cdots z_r) + \Sigma_{1 \le j < k \le r} N(z_1 \cdots \hat{z}_j \cdots \hat{z}_k \cdots z_r) E(z_j z_k) +$
$\Sigma_{1 \le j_1 < k_1 \le r, 1 \le j_2 < k_2 \le r} N(z_1 \cdots \hat{z}_{j_1} \cdots \hat{z}_{k_1} \cdots \hat{z}_{j_2} \cdots \hat{z}_{k_2} \cdots z_r) E(z_{j_1} z_{k_1})$
$E(z_{j_2} z_{k_2}) + \cdots + \Sigma_{1 \le j_1 < k_1 \le r, 1 \le j_2 < k_2 \le r, 1 \le j_s < k_s \le r}\, u\, E(z_{j_1} z_{k_1}) \cdots E(z_{j_s} z_{k_s})$

where s *denotes the integral part of* $r/2$, *and if* r *is even* $u = e$ *(the identity of* **E**), *but if* r *is odd* $u = z_t$, *where* t *is the unique integer such that* $1 \le t \le r$ *and* $t \ne j_m$, $t \ne k_m$ *for all* $m = 1, 2, \ldots, s$. *(Note also that all the sums in b) are taken only over* j_m, k_m *such that* $j_m \ne j_n$, $j_m \ne k_n$, *and* $k_m \ne k_n$ *for all* m, n.*)*

PROOF. To prove *a)* by induction, it suffices to show that the expectation values of both sides agree, and that their commutators with an arbitrary element $z' \in$ **H** are the same. In case $r = 1$, *a)* states that $z_1 z_2 = N(z_1 z_2) + E(z_1 z_2)$, which is easily seen to be valid. For $r > 1$, the identity of the expectation values follows from

LEMMA 7.2.1. *For arbitrary* $z_1, \ldots, z_{r+1}$ *in* **H**, *where* $r > 1$,

$$E(N(z_1 \cdots z_r) z_{r+1}) = 0$$

PROOF OF LEMMA. By linearity, it suffices to consider the case in which each z_j is a creator or an annihilator. If z_{r+1} is an annihilator the conclusion is obvious, so assume it is a creator. In view of the symmetry of $N(z_1 \cdots z_r)$ as a function of the $z_1, \ldots, z_r$, it may be assumed that $z_1 \cdots z_r = ab$, where a is a product of creators and b is a product of annihilators, not both of degree 0. Observing that $E(ab) = E(b^* a^*)$ for arbitrary a and b, it may further be assumed that $a = e$. Note that the map $z \mapsto c(z)$ is complex-linear from **H** to **E**, while the map $z \mapsto c^*(z)$ is antilinear. Thus the unitary transformation e^{Jt} on **H** $(t \in \mathbf{R})$ has $\gamma(e^{Jt}) z_{r+1} = e^{it} z_{r+1}$, while $\gamma(e^{Jt}) b = e^{-irt} b$, so that $\gamma(e^{Jt}) b z_{r+1} = e^{-i(r-1)t} b z_{r+1}$. By Scholium 7.4, it follows that

$$E(b z_{r+1}) = e^{-i(r-1)t} E(b z_{r+1}),$$

implying that $E(b z_{r+1}) = 0$. □

COMPLETION OF PROOF. Assume $r > 1$ in equation a), denoting left and right sides of a) as L and R, for arbitrary z in **H**:

$$[L, z] = N(z_1 \cdots z_r)[z_{r+1}, z] + \sum_{j=1}^{r} N(z_1 \cdots \hat{z}_j \cdots z_{r+1})\ [z_j, z];$$

$$[R, z] = N(z_1 \cdots z_r)[z_{r+1}, z] + \sum_{j=1}^{r} N(z_1 \cdots \hat{z}_j \cdots z_r)\ z_{r+1}\ [z_j, z]$$
$$- \sum_{1 \le j \le r, 1 \le k \le r, j \ne k} N(z_1 \cdots \hat{z}_j \cdots \hat{z}_k \cdots z_r)\ E(z_j z_{r+1})\ [z_k, z].$$

Applying the induction hypothesis to the double sum in the expression for $[R, z]$ by summing first over j shows that this double sum equals

$$\sum_{k=1}^{r} (N(z_1 \cdots \hat{z}_k \cdots z_r) z_{r+1} - N(z_1 \cdots \hat{z}_k \cdots z_{r+1}))\ [z_k, z].$$

On changing the index from k to j in this sum, the equality of $[L, z]$ and $[R, z]$ becomes apparent.

To prove equation b) along similar lines, denoting left and right sides of b) as L and R, first note that the identity of $[L, z]$ with $[R, z]$ for arbitrary $z \in \mathbf{H}$ follows directly from the induction hypothesis; $[L, z]$ is a sum of r terms over i, in the ith term z_i being deleted, while the characterization of N in terms of commutators provides a precisely corresponding sum for $[R, z]$.

It remains only to show that $E(z_1 \cdots z_r) = 0$ if r is odd, while if $r = 2s$ where s is integral,

$$E(z_1 \cdots z_r) = \Sigma \prod_{k=1}^{s} E(z_{i_k} z_{j_k}),$$

the sum being taken over the set of all partitions of $\{1, \ldots, r\}$ into two-element subsets.

In the case when r is odd, the required result follows from the invariance of E under the induced action of the unitary transformation $-I$ on $\mathbf{H}$, which evidently transforms $E(z_1 \cdots z_r)$ into its negative. When r is even, first express $z_1 \cdots z_{r-1}$ as a sum of renormalized products in accordance with the induction hypothesis

$$z_1 \cdots z_{r-1} = \sum_p N(u_p) c_p,$$

where the u_p are certain monomials of degree $1, 2, \ldots, r-1$ and the c_p are scalars. Then

$$z_1 \cdots z_{r-1} z_r = \sum_p N(u_p) z_r c_p.$$

By Lemma 7.2.1, $E(N(u)z) = 0$ except when $\deg u_p = 1$. Thus, only the terms of degree 1 contribute to $E(z_1 \cdots z_r)$, and it follows that it has the stated form.

□

The original definition of Wick product is often used in the physics literature. The definition is in terms of the notion of "normal product," which is defined as a product of creation and annihilation operators in which all annihilation operators are on the right and the creation operators on the left. This has the obvious consequence that the normal vacuum expectation value of a normal product vanishes. In the form given in the physics literature, "The normal product $:A_1(x_1)\cdots A_n(x_n):$ may be defined as the result of reducing to the normal form the ordinary product $A_1(x_1)\cdots A_n(x_n)$ provided that during the *process of reduction* the quantized field functions are regarded as strictly commuting . . . and value zero is assigned to all commutation functions that appear" (Bogolioubov and Shirkov, 1959).

In a more explicit mathematical form, this definition may be formulated as follows: the Wick (or normal) product of vectors $z_1, \ldots, z_n$ in **H** is the element of **E(H)**:

$$N_0(z_1\cdots z_n) = 2^{-n/2} \sum_{S\subseteq\{1,\ldots,n\}} \left\{\prod_{k\in S} c(z_k)\right\} \left\{\prod_{k\in\{1,\ldots,n\}-S} c^*(z_k)\right\}.$$

Noting that the products involved in this expression are independent of the order of the factors, it follows that $N_0(z_1\cdots z_n)$ is well defined.

SCHOLIUM 7.5. *If E is the normal vacuum on* **E(H)**, *then* $N(z_1\cdots z_n) = N_0(z_1\cdots z_n)$ *for arbitrary vectors* $z_1, \ldots, z_n$ *in* **H**.

PROOF. Since both sides of the foregoing putative equality have vanishing normal expectation value, it suffices to show that the right side satisfies the recursion relation characterizing the left side. Thus it suffices to show that

$$[N_0(z_1\cdots z_n), z'] = \sum_{j=1}^{n} [z_j, z']\, N_0(z_1\cdots \hat{z}_j\cdots z_n).$$

To this end it suffices to treat the case where all the z_j are equal. It is easily seen that $N_0(z_1\cdots z_n)$ is symmetric as a function of the z_j and real-linear in each variable z_j. Thus there is a unique linear function $\tilde{N}_0$ from the space of symmetric tensors over **H** to **E** such that

$$N_0(z_1\cdots z_n) = \tilde{N}_0(z_1\vee\cdots\vee z_n),$$

and the above property of N_0 is equivalent to

$$[\tilde{N}_0(z_1\vee\cdots\vee z_n), z'] = \sum_{j=1}^{n} [z_j, z']\, \tilde{N}_0(z_1\vee\cdots\vee\hat{z}_j\vee\cdots\vee z_n).$$

Since this equation is linear in $z_1\vee\cdots\vee z_n$, and the space of symmetric n-tensors over **H** is spanned by those of the form $z\vee\cdots\vee z$ (as it is an irreducible representation of the symmetric group, and such tensors span a subrepresentation), it suffices to prove it in the case $z_1,\ldots,z_n = z$. For any $z \in \mathbf{H}$,

$$N_0(z^n) = 2^{-n/2} \sum_{j=0}^{n} n!\,[j!(n-j)!]^{-1} c(z)^j\, c^*(z)^{n-j};$$

hence for any $z' \in \mathbf{H}$ it follows that

$$N_0(z^n), c(z')] = 2^{-n/2}\langle z', z\rangle \sum_{j=0}^{n} n!\,[j!(n-j-1)!]^{-1} c(z)^j\, c^*(z)^{n-j-1},$$

$$[N_0(z^n), c^*(z')] = -2^{-n/2}\langle z, z'\rangle \sum_{j=0}^{n} n!\,[(j-1)!(n-j)!]^{-1} c(z)^{j-1}\, c^*(z)^{n-j},$$

and

$$[N_0(z^n), z'] = [N_0(z^n), 2^{-1/2}(c(z') + C^*(z'))]$$
$$= in \operatorname{Im}\langle z', z\rangle\, 2^{(n-1)/2} \sum_{0\le j\le n-1} (n-1)!\,[j!(n-j-1)!]^{-1}\, c(z)^j\, c^*(z)^{n-j}$$
$$= in \operatorname{Im}\langle z', z\rangle\, N_0(z^{n-1}) = \sum_{j=1}^{n} [z, z']\, N_0(z^{n-1}),$$

as was to be shown. □

Equations a) and b) of Theorem 7.2 are accordingly known as "Wick's Theorem."

Problems

1. Let $\mathbf{H}$ be a complex Hilbert space and N the renormalization map relative to the normal vacuum on the infinitesimal Weyl algebra $\mathbf{E}$ over $\mathbf{H}$. Show that for any $z \in \mathbf{H}$, $N(z^n)$ is a constant multiple of the nth Hermite polynomial (cf. Glossary) applied to z. Determine the constant.

2. a) Compute explicitly the expressions of all Wick products of degree $\le$ 4 in two vectors z_1, z_2 in $\mathbf{H}$, in terms of their ordinary products and expectation values.

b) Express the ordinary products in terms of the Wick products and the expectation values of the former.

7.3. Regularity properties of boson field operators

An important feature of analysis in boson fields is the domination of relevant operators by a Hamiltonian. In this section the free boson field ($\mathbf{K}$, W,

$\Gamma, \nu)$ over a complex Hilbert space **H** is treated in this context. In the simplest form of this approach, there is a given in **H** a selfadjoint operator A representing the single-particle Hamiltonian. This operator A and the corresponding field Hamiltonian $H = \partial\Gamma(A)$ determine domains of regular vectors (differentiable, analytic, etc.) in **H** and in **K** that facilitate the analysis of singular operators. Such operators are often conveniently formulated as sesquilinear forms on domains of regular vectors. These forms are equivalent to mappings from the regular domains to their antiduals, relative to an appropriate topology on the regular domain, and thus represent generalized operators. Among such operators are the local renormalized products of free relativistic fields.

To this end general machinery will now be developed. For any given densely defined operator T in a Hilbert space **H**, having the property that $Tx = 0$ only if $x = 0$, the domain $\mathbf{D}(T)$ as a pre-Hilbert space with the inner product $\langle x, y\rangle' = \langle Tx, Ty\rangle$ is denoted as $\langle\mathbf{D}(T)\rangle$. The Hilbert space completion of $\langle\mathbf{D}(T)\rangle$ will be denoted as $\langle\langle\mathbf{D}(T)\rangle\rangle$. If A is a given strictly positive selfadjoint operator in **H**, the intersection of the domains $\mathbf{D}(A^n)$ $(n = 0, 1, 2, \cdots)$ will be denoted as $\mathbf{D}_\infty(A)$, and when topologized so that convergence of a sequence means convergence in *each* $\langle\mathbf{D}(A^n)\rangle$, will be denoted as $\langle\mathbf{D}_\infty(A)\rangle$. The union of the $\mathbf{D}(A^n)$, as n varies over *all* integers $(0, \pm 1, \pm 2, \cdots)$, will be denoted as $\mathbf{D}_{-\infty}(A)$ and when topologized so that convergence of a sequence means convergence in *some* $\langle\mathbf{D}(A^n)\rangle$, will be denoted as $\langle\mathbf{D}_{-\infty}(A)\rangle$. If $x \in \langle\mathbf{D}(A^m)\rangle$ and $y \in \langle\mathbf{D}(A^n)\rangle$ where $m + n \geq 0$, then $\langle x, y\rangle$ is canonically definable by continuity, in extension of the definition in **H**, and this extended inner product will be denoted in the same way as in **H**.

A convenient concrete representation of the spaces $\mathbf{D}_{\pm\infty}(A)$ derives from spectral theory. Specifically, any selfadjoint operator A in **H** is unitarily equivalent to the operation M_g of multiplication by a real measurable function g on some measure space **M**, in such a way that **H** is correspondingly unitarily equivalent to $L_2(\mathbf{M})$. The domain of such a multiplication operator is defined as the set of all vectors $f \in L_2(\mathbf{M})$ such that the pointwise product gf is again in $L_2(\mathbf{M})$. It is readily deduced that $\mathbf{D}_\infty(A)$ is correspondingly represented by the space of all measurable functions f such that $g^nf \in L_2(\mathbf{M})$ for all $n \geq 0$. Similarly, $\mathbf{D}_{-\infty}(A)$ is represented by the space of all measurable functions f such that $g^nf \in L_2(\mathbf{M})$ for *some* integer n.

EXAMPLE 7.3. Let **H** denote the space of Cauchy data for the Klein-Gordon equation $\Box\varphi + m^2\varphi = 0$ on $\mathbf{M}_0$, where $m > 0$, that are normalizable relative to the Lorentz invariant norm. Let the operators A and B be as in Section 6.3. Then $\mathbf{D}(A^k) = \mathbf{D}(B^{k+1/2})\oplus\mathbf{D}(B^{k-1/2})$, and $\mathbf{D}(B^k)$ is the Sobolev space $L_{2,k}(\mathbf{R}^n)$ of all functions that are square-integrable together with their first k derivatives. It follows that $\mathbf{D}_\infty(A)$ consists of all pairs (f, g) where f and g are C^∞ on $\mathbf{R}^n$, and together with their partial derivatives of all orders, in $L_2(\mathbf{R}^n)$. It follows also that $\mathbf{D}_{-\infty}(A)$ includes all pairs (f, g) of compactly supported distributions.

As in the previous example, the operators A that are relevant typically arise as generators of one-parameter unitary groups. If $U(t) = e^{itA}$ where A is self-adjoint in $\mathbf{H}$ $(t \in \mathbf{R})$, there is a canonical extension of $U(\cdot)$ to all of $\mathbf{D}_{-\infty}(A)$. This extension, which will again be denoted as $U(\cdot)$ to avoid undue circumlocution, acts continuously on $\langle \mathbf{D}_{-\infty}(A) \rangle$.

The strong smoothing effect of time-integration is exemplified by

SCHOLIUM 7.6. *Suppose that A is a strictly positive selfadjoint operator in the Hilbert space $\mathbf{H}$. If $n < 0$, $x \in \mathbf{D}(A^n)$, and $f(\cdot) \in C_0^{|n|}(\mathbf{R})$, then $\int U(t)xf(t)dt$ is in $\mathbf{H}$ and has norm bounded by $\|A^n x\| \, \|f^{(|n|)}\|_{L_2(\mathbf{R})}$.*

PROOF. The case of general n follows from that for $n = -1$, so assume this. Then $x = Ay$ with $y \in \mathbf{H}$, and using Stone's theorem,

$$\int U(t)xf(t)dt = \int U(t)Ayf(t)dt = \lim_{\varepsilon \to 0} \int U(t)\,(i\varepsilon)^{-1}\,[U(\varepsilon) - I)yf(t)dt$$

$$= \lim_{\varepsilon \to 0} (i\varepsilon)^{-1} \int U(t)y[f(t - \varepsilon) - f(t)]dt = i \int U(t)yf'(t)dt.$$

This has the indicated bound. □

A useful related result is

SCHOLIUM 7.7. *If $\mathbf{L}$ is a dense submanifold in $\mathbf{H}$ that is invariant under $U(\cdot)$ and contained in $\mathbf{D}(A^n)$, then $\mathbf{L}$ is also dense in $\langle \mathbf{D}(A^n) \rangle$, where A is as in Scholium 7.6.*

PROOF. Otherwise let z be a nonzero vector in $\langle \mathbf{D}(A^n) \rangle$ that is orthogonal to $\mathbf{L}$ in $\mathbf{D}(A^n)$: $\langle A^n z, A^n w \rangle = 0$ for all $w \in \mathbf{L}$. By the invariance of $\mathbf{L}$,

$$\langle A^n z, A^n U(t)w \rangle = 0 = \langle A^n z, U(t)A^n w \rangle = \langle A^n U(-t)z, A^n w \rangle$$

for all real t. It follows that $\langle A^n z', A^n w \rangle = 0$ if $z' = \int U(t)zf(t)dt$ with $f \in C_0^\infty(\mathbf{R})$, whence $\langle A^{2n} z', w \rangle = 0$, since $z' \in \mathbf{D}_\infty(A)$. Such z' will be nonzero if f approximates δ appropriately, whence $A^{2n} z' \neq 0$, contradicting the density of $\mathbf{L}$ in $\mathbf{H}$. □

In these terms the field and annihilation operators enjoy the regularity properties specified in the conclusion of

THEOREM 7.3. *Given a complex Hilbert space $\mathbf{H}$, let $(\mathbf{K}, W, \Gamma, \nu)$ be the free boson field over $\mathbf{H}$, let A be a strictly positive selfadjoint operator in $\mathbf{H}$, and let H denote $\partial\Gamma(A)$. Then the map $(z, u) \mapsto \phi(z)u$ is continuous from $\langle \mathbf{D}_\infty(A) \rangle \times \langle \mathbf{D}_\infty(H) \rangle$ into $\langle \mathbf{D}_\infty(H) \rangle$.*

Moreover, the map $(z, u) \to C(z)^*u$ *has a unique continuous extension to a map from* $\langle \mathbf{D}_{-\infty}(A)\rangle \times \langle \mathbf{D}_\infty(H)\rangle$ *into* $\langle \mathbf{D}_\infty(H)\rangle$.

PROOF. We establish the inequalities

$$\|H^m\phi(z)u\| \le c\,\|(H + I)^{m+1/2}u\|\,\|A^m z\| \tag{7.4}$$

by induction on m. The case $m = 0$ is Lemma 6.1.1. Now assume that (7.4) is valid for all nonnegative half-integers $m < n$. To show that $\phi(z)u$ is in the domain of H^n is equivalent to showing that $H^{n-1}\phi(z)u$ is in the domain of H. This is in turn equivalent to showing that $y(t) = t^{-1}(e^{itH} - I)H^{n-1}\phi(z)u$ remains bounded as $t \to 0$ (by a corollary to Stone's theorem). $z_t = e^{itA}z$,

$$\begin{aligned} y(t) &= t^{-1}(e^{itH}H^{n-1}\phi(z)u - H^{n-1}\phi(z)u) \\ &= H^{n-1}[t^{-1}(\phi(z_t) - \phi(z))e^{itH}u] + H^{n-1}\phi(z)t^{-1}(e^{itH} - I)u \end{aligned}$$

by the intertwining relation between ϕ and Γ. Applying the induction hypothesis to each of the two addends on the right side, it results that

$$\begin{aligned} \|y(t)\| \le{} & c\,\|(H + I)^{n-1/2}u\|\,\|A^{n-1}t^{-1}(z_t - z)\| + \\ & c\,\|(H + I)^{n-1/2}t^{-1}(e^{itH} - I)u\|\,\|A^{n-1}z\|. \end{aligned}$$

As $t \to 0$, $t^{-1}(z_t - z) \to iAz$, and if $z \in D(A^{m+1})$, $A^m t^{-1}(z_t - z) \to iA^{m+1}z$ for any positive integer m. Similarly, $H^{n-1/2}t^{-1}(e^{itH} - I)u$ remains bounded in norm and approaches $H^{n+1/2}u$ as $t \to 0$. Thus $y(t)$ remains bounded in norm as $t \to 0$, and

$$\|H^n\phi(z)u\| \le c\,\|(H + I)^{n-1/2}u\|\,\|A^n z\| + c\,\|(H + I)^{n+1/2}u\|\,\|A^{n-1}z\|$$

Using the boundedness away from zero of the spectrum of $H + I$ and A, it follows that

$$\|H^n\phi(z)u\| \le c'\,\|(H + I)^{n+1/2}u\|\,\|A^n z\|$$

for a suitable constant. Thus the induction hypothesis remains valid at the next stage and the proof of inequalities 7.4 is complete

The continuity of the map $(z, u) \to \phi(z)u$ asserted by the theorem follows, and implies the same for the map $(z, u) \to C(z)u$ and $(z, u) \to C(z)^*u$. The stronger continuity claimed for $C(z)^*u$ follows from

LEMMA 7.3.1. *For arbitrary* $z \in \mathbf{H}$, $u \in \mathbf{D}_\infty(H)$, *and positive interger* m,

$$\|C(z)^*u\| \le c\,\|A^{-m}z\|\,\|H^m u\|.$$

PROOF. Lemma 6.1.1 implies that u is in fact in the domain of $C(z)^*$. Now let P denote the projection of $\mathbf{H}$ onto the one-dimensional subspace spanned by z (which may be assumed $\neq 0$). Then for arbitrary $z' \in \mathbf{D}_\infty(A)$,

$$\|z\|^2\langle Pz', z'\rangle = |\langle z, z'\rangle|^2 \leq \|A^{-m}z\|^2 \, \|A^m z'\|^2 = \|A^{-m}z\|^2 \, \langle A^{2m}z', z'\rangle.$$

It follows that $\|z\|^2 P \leq \|A^{-m}z\|^2 \, A^{2m}$, whence

$$\|z\|^2 \, \partial\Gamma(P) \leq \|A^{-m}z\|^2 \, \partial\Gamma(A^{2m}).$$

On the other hand, $\|z\|^2 \, \partial\Gamma(P) = C(z)C(z)^*$, and it follows from the particle representation and the positivity of A that

$$\partial\Gamma(A^{2m}) \leq \partial\Gamma(A)^{2m} = H^{2m},$$

which is equivalent to the conclusion of the lemma. □

PROOF OF THEOREM, CONTINUED. The intertwining relation $e^{itH}W(z)e^{-itH} = W(e^{itA}z)$ leads by analysis similar to that of Section 1.2 to the inclusion $[H, \phi(z)] \subset -i\,\phi(iAz)$ for arbitrary $z \in \mathbf{D}(A)$, whence also $[H, C(z)^*] \subset C(iAz)^*$. By repeated application of this inclusion it follows that if $z \in \mathbf{D}_\infty(A)$, then

$$H^n C(z)^* \subset \sum_{k=0}^{n} c_{k,n} C(A^k z)^* \, H^{n-k}$$

for all nonnegative integers n, where the coefficients $c_{k,n}$ depend only on k and n. It follows in turn that if $u \in \mathbf{D}_\infty(H)$, then for all integers m and $n \geq 1$,

$$\|(H + I)^n C(z)^* u\| \leq c \sum_{k=0}^{n} \|A^{k-m}z\| \, \|H^{n-k+m}u\|,$$

where c depends only on n. This inequality implies in particular the conclusion of the theorem. □

The expression for the Wick product in terms of creation and annihilation operators will now extend directly to one in which the factors are generalized vectors, in $\mathbf{D}_{-\infty}(A)$ rather than in $\mathbf{H}$. In this and later connections, it will be useful to develop some conventions and notation regarding the relation between sesquilinear forms and operators.

Let D denote a dense subspace of a Hilbert space $\mathbf{H}$. If T is a linear operator in $\mathbf{H}$ whose domain includes D, a sesquilinear form F on D is defined by the equation $F(x, y) = \langle Tx, y\rangle$ $(x, y \in D)$. Evidently, $|F(x, y)| \leq \|Tx\| \, \|y\|$, so that for any fixed x, $F(x, y)$ is an antilinear function of y that is bounded relative to the norm in $\mathbf{H}$. Now suppose there is given in D an intrinsic topology stronger than that in $\mathbf{H}$, in terms of which it forms a topological vector space $\mathbf{D}$, and that F is a given continuous sesquilinear form on $\mathbf{D}$. There is then a continuous linear operator T from $\mathbf{D}$ to the antidual $^*\mathbf{D}$ of $\mathbf{D}$ defined by the equation $(Tx)(y) = F(x, y)$. $\mathbf{H}$ is continuously embedded in $^*\mathbf{D}$ by the map $y \to {}^*y$ where $^*y(x) = \langle y, x\rangle$ for $x \in \mathbf{D}$. In this sense, a continuous sesquilinear form F determines a linear operator T from $\mathbf{D}$ into an overspace of $\mathbf{H}$, namely $^*\mathbf{D}$, so that such a form can be considered as a generalized operator. It has a strict Hilbert space operator T_0 associated with it, defined on the domain D_0 of all vectors $x \in D$ such that $|F(x, y)| \leq c\,\|y\|$ for all $y \in \mathbf{D}$, by the equation $F(x, y) = \langle T_0 x, y\rangle$, but in general D_0 may consist only of 0. It will be convenient on occasion in the following to identify forms with operators, where possible without ambiguity, as indicated by the context, in accordance with these considerations.

In these general terms, the extended Wick product of vectors in $\mathbf{D}_{-\infty}(A)$ is a form on $\langle \mathbf{D}_\infty(A)\rangle$, and thus a generalized operator, in accordance with

COROLLARY 7.3.1. *There exists a unique mapping R from monomials in the free associative algebra over* $\mathbf{D}_{-\infty}(A)$ *to continuous sesquilinear forms on* $\langle \mathbf{D}_\infty(H)\rangle$ *that extends the mapping* R_0 *on the monomials in the subalgebra generated by* **H**, *given by the equation*

$$R_0(z_1\cdots z_n)(u, u') = \langle w(N_0(z_1\cdots z_n))u, u'\rangle \; (u, u' \in \mathbf{D}_\infty(H)),$$

and such that $R(z_1\cdots z_n)(u, u')$ *is jointly continuous in* $z_1, \ldots, z_n$, u, *and* u'.

PROOF. This follows from Theorem 7.3 and the equation

$$R_0(z_1\cdots z_n)(u, u') = 2^{-n/2} \sum_{S\subseteq\{1,\ldots,n\}}$$

$$\langle \{\prod_{k\in S} C^*(z_k)\}\, u, \{\prod_{k\in\{1,\ldots,n\}-S} C^*(z_k)\}u'\rangle,$$

which is implied by Scholium 7.5. □

DEFINITION. The *Wick product* of vectors $z_1, \ldots, z_n$ in $\mathbf{D}_{-\infty}(A)$ is the sesquilinear form on $\mathbf{D}_\infty(H)$ given by Corollary 7.3.1. It will be denoted as $:z_1\cdots z_n:$.

It should be mentioned that the domains $\mathbf{D}_{-\infty}(A)$ and $\mathbf{D}_\infty(H)$ used here are not really definitive. Domains of analytic, entire, and other classes of smooth vectors, and their duals, may also be used in a generally similar way (cf. Chap. 8). The present domains are, however, relatively large, simple, and invariant, and will suffice for the purposes of this chapter.

The extended Wick product inherits by continuity many of the properties of the Wick product of vectors in **H**. Like the Wick product of vectors in **H**, the extended Wick product can be characterized by its recursive commutation relations with arbitrary vectors and the vanishing of its vacuum expectation values. More specifically, using operator notation for forms to clarify the underlying algebra:

SCHOLIUM 7.8. *Let* R' *be a mapping from monomials in the free linear associative algebra over* $\mathbf{D}_{-\infty}(A)$ *to continuous sesquilinear forms on* $\langle \mathbf{D}_\infty(H)\rangle$ *such that*

i) For arbitrary $z_1, \ldots, z_n$ *in* $\mathbf{D}_{-\infty}(A)$, $y \in \mathbf{D}_\infty(A)$,

$$[R'(z_1 \cdots z_n), \phi(y)] = i \sum_j R'(z_1 \cdots \hat{z}_j \cdots z_n)\, A(z_j, y),$$

where $A(z, y)$ *denotes* $\mathrm{Im}(\langle z, y\rangle)$.

ii) $\langle R'(z_1 \cdots z_n)v, v\rangle = 0$.

Then R' *coincides with the mapping* R *given by Corollary 7.3.1.*

PROOF. Let $\mathbf{D}_0$ denote the linear submanifold in $\mathbf{K}$ consisting of all finite linear combinations of the $\phi(z_1)\cdots\phi(z_n)v$, with z_j in $\mathbf{D}_\infty(A)$. Then $\mathbf{D}_0$ is dense in $\langle \mathbf{D}_\infty(H)\rangle$ and contained in the domains of all products of the form $\phi(y_1)\cdots\phi(y_m)$ with the y_k in $\mathbf{H}$. This follows by analysis similar to that in Chapter 1, together with the density of $\mathbf{D}_0$ in $\mathbf{K}$. The density follows from Scholium 7.7 and the observation that $\mathbf{D}_0$ is invariant under the e^{itH}. The restriction of $R(z_1 \cdots z_n)$ and $R'(z_1 \cdots z_n)$ from forms on $\mathbf{D}_\infty(H)$ to forms on $\mathbf{D}_0$ agree by algebraic considerations that are similar to the unicity argument in Section 7.1, and will be omitted. The density of $\mathbf{D}_0$ in $\mathbf{D}_\infty(H)$ completes the proof. □

The covariance of Wick products in appropriate contexts, such as in relativistic free fields (cf. the next section) is shown by

THEOREM 7.4. *Let* A *be a strictly positive selfadjoint operator in the Hilbert space* $\mathbf{H}$. *Let* U *be a unitary operator on* $\mathbf{H}$ *such that* U *and* U^{-1} *leave* $\mathbf{D}_\infty(A)$ *invariant and act continuously on* $\langle \mathbf{D}_\infty(A)\rangle$. *Let* $(\mathbf{K}, W, \Gamma, v)$ *denote the free boson field over* $\mathbf{H}$, *and set* $H = \partial\Gamma(A)$. *Then*

i) $\Gamma(U)$ *leaves* $\mathbf{D}_\infty(H)$ *invariant, and acts continuously on* $\mathbf{D}_\infty(H)$;

ii) there is a unique continuous linear operator $\overline{U}$ *on* $\mathbf{D}_{-\infty}(A)$ *that extends* U; *and*

iii) Wick products of vectors in $\mathbf{D}_{-\infty}(A)$, *as sesquilinear forms on* $\mathbf{D}_\infty(H)$, *transform as follows under the induced actions of* U:

$$:(\overline{U}z_1)\cdots(\overline{U}z_1): \;=\; \Gamma(U) : z_1 \cdots z_n : \Gamma(U)^{-1}.$$

PROOF. We assume $A \geq I$, which is no essential loss of generality since A may otherwise be replaced by an appropriate multiple. If $x \in \mathbf{D}_0 \cap \mathbf{K}_n$, $\Gamma(U)x = (U \otimes \cdots \otimes U)x$ and $Hx = \sum_{1 \leq i \leq n} A_{i,n}x$, where $A_{i,n}$ is the operator on $\mathbf{K}_n$ given by $I \otimes \cdots \otimes A \otimes \cdots \otimes I$, with the factor A in the ith place. For an arbitrary integer $p \neq 0$,

$$H^p x = \sum_{i_1, \ldots, i_p = 1}^{n} A_{i_1,n} \cdots A_{i_p,n} x,$$

so that

$$H^p\Gamma(U)x = \sum_{i_1,\dots,i_p=1}^{n} A_{i_1,n} \cdots A_{i_p,n}(U\otimes\cdots\otimes U)x.$$

Since U is continuous from $\langle \mathbf{D}_\infty(A)\rangle$ to itself, there exists a positive integer m and a constant $c > 1$ (depending on p and U) such that for all positive integers $q \le p$, $\|A^q Uz\| \le c\,\|A^m z\|$ for arbitrary $z \in D_\infty(A)$. It follows that if $x \in \mathbf{D}_0 \cap \mathbf{K}_n$,

$$\|A_{i_1,n} \cdots A_{i_p,n}(U\otimes\cdots\otimes U)x\,\| \le c^p\,\|F_1\otimes\cdots\otimes F_n x\|,$$

where the $F_j = A^{mk_j}(0 \le k_j \le p)$ and $\Sigma_j k_j = p$. But such a term F_j is dominated by one in the expansion of H^{mp} in terms of the $A_{i,n}$, whence

$$F_1\otimes\cdots\otimes F_n \le H^{mp}$$

(all of the operators involved here being simultaneously diagonalizable). It follows that for arbitrary $x \in \mathbf{D}_0$, if x_n is the component of x in $\mathbf{K}_n$, then

$$\|H^p\Gamma(U)x\|^2 \le c^{2p} \sum_{n\ge 0} n^{2p}\|H^{mp}x_n\|^2. \tag{7.5}$$

Since $N \le H$, it results that

$$\|H^p\Gamma(U)x\|^2 \le c^{2p} \sum_{n\ge 0} \|H^{(m+1)p}x_n\|^2 = c^{2p}\,\|H^{(m+1)p}x\|^2. \tag{7.6}$$

Turning to ii) of Theorem 7.4, since U^* is continuous from $\langle \mathbf{D}_\infty(A)\rangle$ to itself, for any integer $m \ge 0$, there exists an integer $n(m)$, which may be assumed $\ge m$, such that $\|A^m U^* y\| \le c\,\|A^{n(m)}y\|$, for all $y \in \mathbf{D}_\infty(A)$. Now if $x \in \mathbf{D}_{-\infty}(A)$, say $x \in \mathbf{D}(A^{-m})$, then $|\langle x, U^*y\rangle| \le c\,\|A^{-m}x\|\,\|A^{n(m)}y\|$. Thus $\langle x, U^*y\rangle$ is a continuous antilinear functional of y in the space $\langle \mathbf{D}(A^{n(m)})\rangle$, and hence of the form $\langle x', y\rangle$ for some vector $x' \in \mathbf{D}(A^{-n(m)})$. Defining $\overline{U}$ as the map $x \to x'$, it is straightforward to verify that $\overline{U}$ extends U and is continuous on $\langle \mathbf{D}_{-\infty}(A)\rangle$.

Finally, the unicity of the Wick product and the invariance of ν under all $\Gamma(U)$ imply that for arbitrary $z_1,\dots,z_n$ in $\mathbf{H}$,

$$:(Uz_1)\cdots(Uz_1): = \Gamma(U) : z_1\cdots z_n : \Gamma(U)^{-1},$$

whence iii) follows by continuity. □

Theorem 7.5, which is an analog to Scholium 7.6 for the case of forms rather than vectors, further exemplifies the relations between forms and operators and will be useful later. As noted, if F is a continuous sesquilinear form on a locally convex topological vector space $\mathbf{L}$, then there is a unique linear operator T from $\mathbf{L}$ into $^*\mathbf{L}$ such that $F(x, y) = (Tx)(y)$ for all $x, y \in \mathbf{L}$. We call T the *kernel* of the form F. In the case of the space $\mathbf{L} = \langle \mathbf{D}_\infty(A)\rangle$, where A is as earlier, $^*\mathbf{L}$ is canonically identifiable with $\langle \mathbf{D}_{-\infty}(A)\rangle$ by the correspondence given earlier, and the mapping T is continuous from $\langle \mathbf{D}_\infty(A)\rangle$ into $\langle \mathbf{D}_{-\infty}(A)\rangle$.

The term "form" on $\langle \mathbf{D}_\infty(A)\rangle$ will mean "continuous sesquilinear form on $\langle \mathbf{D}_\infty(A)\rangle$" in the following.

THEOREM 7.5. *Let A be a strictly positive selfadjoint operator in the Hilbert space $\mathbf{H}$. Let F be a continuous sesquilinear form on $\langle \mathbf{D}_\infty(A)\rangle$. Let f be arbitrary in $C_0^\infty(\mathbf{R})$. Then the kernel of the form*

$$G(u, u') = \int F(e^{itA}u, e^{itA}u')f(t)dt$$

is a continuous linear operator from $\langle \mathbf{D}_\infty(A)\rangle$ into itself.

PROOF. This is based on the following estimate.

LEMMA 7.5.1. *Let A be a selfadjoint operator $\geq \varepsilon$ (where $\varepsilon > 0$) in the Hilbert space $\mathbf{H}$. Let B be a bounded linear operator on $\mathbf{H}$. Let f be a given function in $L_1(\mathbf{R})$. Let u and u' be arbitrary in $\mathbf{D}(A^a)$ and $\mathbf{D}(A^b)$ and*

$$G(u, u') = \int \langle B\, e^{itA} A^a u, e^{itA}A^b u'\rangle f(t)dt.$$

Then for arbitrary $k > 0$ such that $u \in \mathbf{D}(A^k)$,

$$|G(u, u')| \leq c_k \|B\|\, \|A^k u\|\, \|u'\|,$$

where $c_k = \sup\{r^{-k}|\hat{f}(r-s)|r^a s^b \colon r, s > \varepsilon\}$.

PROOF. Let E_r denote the spectral resolution of A, i.e., for any Borel function f on $\mathbf{R}$, $f(A) = \int f(r)\, dE_r$ in the usual notation. Similarly, $Bf(A)u = \int f(r)\, d(BE_r u)$. Setting $m(r, s) = \langle BE_r u, E_s u'\rangle$, note that m is a function of bounded variation in the plane. For defining the auxiliary set function $\wedge_m(P, Q)$ for arbitrary finite unions of intervals P and Q by the equation

$$\wedge_m(P, Q) = \langle BE(P)u, E(Q)u'\rangle,$$

then $|\wedge_m(P, Q)| \leq \|B\|\, \|E(P)u\|\, \|E(Q)u'\|$. It follows that if the $\{P_j\}$ and $\{Q_j\}$ are respectively finite sets of disjoint intervals, then

$$\sum_j |\wedge_m(P_j, Q_j)| \leq \|B\| \sum_j \|E(P_j)u\|\, \|E(Q_j)u'\|,$$

which, by the Schwarz inequality and the orthogonality of the $E(P_j)$ and the orthogonality of the $E(Q_j)$, is in turn bounded by

$$h(P, Q) = \|B\|\, \|E(P)u\|\, \|E(Q)u'\|,$$

where P and Q are the unions of the P_j and Q_j. It follows that

$$|\langle Bf(A)u, g(A)u'\rangle| = \left|\int\!\!\int f(r)\, \overline{g(s)} dm(r, s)\right| \leq \int |f(r)|\, |g(s)|\, dh(r, s),$$

where f and g are Borel functions. It follows in particular that

$$\langle Be^{itA}A^a u, e^{itA}A^b u'\rangle = \int\!\!\int e^{it(r-s)} r^a s^b\, dm(r, s).$$

Multiplying by $f(t)$ and integrating with respect to t, it follows in turn that

$$G(u, u') = \int\int \hat{f}(r-s)\, r^a s^b\, dm(r, s).$$

The integral over the plane of $r^k dh(r, s)$ is $\|B\|\,\|A^k u\|\,\|u'\|$, which implies, by the definition of c_k, that

$$\begin{aligned} |G(u, u')| &\le \int\int |\hat{f}(r - s)|\, r^a s^b\, dh(r, s) \\ &\le c_k \int\int r^k\, dh(r, s) = c_k \|B\|\,\|A^k u\|\,\|u'\|. \end{aligned}$$

□

PROOF OF THEOREM 7.5, CONTINUED. F may be represented in the following form, for sufficiently large a:

$$F(u, u') = \langle BA^a u, A^a u'\rangle,$$

where B is bounded. In particular, if $b = a + c$, where $c \ge 0$,

$$F(u, A^c u') = \langle BA^a u, A^b u'\rangle.$$

It follows that

$$|G(u, A^c u')| \le C_k \|B\|\,\|A^k u\|\,\|u'\|. \tag{7.7}$$

If $C_k < \infty$ for $c = 0, 1, 2, \ldots$, this inequality shows that the kernel K of the form G has the properties that $\|A^c Ku\| \le g(c)\,\|A^{m(c)}u\|$, which is equivalent to the continuity of K as an operator from $\langle \mathbf{D}_\infty(A)$ into itself. For $c = 0$, this is true by definition: Ku is defined as a vector such that $\langle Ku, u'\rangle = G(u, u')$. For $c > 0$, this follows from the fact that the restriction of A^c to $\mathbf{D}_\infty(A)$ is essentially selfadjoint, which is easily seen by an approximation argument (see Prob. 1 following this section). Indeed, the inequality 7.7 shows that there exists a vector w in $\mathbf{H}$ such that $G(u, A^c u') = \langle w, A^c u'\rangle$, while on the other hand $G(u, A^c u') = \langle Ku, A^c u'\rangle$. The equality $\langle w, A^c u'\rangle = \langle Ku, A^c u'\rangle$ for all $u' \in \mathbf{D}_\infty(A)$ shows that w is in the domain of $(A^c|\, \mathbf{D}_\infty(A))^*$. By the cited essential selfadjointness, this means that $A^c Ku$ exists, and that $\|A^c Ku\| \le$ const. $\|A^k u\|$ for sufficiently large k. Thus, to conclude the proof, it suffices to show that C_k is finite for all values of c if k is sufficiently large. To this end it is no essential loss of generality to assume $\varepsilon \ge 1$. Taking $k = a + b$, and noting that

$$|\hat{f}(y)| \le C\,(1 + |y|^b)^{-1}$$

since $f \in C_0^\infty(\mathbf{R})$, it follows that

$$C_k \le \sup\{F(r, s);\, r \ge 1 \text{ and } s \ge 1\},$$

where $F(r, s) = C\, r^{-b} s^b (1 + |r - s|^b)^{-1}$. If $s \le 2r$, then $F(r, s) \le 2^b C$, while if $s \ge 2r$, then $|r - s| \ge s/2$, so that

$$F(r, s) \le C\, s^b\, (1 + (s/2)^b)^{-1} \le 2^b C,$$

implying that $c_k \le 2^b C < \infty$. □

Problems

1. Show that if A is a strictly positive selfadjoint operator in the Hilbert space $\mathbf{H}$, then for any positive integer j, the restriction to $\mathbf{D}_\infty(A)$ of A^j is essentially selfadjoint. (Hint: if $x \in \mathbf{D}(A^j)$, show that $x_n = \exp(-A/n)x \to x$ and $A^j x_n \to A^j x$.)

2. Show that Wick's theorem holds for any symplectic transform of the free vacuum, i.e., for renormalized products with respect to the functional E^T, where T is an arbitrary symplectic transformation on the Hilbert space $\mathbf{H}$, and $E^T(u) = E(\Gamma(T)u)$, $\Gamma(T)$ denoting the unique automorphism of the infinitesimal Weyl algebra that carries z into Tz for arbitrary z in $\mathbf{H}$. (Hint: show the invariance of the defining properties of the renormalization map under the induced action on the infinitesimal Weyl algebra of symplectic transformations on $\mathbf{H}$.)

3. With the notation of Section 7.3, compute the result of applying $C(z)$ and $C(z)^*$, where z is arbitrary in $\mathbf{H}$, to $\prod_{j=1}^{n} \phi(z_j)v$, where z_j are in $\mathbf{D}_\infty(A)$. Show that $C(z)^*$ maps these vectors into $\mathbf{D}_\infty(H)$, while $C(z)$ maps $\mathbf{D}_\infty(H)$ into itself only if $z \in \mathbf{D}_\infty(A)$.

7.4. Renormalized local products of field operators

The formation of interaction Hamiltonians and Lagrangians involves local products of fields. This represents the special case of the foregoing theory in which points X in the underlying space-time manifold are mapped into vectors u_X in the space $\mathbf{D}_{-\infty}(A)$, in such a way that the value of the quantized field at the point X may be correlated with u_X. The specifics of the mapping $X \to u_X$ naturally depend somewhat on the particular field in question, but the general procedure is well represented by the case of a scalar field. This section develops the formulation of renormalized local products of quantized scalar fields.

Let M be an arbitrary complete C^∞ Riemannian manifold, and let $L_2(M, \mathbf{R})$, where the measure involved is the canonical one derived from the Riemannian structure, be denoted as $\mathbf{L}$. The Laplacian Δ_0 is nonpositive and essentially selfadjoint on the domain $C_0^\infty(M)$ (cf. Gaffney, 1955). The selfadjoint closure of Δ_0 will be denoted as Δ. The Klein-Gordon equation on $\mathbf{R} \times M$ takes the form

$$(\partial_t^2 - \Delta + m^2)\varphi = 0, \tag{7.8}$$

where $\varphi(t, x)$ is a real function of the variables $t \in \mathbf{R}$ and $x \in M$. We restrict considerations to the case in which the real constant m, the "mass," is positive, to avoid nongeneric technical complications. Setting $B = (m^2 - \Delta)^{1/2}$, the treatment can be consolidated and generalized by consideration of the equation

$$\partial_t^2\phi + B^2\phi = 0, \tag{7.9}$$

where $\phi = \phi(t)$ is a function of $t \in \mathbf{R}$ whose values are in a real Hilbert space $\mathbf{L}$, which in the case of the Klein-Gordon equation consists of functions on space.

This equation is readily quantized by the procedures of Chapter 1. Setting $C = B^{1/2}$, the underlying single-particle (complex Hilbert) space $\mathbf{H}$ is the direct sum $\langle \mathbf{D}(C)\rangle \oplus \langle\langle \mathbf{D}(C^{-1})\rangle\rangle$, with the complex structure in $\mathbf{H}$ defined by the matrix

$$\begin{pmatrix} 0 & -B \\ B^{-1} & 0 \end{pmatrix}$$

relative to the given direct sum decomposition of $\mathbf{H}$, and the following inner product between the vectors ϕ and ψ in $\mathbf{H}$:

$$\langle \phi, \psi\rangle_{\mathbf{H}} = \langle C\phi(t), C\psi(t)\rangle_{\mathbf{L}} + \langle C^{-1}\partial_t\phi(t), C^{-1}\partial_t\psi(t)\rangle_{\mathbf{L}}$$
$$+ \, i\,(\langle C^{-1}\partial_t\phi(t), C\psi(t)\rangle_{\mathbf{L}} - \langle C\phi(t), C^{-1}\partial_t\psi(t)\rangle_{\mathbf{L}}). \tag{7.10}$$

In the last equation, t is arbitrary in $\mathbf{R}$; the right side of the equation is independent of t by virtue of the underlying differential equation, which is now to be understood in its integrated form:

$$[\phi(t)\oplus\partial_t\phi(t)] = \begin{pmatrix} \cos(tB) & B^{-1}\sin(tB) \\ -B\sin(tB) & \cos(tB) \end{pmatrix} [\phi(0)\oplus\partial_t\phi(0)]. \tag{7.11}$$

The single-particle Hamiltonian A is the selfadjoint generator of the one-parameter unitary group defined by equation 7.11. As earlier, the inner product in $\mathbf{H}$ may be extended to a partially defined one in $\mathbf{D}_{-\infty}(A)$, and the real inner product in $\mathbf{L}$ may similarly be extended to a partially defined one in $\mathbf{D}_{-\infty}(B)$.

Nonlinear local interactions typically involve formal expressions that appear as

$$(D_1\phi)(X)(D_2\phi)(X)\cdots(D_1'\partial_t\phi)(X)(D_2'\partial_t\phi)(X)\cdots, \tag{7.12}$$

where the D_j and D_j' are given linear partial differential operators, as well as integrals of such expressions after multiplication by a smooth function on space or space-time. Thus, the most general local Hamiltonian is formally a sum of such terms. When rigorously defined quantized fields are substituted for the symbolic fields in equation 7.12, the result is a product of distributions, and so is undefined, or at best, singular; and its putative integral is scarcely less so. However, in perturbative scattering theory, the quantized fields (i.e., the so-called incoming and outgoing fields) are free. In this case, a rigorous version of Wick's procedure leads ultimately to interpretation of these singular quantities as forms on $\mathbf{D}_\infty(H)$, where H is the Hamiltonian of the free field.

Basically, the only feature involved in the rigorous formulation and treatment of expression 7.12 is the suitable representation of the point-evaluation vectors Ω_X, as generalized vectors, provided that the differential operators D_j and D_j' are appropriately dominated by powers of B, as is normally the case. The covariance, continuity, and boundedness of the map $X \rightarrow \Omega_X$ is relevant, where a subset of $\mathbf{D}_{-\infty}(A)$ is said to be bounded in case it is bounded as a subset of $\langle \mathbf{D}(A^n) \rangle$ for some integer n. In the case of the Klein-Gordon equation, the basic point evaluation functionals are described in Theorem 7.6. To avoid technical complications, we assume that the manifold M is either $\mathbf{R}^n$ or is compact; but the argument below applies to arbitrary compact regions in a noncompact manifold, to which noncompact regions are largely reducible, by virtue of hyperbolicity considerations. No infinite-dimensional issues are involved in the proofs of the next two results, and only the essentials will be given.

SCHOLIUM 7.9. *Let M be a Riemannian manifold, either compact or $\mathbf{R}^n$. Let $\mathbf{L}$ denote $L_2(M, \mathbf{R})$, and let $B = (m^2 - \Delta)^{1/2}$, where $m > 0$. For arbitrary $x \in M$, there exists a unique vector δ_x in $\mathbf{D}_{-\infty}(B)$ such that for arbitrary f in $\mathbf{D}_\infty(B)$,*

$$f(x) = \langle f, \delta_x \rangle. \tag{7.13}$$

Moreover, the map $x \rightarrow \delta_x$ is bounded and continuous from M into $\mathbf{D}_{-\infty}(B)$, and covariant with respect to the isometry group of M. If T is an arbitrary C^∞ linear differential operator in the case of compact M, or an arbitrary such operator with constant coefficients in the case of $\mathbf{R}^n$, then T extends uniquely to a continuous linear operator on $\mathbf{D}_{-\infty}(B)$, and

$$(Tf)(x) = \langle f, T^*\delta_x \rangle \qquad (f \in \mathbf{D}_\infty(B)). \tag{7.14}$$

PROOF. That $f(x)$ is for fixed x a continuous linear functional of f in a Sobolev space $L_{2,r}$, if $r > \frac{1}{2}\dim(M)$, follows from the Plancherel theorem in the case $M = \mathbf{R}^n$. The Sobolev space involving derivatives of order r is however identical to $\mathbf{D}(B^r)$. This shows the existence of δ_x when $M = \mathbf{R}^n$, and the argument provides an explicit expression for δ_x, from which the bounded continuity of the map $x \rightarrow \delta_x$ follows. The case of a compact manifold is reducible locally to the euclidean case, and with the use of a finite partition of unity, follows globally.

Covariance with respect to arbitrary isometries, meaning that if g is an arbitrary isometry on M, and if $U(g)$ denotes the mapping $f(x) \rightarrow f(g^{-1}x)$ on $\mathbf{D}_\infty(B)$, and also the extended map to $\mathbf{D}_{-\infty}(B)$, that $U(g)$ sends δ_x to $\delta_{g^{-1}x}$, follows from the unicity of δ_x as defined by the equation $f(x) = \langle f, \delta_x \rangle$.

For any linear differential operator T of the type indicated, it is not difficult to show that T is dominated by a power of B in the sense that TB^{-r} is a bounded operator on $L_2(M)$ if r exceeds the order of T. In consequence, T admits a unique continuous extension to all of $\langle \mathbf{D}_{-\infty}(B)\rangle$ and satisfies the indicated equation by virtue of the definition of T^*. □

THEOREM 7.6. *Let M, $\mathbf{L}$, B and δ_x be as in Scholium 7.9. Let* $\mathbf{H}$ *denote the complex Hilbert space of all real normalizable solutions ϕ of the Klein-Gordon equation 7.8. For arbitrary $X \in \mathbf{R} \times M$, there exists a unique vector $\Omega_X \in \mathbf{D}_{-\infty}(A)$ such that for arbitrary $\phi \in \mathbf{D}_\infty(A)$,*

$$\phi(X) = \mathrm{Re}(\langle \phi, \Omega_X\rangle).$$

Moreover, the map $X \to \Omega_X$ is bounded and continuous from $\mathbf{R} \times M$ into $\langle \mathbf{D}_{-\infty}(A)\rangle$, and covariant with respect to the isometry group of M and time-translation. In the case of Minkowski space-time, it is also Lorentz-covariant. Ω_X may be described explicitly as the solution of the Klein-Gordon equation having the Cauchy data $[B^{-1}\delta_x \oplus 0]$ at time t, for $X = (t, x)$.

PROOF. Observe that $\mathbf{H}$ is unitarily equivalent to $L_2(M, \mathbf{C})$: for each t, there is a unitary operator $S(t)$ from $\mathbf{H}$ onto $L_2(M, \mathbf{C})$ that maps the vector ϕ in $\mathbf{H}$ into $C\phi(t) - iC^{-1}\partial_t\phi(t)$. This unitary equivalence carries the operator A in $\mathbf{H}$ into the operator $(m^2 - \Delta)^{1/2}$ in $L_2(M, \mathbf{C})$; we denote this operator as B^c to distinguish it from the operator of the same form in the real Hilbert space $\mathbf{L}$. Since $\mathbf{D}_\infty(A)$ is a unitarily invariant concept, $S(t)$ extends uniquely to an isomorphism of $\langle \mathbf{D}_\infty(A)\rangle$ with $\langle \mathbf{D}_\infty(B^c)\rangle$ and of $\langle \mathbf{D}_{-\infty}(A)\rangle$ with $\langle \mathbf{D}_{-\infty}(B^c)\rangle$. Now

$$\mathrm{Re}(\langle \phi, \psi\rangle_{\mathbf{H}}) = \langle C\phi(t), C\psi(t)\rangle_{\mathbf{L}} + \langle C^{-1}\partial_t\phi(t), C^{-1}\partial_t\psi(t)\rangle_{\mathbf{L}},$$

while $\phi(t, x) = \langle \phi(t), \delta_x\rangle_{\mathbf{L}}$, whence $\phi(t, x) = \langle C\phi(t), C^{-1}\delta_x\rangle_{\mathbf{L}}$. Setting Ω_X for the solution of the Klein-Gordon equation with the Cauchy data at time t, $[B^{-1}\delta_x \oplus 0]$, it follows that $\phi(X) = \mathrm{Re}(\langle \phi, \Omega_X\rangle)$. Unicity is immediate, and boundedness and continuity of Ω_X as a function of X follows from that of δ_x in conjunction with the observation that temporal evolution is e^{itA}, which leaves invariant the norms in $\langle \mathbf{D}(A^n)\rangle$ for arbitrary n. The Lorentz covariance in the case of Minkowski space-time follows from the Lorentz invariance of the norm in $\mathbf{H}$. □

COROLLARY 7.6.1. *With the same notation as earlier, and with linear differential operators D_j and D'_k that are of constant coefficients in the case $M = \mathbf{R}^n$ or C^∞ if M is compact, the (finite) Wick product*

$$V(X) = :(D_1\phi)(X)(D_2\phi)(X)\cdots(D'_1\partial_t\phi)(X)(D'_2\partial_t\phi)(X)\cdots:$$

exists and is a continuous and bounded function of X with values in $\langle \mathbf{D}_{-\infty}(H)\rangle$.

For arbitrary $f \in C_0^\infty(\mathbf{R} \times M)$ (resp. in $C_0^\infty(\mathbf{R})$),

$$\int V(t, x)f(X)dX \qquad (\textit{resp.}\ \int V(t, x)f(t)dt)$$

exists and is a continuous linear operator from $\langle \mathbf{D}_\infty(H)\rangle$ to itself.

PROOF. It follows from Theorem 7.6 that if S is a continuous linear operator on $\mathbf{D}_{-\infty}(A)$, then $(S\phi)(X) = \mathrm{Re}(\langle\phi, S^*\Omega_X\rangle)$. Thus, the factors $D_j\phi(X)$ and $D'_k\partial_t\phi(X)$ in $V(X)$ are all of the form $\phi(z)$ for suitable vectors z in $\mathbf{D}_{-\infty}(A)$, to which the Wick product applies. Continuity results from the continuity of Wick products as a function of the factors involved. For a finite number of differential operators of bounded orders, the $V(X)$ all lie in $\mathbf{D}(A^n)$ for values of n bounded away from $-\infty$. Continuity as a function of X, together with invariance of the norm in $\langle\mathbf{D}(A^n)\rangle$ under temporal evolution, then implies that the $V(X)$ form a bounded function of X within any compact X-region.

It follows that the indicated integrals exist as forms. That the temporal integral represents a continuous linear operator on $\langle\mathbf{D}_\infty(A)\rangle$ to itself follows from Scholium 7.8. By virtue of the boundedness of the integrand and the estimates of Lemma 7.5.1, the further integral over X involved in the space-time integral remains a continuous linear operator on $\langle\mathbf{D}_\infty(H)\rangle$. □

COROLLARY 7.6.2. *Let φ denote the free Klein-Gordon field over Minkowski space $\mathbf{M}_0$. The map $f \to \int :\varphi(X)^n: f(X)dX$ from $C_0^\infty(\mathbf{M}_0)$ to forms on $\langle\mathbf{D}_\infty(H)\rangle$ in fact maps into the continuous linear operators on $\langle\mathbf{D}_\infty(H)\rangle$ and is Poincaré-covariant.*

PROOF. This is immediate from the preceding corollary and the Lorentz-covariance of the Wick product. □

LEXICON. In the early days of quantum field theory it was noted by Kramers and emphasized by Bohr and Rosenfeld (1933) that the value of a quantum field at a point was a purely ideational object, physically speaking. The view was that any physically realizable probe of the field could at best determine the field average, in some region, or with respect to an appropriate averaging function. In this connection, Bohr and Rosenfeld computed the commutation relations for the appropriately averaged quantized free electromagnetic field in the neighborhood of a point. Since that time it has been recognized that, in mathematical terms, appropriate local smoothing of the quantized field was required to obtain a bona fide (densely defined) operator in Hilbert space. In the interest of manifest relativistic invariance, smoothing in both time and space is indicated, and is described above in the technically adaptable space $C_0^\infty(\mathbf{M}_0)$.

For dynamical purposes, averaging in space at a fixed time is more effective, particularly where nonlinear expressions are involved. For example, the Hamiltonian H of the free Klein-Gordon field may be expressed in terms of the field $\varphi(t, x)$ at the fixed time t as

$$H = \int [:(\nabla\varphi(X))^2: + [(\partial_t\varphi(X))^2: + m^2 :\varphi(X)^2:]dx$$

This symbolic expression, obtained simply by substitution of the quantized field into the expression for the classical Hamiltonian (i.e., the correspondence principle) is much more complicated from a mathematical position than the expression of H as $\partial\Gamma(A)$. In a way it is remarkable that it can be validated at all; each of the three summands under the integral sign is only a form-valued distribution on space. But this integral expression for H is important because it is essential to justify the use of the correspondence principle for the quantization of nonlinear wave equations. In the latter case there is no established representation for H in a form analogous to $\partial\Gamma(A)$. We leave to the reader the problem of inserting an appropriate spatial cutoff $f \in C_0^\infty(\mathbf{R}^n)$ into the above expression for H, obtaining an essentially selfadjoint operator by virtue of cancellations in the sum of the three forms. Following this, it is necessary to pass to the limit $f \to 1$ in the space of selfadjoint operators. Although lengthy, no essential difficulty is involved.

This is also no difficulty in expressing the underlying equation and commutation relations in terms of the quantized field as a form-valued function. The equation is simply

$$(\Box + m^2)\varphi = 0,$$

which in terms of space-time averages takes the form $\Phi((\Box + m^2)f) = 0$ for arbitrary $f \in C_0^\infty(\mathbf{M}_0)$. Forms in general have no well-defined commutator (which is again a form), but if one of the forms is an operator, the commutator may be defined. Thus in relativistic terms,

$$W(f)^{-1}\varphi(X)W(f) = \varphi(X) + \int D(X - Y)f(Y) \quad (f \in C_0^\infty(\mathbf{M}_0)),$$

where $W(f)$ is defined as $W(Tf)$, T being the Poincaré-covariant projection of $C_0^\infty(\mathbf{M}_0)$ into the solution manifold $\mathbf{H}$ of the Klein-Gordon equation. We again leave the details of these alternative formulations to the reader.

Problems

1. a) For the free Klein-Gordon field φ on Minkowski space of dimension $n + 1$, show that $\int_{\mathbf{R}^n} :\varphi(X)^r: f(x)dx$ exists as a continuous sesquilinear form F on $\langle \mathbf{D}_\infty(H)\rangle$, if $f \in C_0^\infty(\mathbf{R}^n)$ and r is a positive integer.

b) Show that if $n = 1$, the kernel of F maps from $\mathbf{D}_\infty(H)$ into the Hilbert space $\mathbf{K}$ (i.e., $|F(u, u')| \leq c(u)\,\|u'\|$ for all $u \in \mathbf{D}_\infty(H)$).

c) Show that if $n > 1$ and $r > 1$, then in general the Hilbert space operator corresponding to F has only 0 in its domain.

2. Let M denote the n-torus T^n, where $n > 1$, and let φ denote the free Klein-Gordon field over $\mathbf{R} \times M$. Show that if $p > 1$, then the domain of $\int :\varphi(0, x)^p: dx$ as a Hilbert space operator in $\mathbf{K}$ has domain consisting only of 0.

3. Show that if $f \in C_0^\infty(\mathbf{M}_0)$ and if φ is the free Klein-Gordon field over $\mathbf{M}_0$, then $\int :\varphi(X)^2: f(X)\, dX$ is essentially selfadjoint on $\mathbf{D}_\infty(H)$. (The analog for higher, even powers is an open question even with the restriction that f is nonnegative.)

4. Extend the treatment of the Wick products of free Klein-Gordon field operators to the case $m = 0$.

5. Show that Theorem 7.6 remains valid if the conclusion is changed to the existence of a unique vector $\Omega'_X \in \mathbf{D}_{-\infty}(A)$ such that $\phi(X) = \mathrm{Im}(\langle \phi, \Omega'_X \rangle_{\mathbf{H}})$. Determine the Cauchy data for Ω'_X at time t.

6. Show that $\langle \mathbf{D}_\infty(A) \rangle$ is irreducible under the action of the Poincaré group in the case of the Klein-Gordon equation in $\mathbf{M}_0$. Show also that the most general Poincaré-invariant continuous sesquilinear form on $\langle \mathbf{D}_\infty(A) \rangle$ is the given inner product in $\mathbf{H}$, within a constant factor. Establish the similar irreducibility of $\langle \mathbf{D}_{-\infty}(A) \rangle$. Derive from this the most general continuous Poincaré covariant map from $\mathbf{M}_0$ into $\langle \mathbf{D}_{-\infty}(A) \rangle$.

7. Let $x(t)$ denote the Brownian motion as formulated by Wiener, and let $y(t)$ denote its fractional derivative of order ½, as a random-variable-valued Schwartz distribution (defined by Fourier transformation). Show that there exist similar Schwartz distributions, which may be denoted as $:y(t)^n:$, that are uniquely determined by the properties of having vanishing expectation values, and satisfying the relations

$$:(y(t) + f(t))^n: \;=\; :y(t)^n: + nf(t):y(t)^{n-1}: + \tfrac{1}{2}n(n-1)f(t)^2:y(t)^{n-2}: + \cdots,$$

where f is an arbitrary real function in $L_\infty(0, 1)$, and $:(y(t) + f(t))^n:$ is defined as the transform of $:y(t)^n:$ under the transformation $y(t) \to y(t) + f(t)$, in the following sense: let U denote the unitary operator on $L_2(\mathbf{C})$, where $\mathbf{C}$ denotes Wiener space, induced canonically from the absolutely continuous transformation that carries $\int h(t)dx(t)$ into $\int h(t)d(x(t) + F(t))$, where $F(t) = \int_0^t f(\tau)d\tau$, for arbitrary $h \in L_2(0, 1)$. Then, conjugation by U transforms the operation of multiplication by $:y(t)^n:$ into the operation of multiplication by $:(y(t) + f(t))^n:$ (each after integration relative to the same function in $C_0^\infty(0, 1)$). (Hint: show that $y(t)$ differs inessentially from the fixed time free quantum field for the wave equation in two space-time dimensions.)

8. The "white noise" w over a measure space M may be defined as the isonormal distribution over the real Hilbert space $L_2(M, \mathbf{R})$. Let M be a C^∞ Riemannian manifold, either $\mathbf{R}^k$ or compact, let $B = (m^2 - \Delta)^{1/2}$ as earlier, and let ϕ denote the random Schwarz distribution over $C_0^\infty(M)$ (i.e., linear mapping from the latter space to random variables), $f \to w(B^{-1/2}f), f \in C_0^\infty(M)$. Show that this distribution is (probabilistically) equivalent to the Klein-Gordon free field over space at a fixed time, relative to free vacuum expectation values.

9. a) Let $\mathbf{H}$ be the Hilbert space solution manifold of the Klein-Gordon equation in $\mathbf{R} \times M$, as earlier, and suppose that all vectors are periodic in t

with a fixed period 2p. Show that if φ denotes the free quantized Klein-Gordon field over $\mathbf{R} \times M$, then the space-time integral $\int_{(-p,p)\times M} :\varphi(X)^r: dX$ is essentially selfadjoint on $\mathbf{D}_\infty(H)$. (Hint: use the result of Poulsen (1972) on the essential selfadjointness of invariant forms.)

b) Taking $M = S^3$ with its usual Riemannian structure, and $m = 1$, show the periodicity in (*a*) is valid with $p = \pi$.

c) By exploiting conformal invariance of the wave equation, show the unitary equivalence of the free quantized wave equation field over four-dimensional Minkowski space $\mathbf{M}_0$ and the quantization of the Klein-Gordon equation $(\partial_t^2 - \Delta + 1)\varphi = 0$ on $\mathbf{R} \times S^3$.

d) Conclude that if φ is the free quantized wave equation field over $\mathbf{M}_0$, then $\int_{\mathbf{M}_0} :\varphi(X)^4: dX$ has a natural (conformally covariant) interpretation as a selfadjoint operator in $\mathbf{K}$. (Hint: note the conformal invariance of the corresponding classical integral. Do not use the usual relativistic Hamiltonian H, which suffers from "infrared" problems, but work with forms on $\mathbf{D}_\infty(H')$, where H' is the field Hamiltonian corresponding to temporal evolution in $\mathbf{R} \times S^3$.)

10. Develop the notion of power of a quasi-invariant distribution ϕ on C_0^∞ over a manifold M: $\phi(f) \approx \int \varphi(x)f(x)dx$. Show that when $N(\varphi(x)^r)$ exists for all r, where N denotes renormalization with respect to the expectation functional derived from the given probability measure, it is local in the additional sense that $\varphi(x)$ is unaffected by the induced action of the translation $\phi \rightarrow \phi + f$, where $f \in C_0^\infty(M)$, if f vanishes in a neighborhood of the point x.

11. Suppose that f and g are in $C_0^\infty(\mathbf{M}_0)$ and that the region of influence (relative to the Klein-Gordon equation) of the support of f is disjoint from the support of g. Show that $\int :\varphi(X)^r: f(X)dX$ and $\int :\varphi(X)^s: g(X)dX$ commute, as operators on $\mathbf{D}_\infty(H)$, where φ is the free Klein-Gordon field, r and s being arbitrary integers.

Bibliographical Notes on Chapter 7

The standardization of renormalization for products of free fields by Wick (1950) is rigorously applicable to normalizable fields; its adaptation in the physical literature to products of fields at a point is heuristic. The present definition applicable also to interacting fields was given in an implicitly axiomatic form by Segal (1964b) and developed along rigorous constructive lines by Segal (1967; 1969a, b; 1970c; 1971). Wightman and Gårding (1964) showed that the (Schwartz) distributions corresponding to local products of free fields were densely defined operators in Hilbert space, with the aid of combinatorial properties of Wick products established by Caienello. Nelson (1972) and Baez (1989) treated Wick product theory on the basis of scales of the $\mathbf{D}_\infty$ type.

8

Construction of Nonlinear Quantized Fields

8.1. Introduction

The intuitive conceptual simplicity of nonlinear quantized field theory contrasts strikingly with the depth of the complications that arise from attempts at rigorous mathematical implementation of the underlying ideas. A priori, notwithstanding the formal simplicity and physical appeal of the proposed theory—rooted in the work of Dirac, Heisenberg, Pauli, et al.—there is no assurance that this theory actually exists, nor as yet is there a definitive mathematical interpretation for it. In direct formal terms, it involves nonlinear functions of quantized distributions, which are rightly looked on with considerable suspicion from a mathematical standpoint. In practical terms, the situation is not that much better; in the original and most studied case of quantum electrodynamics, there is still no mathematically manifestly rigorous proof of "renormalizability" in formal perturbation theory, although a posteriori and partially subjective computations along these lines have reproduced the results of some of the basic measurements that were indicative of the empirical relevance of quantum field effects.

For these reasons it appears most important that the construction of nonlinear quantum fields be not only technically rigorous but also intuitively convincing, by virtue of close conformity to both basic physical principles and the mathematical ideas in terms of which these fields are formulated and developed. Otherwise there will remain a real possibility that nonlinear quantum field theory as sought for more than a half century may be a specious illusion.

As yet it is only in two space-time dimensions that nonlinear quantum fields have been constructed in conformity with such desiderata, and even in this simplified, and of course unphysical case, there remain some unresolved and significant issues. But the general method is a mathematically interesting and physically convincing one, which extends on a preliminary basis (e.g., construction of "cutoff" theories) to the empirically relevant four-dimensional case. This chapter treats the underlying mathematical theory in the simplest form that displays the essential ideas but also has potential adaptability to an attack on the four-dimensional case.

In formal principle, the Hamiltonian H for the prototypical scalar nonlinear wave equation $\Box\varphi + p'(\varphi) = 0$, where p is a given polynomial, is a functional of the putative "interacting" field φ satisfying this classical equation, given as follows:

$$H(\varphi) = \int [\tfrac{1}{2}((\nabla\varphi)^2 + (\partial_t\varphi)^2) + p(\varphi)]\, dx.$$

The integration here is over space at a fixed time, but is independent of time by virtue of the underlying differential equation. According to standard quantum-mechanical ideas, the corresponding quantized field φ is propagated by the Hamiltonian $H(\varphi)$ obtained by substituting the quantized field for the classical field in the expression for the Hamiltonian. Formally, if the Cauchy data are given for the field φ and its first time derivative $\partial_t\varphi$ at an initial time, say $t = 0$, then $H(\varphi)$ may be evaluated, and the quantized field obtained at all time by conjugation of the initial data with $e^{itH(\varphi)}$. However, $p(\varphi)$ requires renormalization, being infinite if interpreted by a straightforward limiting procedure, and this renormalization depends on the vacuum, as seen in Chapter 7. The "physical" vacuum, represented by the lowest eigenvector of $H(\varphi)$, is the appropriate one here. However, since this is not known a priori, the precise interpretation of $p(\varphi)$ is also not known. In addition, the initial data, i.e., the quantized field φ and its first time derivative $\partial_t\varphi$ at the initial time, are not known a priori, apart from the presumption of irreducibility and the canonical commutation relations, which, as seen in Chapter 4, by no means determine (within unitary equivalence, which is all that matters here) these initial fields.

Some such complications are not unexpected in a truly nonlinear problem. A natural strategy for their resolution is by successive approximation starting from the free field as a first approximation. The next approximation is then defined by a formally *linear* differential equation, or equivalently as the field obtained from the free field by propagation with the Hamiltonian formally given as $H(\varphi_0)$, where φ_0 denotes the free field. However, the rigorous implementation of this formal procedure is not straightforward. The problem is, briefly, that $H(\varphi_0)$ is formally equal to $H_0 + V$, where H_0 is the Hamiltonian for the free quantum field, while the interaction energy $V = \int p(\varphi_0)dx$ is relatively singular, and thereby outside the scope of conventional perturbation theory, e.g., of the Kato-Rellich type. The interpretation of $H_0 + V$ as a self-adjoint operator having a unique lowest eigenvector, etc., will occupy the first part of this chapter.

The strategy of the treatment of $H(\varphi_0)$ is as follows: one needs a common domain on which both H_0 and V are well defined and act appropriately. Rather curiously, one proceeds via analysis of the one-parameter semigroups e^{-tH_0} and e^{-tV} ($t > 0$), to establish a one-parameter selfadjoint semigroup that can be identified with e^{-tH}, where $H = H_0 + V$. A direct construction of the physical time evolution group e^{itH} is not presently known. The treatment of these semigroups is facilitated by relating them to a scale of Banach subspaces $\mathbf{K}_p$ of $\mathbf{K}$, whose norms increase in strength with p. More specifically, e^{-tH} improves the status (relative to the scale) of a given vector at a rate that is more rapid than the rate at which e^{-tV} worsens its status. In a fairly general context of this type it is then possible to establish $H_0 + V$ as a selfadjoint operator, and to develop aspects of its spectrum. The functional-analytic techniques in so doing involve notably the Duhamel and Lie-Trotter formulas for e^{A+B}, where A and B are given operators, and are facilitated by basic interpolation theory for operators.

8.2. The L_p scale

Since the quantized field at a given initial time represents the Cauchy data in the solution of quantized differential equations, there is a corresponding conjugation $\varkappa$ in the single-particle space $\mathbf{H}$, as seen in Chapter 6. The real subspace $\mathbf{H}_\varkappa$ of all vectors x in $\mathbf{H}$ for which $\varkappa x = x$ is associated to the canonical pairing between the field and its first time derivative *at the given time*. Accordingly, $\mathbf{H}_\varkappa$ is not at all invariant under temporal evolution, i.e., under the operators e^{itH_0}, where H_0 is the Hamiltonian for the field. Despite this lack of invariance, it serves as the basis for the establishment of an effective scale of spaces; specifically, the scale $L_p(\mathbf{H}_\varkappa, g)$ for $p \in (1, \infty)$.

Formally, the Hamiltonian H for an interacting field is representable as $H = H_0 + V$, where H_0 is the Hamiltonian for the presumed associated free field and V is the interaction Hamiltonian. It is with the actions of the semigroups e^{-tH_0} and e^{-tV} relative to the indicated scale that we are here concerned. These two semigroups behave quite differently, and we begin with the former. The isonormal distribution g will be understood as the underlying distribution (unless otherwise indicated), with variance parameter 1 in connection with the real wave representation.

THEOREM 8.1. *There exists a universal constant $\varepsilon > 0$ with the following property. Let $\mathbf{H}$ be a complex Hilbert space (of arbitrary dimension), and let $\varkappa$ be a given conjugation on $\mathbf{H}$. Let A be a ($\varkappa$-) real selfadjoint operator on $\mathbf{H}$ that is $\geq I$. Let $(\mathbf{K}, W, \Gamma, \nu)$ denote the free boson field over $\mathbf{H}$ in the real wave representation relative to $\varkappa$. Then $e^{-t\partial\Gamma(A)}$ is a contraction from $L_q(\mathbf{H}_\varkappa, g)$ to $L_p(\mathbf{H}_\varkappa, g)$, where $p = qe^{\varepsilon t}$ and $q \geq 2$, for all $t > 0$.*

We set $H = \partial\Gamma(A)$ and $N = \partial\Gamma(I)$, and drop g where indicated by the context.

LEMMA 8.1.1. *If* $\mathbf{H}$ *is one-dimensional,* e^{-tH} *is bounded from* $L_2(\mathbf{H}_\varkappa)$ *to* $L_4(\mathbf{H}_\varkappa)$, *for all sufficiently large* t, *uniformly in* A.

PROOF. Identifying $\mathbf{H}$ with $\mathbf{C}$ and $\varkappa$ with complex conjugation, A has the form $z \mapsto az$, where $a \geq 1$. Since $e^{-tH} = e^{-tN}e^{-t\partial\Gamma(A-I)}$, and $e^{-t\partial\Gamma(A-I)}$ is a contraction from $L_2(\mathbf{H}_\varkappa)$ to itself, it suffices to show that e^{-tN} is uniformly bounded from $L_2(\mathbf{H}_\varkappa)$ to $L_4(\mathbf{H}_\varkappa)$, for all sufficiently large t. A classic formula of Mehler (rigorously established by Hille, 1926) gives an expression for $\exp(-tN)$ as an integral operator, with kernel

$$K(x, y) = (1 - c^2)^{-\frac{1}{2}}\exp[\{-c^2(x^2 + y^2) + 2cxy\}/2(1 - c^2)],$$

where $c = e^{-t}$. By the Schwarz inequality, and the evaluation of a Gaussian integral, for arbitrary $f \in L_2(\mathbf{R}^1, g)$,

$$|(e^{-tN}f)(x)| \leq (1 - c^4)^{-\frac{1}{4}}\exp[\tfrac{1}{2}c^2x^2(1 + c^2)^{-1}]\,\|f\|_2.$$

It follows that if $c^2 < 1/3$,

$$\|e^{-tN}f\|_4 \leq (1 - c^4)^{-\frac{1}{4}}[(1 - 3c^2)/(1 + c^2)]^{-\frac{1}{8}}\,\|f\|_2,$$

completing the proof.

LEMMA 8.1.2. *If* $\mathbf{H}$ *is one-dimensional,* e^{-tN} *is a contraction from* $L_p(\mathbf{H}_\varkappa)$ *to* $L_p(\mathbf{H}_\varkappa)$ *for all* $p \geq 2$ *and* $t \geq 0$.

PROOF. From $(e^{-tN}f)(x) = \int K(x, y)f(y)dg(y)$ and the factorization

$$K(x, y)f(y) = (K(x, y)^{1/p}f)\,(K(x, y)^{1/p'}),$$

where $p' = p/(p - 1)$, it follows via Hölder's inequality that

$$|(e^{-tN}f)(x)| \leq (\int K(x, y)|f(y)|^p dg(y))^{1/p}(\int K(x, y)dg(y))^{1/p'}$$

But $e^{-tN}1 = 1$, so $\int K(x, y)dg(y) = 1$ (equally obtainable by direct integration), and it results that $|(e^{-tN}f)(x)|^p \leq \int K(k, y)|f(y)|^p dg(y)$. Integrating with respect to x, the lemma follows. □

LEMMA 8.1.3. *If* $\mathbf{H}$ *is one-dimensional, then* e^{-tN} *is a contraction from* $L_2(\mathbf{H}_\varkappa)$ *to* $L_4(\mathbf{H}_\varkappa)$ *for sufficiently large* t.

PROOF. It suffices to show that $\|e^{-tN}f\|_4 \leq \|f\|_2$ for all $f = 1 + h$, where $\langle h, 1\rangle = 0$, and all sufficiently large t. For such f, $e^{-tN}f = 1 + k$, where $k = e^{-tN}h$, and

$$|e^{-tN}f|^4 = (1 + k)^2(1 + \bar{k})^2$$
$$= 1 + (k + \bar{k})^2 + (k\bar{k})^2 + 2k\bar{k} + 2(k + \bar{k})k\bar{k} + 2(k + \bar{k}).$$

Using the Schwarz inequality, it follows that

$$\|1 + k\|_4^4 \le 1 + C\|k\|_2^2 + C'\|k\|_4^4.$$

By Lemma 8.1.1, $C'\|k\|_4 \le C''\|h\|_2$ if $t \ge$ some t_0, and evidently $\|k\|_2 \le e^{-t}\|h\|_2$. For arbitrary $t_1 > 0$ and $t > t_0 + t_1$,

$$\|\exp(-tN)h\|_4 = \|\exp(-(t - t_0 - t_1)N)\exp(-t_0N)\exp(-t_1N)h\|_4.$$

Now $\exp(-(t - t_0 - t_1)N)$ is a contraction on L_4. Hence if t_1 is chosen so that $\exp(-4t_1)C'' < 1$ and $\exp(-2t_1)C < 2$, then

$$\|1 + k\|_4^4 \le 1 + 2\|h\|_2^2 + \|h\|_2^4 = (1 + \|h\|_2^2)^2 = \|f\|_2^4,$$

completing the proof. □

LEMMA 8.1.4. *Let M_j ($j = 1, 2, \ldots, n$) be separable measure spaces, and let T_j be an integral operator on $L_1(M_j)$ that is a contraction from $L_p(M_j)$ to $L_q(M_j)$ (p and q being fixed, finite, and independent of j). Suppose also that the kernels $K_j(x, y)$ of the T_j are nonnegative. Then the algebraic tensor product $T_1 \times T_2 \times \cdots \times T_n$ is a contraction from $L_p(M_1 \times M_2 \times \cdots \times M_n)$ to $L_q(M_1 \times M_2 \times \cdots \times M_n)$.*

PROOF. It suffices by associativity to treat the case $n = 2$. To this end, let $\mathbf{B}$ be an arbitrary separable Banach space, M an arbitrary measure space, and let $L_p(M, \mathbf{B})$ denote the space of all strongly measurable $\mathbf{B}$-valued functions F on M for which the norm $\|F\|_p = (\int \|F(x)\|^p dx)^{1/p}$ is finite. Suppose T is an integral operator that is a contraction from $L_p(M)$ to $L_q(M)$ whose kernel $K(x, y)$ is nonnegative. Then the operator T' from $L_p(M, \mathbf{B})$ to $L_q(M, \mathbf{B})$ defined by $F \mapsto G$, where $G(x) = \int K(x, y)F(y)dy$ exists and is a contraction. For the mapping $y \to K(x, y)F(y)$ is easily seen to be strongly measurable from M to $\mathbf{B}$, for each x; and $\|G(x)\|_{\mathbf{B}} \le \int K(x, y)\, \|F(y)\|_{\mathbf{B}}dy$, which imply that $\|\, \|G(\cdot)\|\, \|_q \le \|\, \|F(\cdot)\|\, \|_p$. This shows the absolute integrability of the integral defining $G(x)$ almost everywhere, and yields the estimate $\|T'\| \le 1$.

Similarly, if $\mathbf{B}'$ is another separable Banach space, and if T is a contraction from $\mathbf{B}$ to $\mathbf{B}'$, then the operator T'' from $L_q(M, \mathbf{B})$ to $L_q(M, \mathbf{B}')$, defined by the equation $(T''F)(x) = TF(x)$, for $F \in L_q(M, \mathbf{B})$, is easily seen to be a contraction. Taking $\mathbf{B}$ as $L_p(M_2)$ and $\mathbf{B}'$ as $L_q(M_2)$, and making the natural identifications of $L_p(M_1, \mathbf{B})$ with $L_p(M_1 \times M_2)$ and of $L_q(M_1, \mathbf{B}')$ with $L_q(M_1 \times M_2)$ that are justified by the Fubini theorem, it follows that the contraction $T_2''T_1'$ extends the algebraic tensor product $T_1 \times T_2$, which implies that the latter is also a contraction. □

LEMMA 8.1.5. *For all* $p \in (2, \infty)$, $t > 0$ *and any real selfadjoint operator* B *such that* $B \geq 0$, $\exp(-t\partial\Gamma(B))$ *is a contraction from* $L_p(\mathbf{H}_\times)$ *to* $L_p(\mathbf{H}_\times)$.

PROOF. If $\mathbf{H}$ is n-dimensional ($n < \infty$) and B is taken in diagonal form, the kernel for $\exp(-t\partial\Gamma(B))$ is a product of kernels for the one-dimensional case, and hence positive. The same argument as in the one-dimensional case then applies by virtue of Lemma 8.1.4. Letting $n \to \infty$, the case of an arbitrary B of pure point spectrum follows. To treat the general B, let $\{f_n\}$ be a sequence of countably-valued Borel functions on $(0,\infty)$ such that $0 \leq f_n(x) \leq x$ and $f_n(x) \to x$ as $n \to \infty$. Set $B_n = f_n(B)$, and let $H_n = \partial\Gamma(B_n)$ and $H = \partial\Gamma(B)$. Then $\exp(itB_n) \to \exp(itB)$ (in the strong operator topology) implying that $\Gamma(\exp(itB_n)) \to \Gamma(\exp(itB))$, i.e., $\exp(itH_n) \to \exp(itH)$. But this implies that $H_n \to H$ in the strong topology for unbounded selfadjoint operators, by virtue of the uniform semiboundedness of the H_n and H. This implies in turn that $\exp(-tH_n) \to \exp(-tH)$ as $n \to \infty$. The proof is completed by applying Fatou's lemma to the inequality $\|\exp(-tH_n)f\|_p \leq \|f\|_p$.

PROOF OF THEOREM. Since $A - I \geq 0$ and

$$\exp(-t\partial\Gamma(A)) = \exp(-t\partial\Gamma(I))\exp(-t\partial\Gamma(A - I)),$$

it follows from Lemma 8.1.5 that it suffices to establish the conclusion of the theorem for the case $A = I$. The operator $\exp(-(t + is)N)$, where $N = \partial\Gamma(I)$, is holomorphic as a function of $t + is$ for $t > 0$, continuous as a function of s when $t = 0$, and unitary on $L_2(\mathbf{H}_\times)$; for $t = t_0$ it is a contraction from $L_2(\mathbf{H}_\times)$ to $L_4(\mathbf{H}_\times)$. According to the operator-interpolation theorem of Stein (1956), $\exp(-(t + is)N)$ is then a contraction from $L_2(\mathbf{H}_\times)$ to $L_{p(t)}(\mathbf{H}_\times)$, where $p(t)^{-1} = \frac{1}{2}(1 - t/t_0) + \frac{1}{4}(t/t_0)$. By an elementary estimate, $p(t) > 2e^{\varepsilon t}$ for some constant $\varepsilon > 0$. The convexity theorem of M. Riesz then implies that for arbitrary $q > 2$, e^{-tN} is a contraction from $L_q(\mathbf{H}_\times)$ to $L_{q\exp(\varepsilon t)}(\mathbf{H}_\times)$. It follows by induction that $\exp(-nt_0N)$ is a contraction from $L_2(\mathbf{H}_\times)$ to $L_{2\exp(n\varepsilon t_0)}(\mathbf{H}_\times)$, from which it follows that $\exp(-tN)$ is a contraction from $L_2(\mathbf{H}_\times)$ to $L_{2\exp(\varepsilon t)}(\mathbf{H}_\times)$ for all t. Applying the theorem of Riesz once again, it follows that e^{-tN} is a contraction from $L_q(\mathbf{H}_\times)$ to $L_{q\exp(\varepsilon t)}(\mathbf{H}_\times)$ for all $q \geq 2$. □

Theorem 8.1 implies the following estimate for quasi-polynomials in an infinite number of Gaussian random variables.

COROLLARY 8.1.1. *There exists a universal constant* $\varepsilon > 0$ *such that if* $V \in L_2(\mathbf{H}_\times)$ *and if* V *is in the closed linear span of the polynomials of degree* $\leq d$ *on* $\mathbf{H}_\times$, *then* $\|V\|_p \leq (p/2)^{d/\varepsilon}\|V\|_2$. *Moreover,* $\exp(|V|^a) \in L_1(\mathbf{H}_\times)$ *for some* $a > 0$.

Proof. Let $\mathbf{M}$ denote the closed linear span of the polynomials of degree $\leq d$. Then $\mathbf{M}$ is invariant under the e^{itN}, and N is bounded by d on $\mathbf{M}$. Thus if $p = 2\exp(\varepsilon t)$,

$$\|V\|_p = \|e^{-tN}(e^{tN}V)\|_p \leq \|e^{tN}V\|_2 \leq (p/2)^{d/\varepsilon}\,\|V\|_2.$$

Now

$$\|\exp(|V|^a)\|_1 \leq \sum_n (n!)^{-1}\|\,|V|^{an}\|_1 \leq \sum_n (n!)^{-1}(\|V\|_{an})^{an}.$$

There are only a finite number of n for which $an < 2$, and the contribution from these terms to the foregoing sum is clearly finite. For the remaining values of n, $(\|V\|_{an})^{an} \leq (\tfrac{1}{2}an)^{adn/\varepsilon}\|V\|_2^{an}$ and choosing a sufficiently small, the series $\Sigma_n (n!)^{-1}(\tfrac{1}{2}an)^{adn/\varepsilon}\|V\|_2^{an} < \infty$. □

8.3. Renormalized products at fixed times

The interaction Hamiltonian for a nonlinear quantized field has formally the expression $V = \int p(\varphi(t, x))dx$, where p is a given polynomial, and $\varphi(t, x)$ denotes the putative interacting field. Even the free field at a fixed time is a distribution or other generalized function, as a function of the space variable x, and there is no reason to expect the interacting field to be any more regular. And in the case of the free field, explicit analysis shows that if p is a monomial of even degree, $p(s) = s^{2r}$ for some integer r, then V is identically infinite, if interpreted as a random variable relative to the Gaussian distribution associated with the free field vacuum. There is thus no question that some species of renormalization is required to give clear meaning to the foregoing expression. A natural—local, invariant, etc.—such renormalization was treated in Chapter 7, which in the physical context should be defined relative to the *physical* vacuum, which is usually described as the putative lowest eigenvector of the *total* Hamiltonian. But at this point an essential complication intervenes: the physical vacuum is not known at this stage, or even known to exist. It is therefore much simpler to treat renormalization with respect to the *free* vacuum, which of course is known; and as a step toward the establishment of the physical vacuum, this is what is done, initially.

Some properties of renormalized products relative to the free vacuum have already been developed in Chapter 7. In particular, when averaged relative to a smooth function vanishing outside a compact subset of space-time, the result is a densely defined operator on the underlying free field state vector space $\mathbf{K}$. On the other hand, for the treatment of a nonlinear evolutionary differential equation, the integration over time must be eliminated. Such an equation has the form $du/dt = F(u, t)$, where F is a given function of the unknown function

u, whose dependence on t is under consideration. When F is linear, this equation is readily reformulated in terms of temporal averages. Thus, if $F(u, t)$ takes the form $F(t)u$, where $F(t)$ is a given linear transformation, and if $U(g) = \int u(t)g(t)dt$, where g is arbitrary in $C_0^\infty(\mathbf{R})$, then the differential equation may be given an equivalent formulation in terms of $U(\cdot)$ as the equation $U(g') = -U(Fg)$ for arbitrary such g. But if F is truly nonlinear, this device is ineffective.

Thus, without some wholesale reformulation of the underlying differential equations of quantum field theory, it appears essential that the interaction Hamiltonian be analyzed *at a fixed time*, without smoothing with respect to time. Since the latter type of smoothing has a much stronger regularizing effect than smoothing over space, the fixed-time interaction Hamiltonian must be expected to be considerably more singular than the densely defined operators obtained from smooth averages over space-time. Indeed, when the number of space-time dimensions of a relativistic free field is greater than two, the Wick powers of the field do not become densely defined operators on averaging only over space with respect to a function in C_0^∞. On the other hand, the analysis of local nonlinear functions of a relativistic quantum field is greatly facilitated by relativistic causality (in one of its interpretations), according to which the field operators at different *relatively spacelike* points (e.g., points at a fixed time) commute with each other.

The next sections are devoted to the establishment of the fixed-time properties of fields that are relevant to the construction of interacting fields. In these sections we are concerned with a quantum field in space, and time will not enter into the considerations explicitly. It will be advantageous for logical clarity and generality to treat successively more restricted classes of spaces, arriving eventually at the cases of $\mathbf{R}$ and S^1 for the presentation of the basic theory of quantized nonlinear equations.

Renormalized products of theoretically observable quantum field operators at a fixed time can be effectively characterized by a nonlinear variant of the Weyl relations involving only bounded operators. This serves to suppress irrelevant pathology that derives from formal manipulations with unbounded operators. It will suffice here to treat the case of powers, or, more generally, polynomials, in a scalar field. Initially, we make no assumption concerning the space on which this field is defined, other than that it is a measure space (S, s). Here S denotes the underlying set, and s the given nonnegative countably-additive measure on a given σ-ring of subsets of S. We denote $L_2(S, s)$ as $\mathbf{H}$, and assume given a dual couple $(\mathbf{M}, \mathbf{N}, \langle\cdot,\cdot\rangle)$ consisting of linear subspaces $\mathbf{M}$ and $\mathbf{N}$ of real functions in $\mathbf{L}_2(S, s)$, with the pairing $\langle\cdot,\cdot\rangle$ defined as the usual inner product in $\mathbf{L}_2$. We assume given a Weyl pair $(U, V, \mathbf{K})$ over this dual couple, and make the special assumption appropriate to the consideration of local products of fields that if $f \in \mathbf{M}$ and $g \in \mathbf{N}$, then $gf \in \mathbf{M}$. We call such a dual couple *multiplicative*, and are concerned with a Weyl pair over a multiplicative dual couple that is grounded by the designation of a unit vector v in $\mathbf{K}$.

A *renormalized power system* for the grounded pair $(U, V, \mathbf{K}, v)$ over a multiplicative dual couple is defined to consist of continuous unitary representations U_n of the additive group of $\mathbf{M}$ on $\mathbf{K}$ $(n = 0, 1, 2, \ldots)$ having the following properties: denoting $\exp(i\int h(x)dx)$ as $U_0(h)$ for $h \in L_1(S, s)$, and $\exp(i\varphi(f))$ as $U_1(f)$, then for arbitrary $f \in \mathbf{M}$ and $g \in \mathbf{N}$,

(1) $V(g)^{-1}U_n(f)V(g) = U_n(f)U_{n-1}(\binom{n}{1}fg)U_{n-2}(\binom{n}{2}fg^2)\cdots U_0(fg^n)$; and

(2) $\langle U_n(f)v, v\rangle = 1$ if $n \geq 1$.

The selfadjoint generator of the one-parameter unitary group $U_n(tf)$, t real, will be denoted as $\phi^n(f)$ and symbolically as $\int :\varphi(x)^n: f(x)dx$.

LEXICON. At first glance, the significance of the *nonlinear Weyl relations* 1) may not be apparent. To explain how they arise naturally from the problem of the appropriate definition of nonlinear local functions of a quantum field, we interpolate a heuristic explanation.

Let φ and π denote point functions representing canonically conjugate fields over S. That is, $\varphi(x)$ and $\pi(x)$ are hermitian fields, and they satisfy the commutation relations

$$[\varphi(x), \pi(y)] = -i\,\delta(x-y); \quad [\varphi(x), \varphi(y)] = 0 = [\pi(x), \pi(y)] \quad (x, y \in S).$$

Consider the problem of defining $\varphi(x)^2$. Whatever this may be, one would at least expect that it commutes with all $\varphi(y)$, since formally

$$[\varphi(y), \varphi(x)^2] = [\varphi(y), \varphi(x)]\varphi(x) + \varphi(x)[\varphi(y), \varphi(x)] = 0.$$

Similarly, one would expect that

$$[\pi(y), \varphi(x)^2] = [\pi(y), \varphi(x)]\varphi(x) + \varphi(x)[\pi(y), \varphi(x)] = 2i\,\delta(x-y)\varphi(x).$$

In terms of the corresponding distributions ϕ, Π and ϕ^2, where

$$\phi^2(f) = \int_S \varphi(x)^2 f(x)dx,$$

these equations mean that for arbitrary functions f and h,

$$[\phi(f), \phi^2(h)] = 0, \qquad [\Pi(f), \phi^2(h)] = 2i\,\phi(fh). \qquad (*)$$

But as equations for $\phi^2(h)$, equations * are fairly regular; they state that the a priori undefined object $X = \phi^2(h)$ satisfies the mathematically meaningful equations $[\phi(f), X] = 0$, $[\Pi(f), X] = 2i\phi(fh)$, in which all terms other than X are well-defined operators. Thus, either such operators X exist, in which case we say $\varphi(x)^2$ exists and defines the distribution $\int \varphi(x)^2 f(x)dx = X$; or they do not exist, in which case we may conclude that the intuitive concept of the operator-valued distribution representing the square of the field was an illusion. Moreover, when X exists, it is unique within an additive constant by virtue of the irreducibility of the $\varphi(x)$ and $\pi(y)$ in their totality, in the cases of the free and other fields normally contemplated. This additive constant may be fixed in a natural invariant way by requiring that its vacuum expectation value vanish.

Suppose we have resolved the question of the existence of $\varphi(x)^2$ positively; we may then proceed to treat $\varphi(x)^3$ in a similar way. That is, formally it is naturally constrained by the relations

$$[\varphi(y), \varphi^3(x)] = 0, \qquad [\pi(y), \varphi^3(x)] = 3i\,\delta(x - y)\varphi(x)^2,$$

providing a mathematically meaningful equation as earlier. Next, in case $\varphi(x)^3$ exists, we may proceed to the treatment of $\varphi(x)^4$; and so on by recursion for arbitrary powers.

Rigorous mathematical implementation of the foregoing idea would run into ambiguities connected with the unboundedness of the operators involved, and technical issues concerning domains, for which mere density in an underlying Hilbert space, together with invariance under the operators in question, would be quite insufficient, even though the appropriateness of such a technical assumption is arguable. Some exponentiation to unitary operators, which are then free of technical domain requirements, is indicated, and there is no problem exponentiating the $\Pi(g)$ to obtain the $V(g) = \exp(i\Pi(g))$. By formal power series manipulations it follows from the relations $[\pi(y), \varphi(x)^n] = in\,\delta(x - y)\varphi(x)^{n-1}$ that

$$V(g)^{-1}\phi^n(f)V(g) = \phi^n(f) + \cdots + \binom{n}{m}\phi^{n-m}(fg^m) + \cdots + \phi^0(fg^n).$$

This relation has been developed in a purely formal way; if, however, the $\phi^r(h)$ are all selfadjoint and strongly commutative, then there is a natural way to rigorize the relation, by forming the closure of the right-hand side, which necessarily exists and is selfadjoint (by spectral theory). Finally, exponentiation of this equation produces the nonlinear Weyl relations. Conversely, these relations rigorously imply the indicated partially infinitesimal relations.

Before resuming the rigorous development, beginning with some general properties of solutions of the nonlinear Weyl relations, including the property just described, one last point about the physical justification for the normalization via vanishing vacuum expectation values should be noted. Mathematically, it would be possible to give arbitrary values to these vacuum expectation values. However, in the putative eventual application to quantized nonlinear wave equations, the vacuum expectation values are time independent, since the vacuum state vector is eigenvector of the Hamiltonian. This means the vacuum expectation value would be the same in the infinite time limit, but the physical expectation is that in this limit the quantized field is asymptotic to a free field, and the nonlinear interaction term in the wave equation should tend to 0, and so have vacuum expectation value tending to 0.

Recapitulating in rigorous terms,

SCHOLIUM 8.1. *The nonlinear Weyl relations imply the equations*

$$V(g)^{-1}\phi^n(f)V(g) = \phi^n(f) \mathbf{+} \cdots \mathbf{+} \binom{n}{m}\phi^{n-m}(fg^m) \mathbf{+} \cdots \mathbf{+} \phi^0(fg^n), \qquad (*)$$

where boldface "$\mathbf{+}$" indicates that the closure of the indicated sum is formed. Conversely, suppose given a Weyl pair ($\mathbf{K}$, $U(\cdot)$, $V(\cdot)$) *over the multiplicative dual couple* ($\mathbf{M}$, $\mathbf{N}$) *and mappings* ϕ^n, $n = 1, 2, \ldots$, *from* $\mathbf{M}$ *to the selfadjoint operators on* $\mathbf{K}$, *that are linear and continuous (with respect to the strong addition, etc., of operators),*

$$\phi^n(f_1 + f_2) = \phi^n(f_1) \mathbf{+} \phi^n(f_2),$$

$$\phi^n(\alpha f_1) = \alpha\,\phi^n(f_1), \qquad \alpha \in \mathbf{R}; \qquad f_1, f_2 \in \mathbf{M}.$$

Suppose also that the $\phi^n(f)$ ($f \in \mathbf{M}$) *are affiliated with the* W^**-algebra* $\mathbf{R}$ *generated by the* $U(h)$, $h \in \mathbf{M}$. *Then the nonlinear Weyl relations hold with* $U_n(f) = \exp[i\phi^n(f)]$ *if* (*) *holds for* $f \in \mathbf{M}$ *and* $g \in \mathbf{N}$.

The formal indications of unicity for the renormalized powers may be rigorously confirmed along the following lines: we define a grounded Weyl pair to be *simple* in case the W^*-algebras $\mathbf{R}$ and $\mathbf{S}$ generated by the totality of the $U(f)$ and the totality of the $V(g)$, respectively, are maximal abelian in the algebra of all bounded linear operators on $\mathbf{K}$; and if, in addition, $\mathbf{R}$ together with $\mathbf{S}$ forms an irreducible set of operators on $\mathbf{K}$. We recall that in a separable Hilbert space an abelian algebra is maximal abelian if and only if it has a cyclic vector, in which case it is sometimes said to have a "simple spectrum," since any selfadjoint operator that generates the algebra will then have (generalized) eigenvalues of simple (unit) multiplicity.

SCHOLIUM 8.2. *If a renormalized power system exists for a simple grounded Weyl pair over a multiplicative dual couple, then it is unique.*

PROOF. As the start of an induction argument, observe that U_0 is unique by definition, and suppose that two renormalized power systems, denoted U_m and U'_m coincide for $m < n$. It follows that for all f and g,

$$V(g)^{-1}U_n(f)V(g)U_n(f)^{-1} = V(g)^{-1}U'_n(f)V(g)U'_n(f).$$

From this it follows that

$$U_n'(f)^{-1}U_n(f)V(g) = V(g)U_n'(f)^{-1}U_n(f).$$

Let **S** denote the maximal abelian algebra generated by the $V(g)$. The last equation shows that $U_n'(f)^{-1}U_n(f)$ is in **S**. However, it is also in the W^*-algebra **R** generated by the $e^{i\phi(h)}$ as h varies, since it commutes with every element of **R**, and **R** also is maximal abelian. By the assumed irreducibility, the common part of **R** and **S** consists only of scalars, so that $U_n'(f) = c(f)U_n(f)$ for some scalar $c(f)$. Hence $\langle U_n'(f)v, v\rangle = c(f)\langle U_n(f)v, v\rangle = 1$, showing that $c(f) = 1$. □

Renormalized power systems could now be treated over arbitrary complete Riemannian manifolds, but in practice symmetries lacking in the general case are important, and it will suffice here to treat the case in which the underlying space S is a group G. The logical context is clarified by permitting G to be an arbitrary locally compact abelian group, with the further assumption that the energy operator is group invariant. Denoting this operator as B, this means that $BL(a) = L(a)B$, where $L(a)$ denotes the operator $f(a) \to f(x - a)$, $f \in L_2(G)$, and the measure in G is taken to be the essentially unique invariant (Haar) measure. Any such operator has the form $B = \mathbf{F}^{-1}M_b\mathbf{F}$, where $\mathbf{F}$ denotes the Fourier transform operation, and M_b denotes the operation of multiplication by the fixed measurable function b on the dual (character) group G^* of G. We call b the *spectral function* of B, and denote it on occasion as $B(\cdot)$. In physical terms, G^* is the momentum space and $b(g^*)$ for $g^* \in G^*$ is the energy corresponding to the momentum g^*.

In this context, a basic existence theorem for renormalized powers at a fixed time is as follows: here and elsewhere, if I is a subinterval of $[1, \infty]$, $L_I(M)$ denotes the intersection of the $L_p(M)$, $p \in$ I.

THEOREM 8.2. *Let B be a given real G-invariant positive selfadjoint operator in $L_2(G)$, where G is a given separable locally compact abelian group, such that $B(\cdot)^{-1} \in L_{(1,\infty]}(G^*)$, where $B(\cdot)$ denotes the spectral function of B. Let* **M** *denote all real functions in $L_{[2,\infty)}(G)$, and let* **N** *denote all real functions in the domain of $C = B^{1/2}$. Let $(\mathbf{K}, W, \Gamma, v)$ denote the free boson field over* **H** *$= L_2(G)$, and set $U(f) = W(C^{-1}f), f \in \mathbf{M}$, and $V(g) = W(iCg), g \in \mathbf{N}$. Then there exists a unique renormalized power system for this Weyl pair with ground state vector v.*

PROOF. Observe first that $(\mathbf{M}, \mathbf{N}, \langle\cdot,\cdot\rangle)$ is a multiplicative dual couple. For if $g \in \mathbf{N}$, then $\hat{g}C(\cdot) \in L_2(G^*)$, whence $\hat{g} \in L_q(G^*)$ for all $q \in (1, 2]$. Accordingly, $g \in L_r(G)$ for all $r \in [2, \infty)$. Therefore if $f \in \mathbf{M}$ and $g \in \mathbf{N}$, then $fg \in \mathbf{M}$. Moreover, the $C^{-1}f$ with f in **M** and the Cg with g in **N** are dense in $\mathbf{H}_\kappa = L_2(G, \mathbf{R})$, so that the W^*-algebra **R** generated by the $U(f)$ with $f \in \mathbf{M}$ is the same as that generated by the $W(f')$ with $f' \in \mathbf{H}_\kappa$, and hence maximal abelian; and similarly for the W^*-algebra $\mathbf{S}$ generated by the $V(g)$ with $g \in \mathbf{N}$. Moreover, irreducibility for (U, V) follows from the corresponding property of the free field over $\mathbf{H}$, so that the present Weyl system is simple.

To show existence of a renormalized power system, $\phi^n(f)$ will be constructed as a limit in $L_2(\mathbf{R}, v)$ of approximations to $\phi^n(f)$ obtained by smoothing the underlying quantized field $\varphi(x)$. Let g be an arbitrary even element of $\mathbf{M}$ for which $\int g(x)dx = 1$ and $|\hat{g}(x)| \leq 1$; let $\mathbf{M}_0$ denote the class of such g. To simplify the notation we write $\Gamma(a)$ for $\Gamma(L(a))$, and note that $\Gamma(a)^{-1}\phi(g)\Gamma(a) = \phi(g_a)$, where $g_a(x) = g(x - a)$, writing the group G additively, and similarly for the $\Pi(h)$, by virtue of the G-invariance of B. Moreover, conjugation by $\Gamma(a)$ leaves $\mathbf{R}$ and $\mathbf{S}$ invariant (as a set), and also leaves invariant the expectation functional E, $E(S) = \langle Sv, v\rangle$, having state vector v, defined on all bounded linear operators S on $\mathbf{K}$. Accordingly, if S is a given operator in $L_2(\mathbf{R}, v)$, the mapping $a \to \Gamma(a)^{-1}S\Gamma(a)$ is continuous from G into $L_2(\mathbf{R}, v)$. It is clear that v is in the domain of all polynomials in the $\phi(g)$, and we denote by $:\cdots:$ the operation of renormalization with respect to E.

Now for arbitrary $g \in \mathbf{M}$ and integer $r > 0$, $:\phi(g)^r:$ is a polynomial in $\phi(g)$ of degree r with real coefficients, whose term of highest degree is $\phi(g)^r$. It follows by a simple estimate based on this observation, together with spectral theory, that $:\phi(g)^r:$ is selfadjoint and in $L_2(\mathbf{R}, v)$. The same is true of $\Gamma(a)^{-1}:\phi(g)^r:\Gamma(a)$, and it follows that the map $a \to :\phi(g_a)^r:$ is continuous and bounded from G into $L_2(\mathbf{R}, v)$.

It follows that $\int :\phi(g_y)^r:f(y)dy$ exists as a Banach-space valued integral if $f \in L_1(G)\cap\mathbf{M}$, where the Banach space in question is $L_2(\mathbf{R}, v)$. We denote the value of this integral as $u(g)$. It will next be shown that if $\{g_n\}$ is a sequence in $\mathbf{M}_0$ such that $\hat{g}_n(y) \to 1$ as $n \to \infty$ (pointwise on G^*), then the sequence $\{u(g_n)\}$ is convergent in $L_2(\mathbf{R}, v)$. To this end, we compute the inner product $\langle u(g), u(h)\rangle$, in the space $L_2(\mathbf{R}, v)$, which will henceforth be denoted as $\mathbf{K}'$. (Note that $\mathbf{K}'$ is unitarily equivalent to $\mathbf{K}$ via the mapping $T \to Tv$, since v is a cyclic vector for $\mathbf{R}$.) Applying Wick's theorem, $E(:z^r: :z'^r:) = r!E(zz')^r$ for arbitrary z and z' of the form $g, h \in \mathbf{M}$, whence

$$\begin{aligned}\langle u(g), u(h)\rangle &= \int\int \langle :\phi(g_a)^r:, :\phi(h_b)^r:\rangle f(a)\overline{f(b)}dadb \\ &= r! \int\int \langle C^{-1}g_a, C^{-1}h_b\rangle^r f(a)\overline{f(b)}dadb.\end{aligned}$$

By Fourier analysis, $\langle C^{-1}g_a, C^{-1}h_b\rangle$ may also be expressed as $K(a - b)$, K being the inverse transform of $k = B(\cdot)^{-1}\hat{g}\hat{h}$, where $\hat{g}$ and $\hat{h}$ are the Fourier transforms of g and h. Since $B(\cdot)^{-1} \in L_{(1,\infty]}(G^*)$, k is in $L_{(1,\infty]}(G^*)$, implying that K is in $L_{[2,\infty)}(G)$. Hence K^r is in $L_{[2,\infty)}(G)$. The convolution $\int K^r(a - b)f(b)db$ is again in $L_{[2,\infty)}(G)$, since $f \in L_1(G)$. Now $\langle u(g), u(h)\rangle$ is the inner product of the foregoing convolution with f, which by Parseval's theorem for the L_p-Fourier transform is the same as the inner product of the respective Fourier transforms. Accordingly,

$$\langle u(g), u(h)\rangle = r! \int_{G^*} k^{(*r)}(y)|\hat{f}(y)|^2 dy,$$

where $\hat{f}$ is the Fourier transform of f and $k^{(*r)}$ is the r-fold convolution of k with itself. Applying this result to the evaluation of $\|u(g) - u(h)\|$, this may be expressed as

$$\int_{G^*}[(\hat{g}^2B(\cdot)^{-1})^{(*r)} + (\hat{h}^2B(\cdot)^{-1})^{(*r)} - 2(\hat{h}\hat{g}B(\cdot)^{-1})^{(*r)}](y)|\hat{f}(y)|^2 dy$$

(Note that since g and h are taken invariant under the map $x \to -x$, their Fourier transforms are real.) Setting $g = g_m$ and $h = g_n$, the Fourier transforms of g_n and g_m are uniformly bounded by 1 and converge pointwise to 1. The expression

$$[(\hat{g}_n^2B(\cdot)^{-1})^{(*r)} + (\hat{g}_mB(\cdot)^{-1})^{(*r)} - 2(\hat{g}_n\hat{g}_mB(\cdot)^{-1})^{(*r)}]$$

is dominated by $4(B(\cdot)^{-1})^{(*r)}$ and converges pointwise to 0 by the Hausdorff-Young inequality. As $n, m \to \infty$, the entire integrand over G^* converges to 0 and is dominated by the function

$$4[B(\cdot)^{-1}]^{(*r)}(y)\ |\hat{f}(y)|^2,$$

which is integrable since $[B(\cdot)^{-1}]^{(*r)}$ is in $L_p(G^*)$ for all $p > 1$ and $\hat{f}(y) \in L_2(G^*)$. Hence $\{u(g_n)\}$ has a limit in $\mathbf{K}'$, which we denote as $\phi^r(f)$. By continuity, $\phi^r(f)$ may be extended from $L_1(G)\cap\mathbf{M}$ to all of $\mathbf{M}$, and the extension will also be denoted as $\phi^r(f)$.

It is clear from this construction that $e^{-i\phi^r(f)}$ is in $\mathbf{R}$. To complete the proof of the theorem it remains only to show the nonlinear Weyl relations. To this end, note that

$$V(h)^{-1}\phi(f)^rV(h) = (\phi(f) + \langle f, h\rangle)^r \qquad (r = 1, 2, \ldots).$$

In order to establish the nonlinear Weyl relation, it suffices to establish the following half-infinitesimal form:

$$V(h)^{-1}\phi^n(f)V(h) = \phi^n(f) + n\phi^{n-1}(fh) + \cdots + b_{n,m}\phi^{n-m}(fh^m) + \cdots$$

provided the $\phi^k(g)$ are all known to be affiliated with the same abelian ring $\mathbf{R}$, as is the case for the $\phi^r(f)$. To establish the half-infinitesimal form of the Weyl relations, observe first that the corresponding relations hold when $\varphi(x)$ is replaced by $\phi(g_x)$, where g is as earlier: $V(h)^{-1}\phi(k)^nV(h) = [\phi(k) + \langle k, h\rangle]^n$ for $n = 1, 2, \ldots$ (as a consequence of the case $n = 1$). Observe next the equation $V(h)^{-1}{:}\phi(k)^n{:}V(h) = {:}\phi(k)^n{:} + n{:}\phi(k)^{n-1}{:}\langle k, h\rangle + \cdots + \binom{n}{m}{:}\phi(k)^m{:}\langle k, h\rangle^m + \cdots$. This follows by induction on n. More specifically, substitution of th for h followed by differentiation with respect to t reduces it to the same equation with n replaced by $n - 1$. In this reduction use is made of the equation for real h and k

$$[:\phi(k)^n:, \phi(ih)] = in\langle k, h\rangle :\phi(k)^{n-1}:,$$

representing a special case of the defining relation for Wick products.

On replacing k by g_x and integrating over G after multiplication by $f(x)$, this implies a relation of the form

$$V(h)^{-1}\phi_n(f, g)V(h) = \phi_n(f, g) + n\phi_{n-1}(f, g, h) + \cdots + \binom{n}{m}\phi_{n-m}(f, g, h) + \cdots + \int \langle g_x, h\rangle^n f(x)dx,$$

where

$$\phi_n(f, g) = \int :\phi(g_x)^n: f(x)dx, \qquad \phi_r(f, g, h) = \int :\phi(g_x)^r:\langle g_x, h\rangle^r f(x)dx$$

are operators in $L_2(\mathbf{R}, \nu)$ that converge in this space to corresponding operators in the putative half-infinitesimal form of the nonlinear Weyl relations. To complete the proof, it therefore suffices to show that if the $S_{n,j}$ are selfadjoint operators in $\mathbf{K}'$ such that $S_{n,j} \to S_n$ as $j \to \infty$, and such that for some unitary operator V for which $V^{-1}RV \in \mathbf{R}$ for all $R \in \mathbf{R}$,

$$V^{-1}S_{n,j}V = S_{n,j} + S_{n-1,j} + \cdots + S_{n-m,j} + \cdots,$$

then the same equation holds with each $S_{n,j}$ replaced by its limit S_n. This follows from the use of spectral theory to represent all the $S_{n,j}$ by multiplication operators on a probability measure space. It follows from this representation that if a sequence of selfadjoint operators converges in $\mathbf{K}'$, then it converges in the strong operator topology. For such convergence of a sequence $\{H_n\}$ of selfadjoint operators to another selfadjoint operator H is equivalent to the convergence of $\exp(itH_n)$ to $\exp(itH)$ in the strong operator topology for bounded operators, for all real t. Lebesgue integration theory shows that if $\{h_n\}$ is a sequence of real measurable functions on a probability measure space P that converges in $L_2(P)$ to h, then the operation M_{h_n} of multiplication by h_n, acting on $L_2(P)$, converges in the strong operator topology to M_h. The unitary invariance of this type of convergence for unbounded selfadjoint operators then implies that $V^{-1}S_{n,j}V$ converges to $V^{-1}S_nV$ in this topology.

Problems

1. a) Show that the $\mathbf{M}$ of Theorem 8.2 is a topological algebra in the topology of convergence in each $L_p(G)$, $p \in [2, \infty)$.

   b) Show that $\phi^n(f)$ is continuous from $\mathbf{M}$ in this topology to $L_{[2,\infty)}(\mathbf{R}, \nu)$.

2. Show that Theorem 8.2 remains valid with the following choices for $\mathbf{M}$

and **N**: **M** consists of all real functions in $L_2(G)$; **N** consists of all real such functions whose Fourier transforms have compact supports.

8.4. Properties of fixed-time renormalization

This section develops properties of the fixed-time polynomials in a quantum field. It treats function-space aspects, the applicability of renormalization to interacting fields rather than the free field, and locality features. All of these are involved in the solution of nonlinear quantized wave equations. For an arbitrary polynomial of the form $p(x) = a_0x^n + a_1x^{n-1} + \cdots a_{n-1}x$, we define $\phi_p(f)$ as closure of $a_0\phi^n(f) + a_1\phi^{n-1}(f) + \cdots a_{n-1}\phi^1(f)$ when this exists. With the hypotheses of Theorem 8.2, this is the case for all f in **M**, by spectral theory.

COROLLARY 8.2.1. *With the hypotheses of Theorem 8.2, $\phi_p(f)$ is in $L_q(\mathbf{R}, v)$ for all $2 \leq q < \infty$, p being an arbitrary given polynomial.*

PROOF. This could be shown by direct computation of the vacuum expectation value of $\phi_p(f)^q$, for even integers q, using Wick's theorem. However, it is easy to deduce the result from Theorem 8.2, which also yields a simple estimate for $\|\phi_p(f)\|_q$. Let N denote the number (of particles) operator in the free boson field over $\mathbf{H} = L_2(G)$. It is clear that $\phi_p(f)v$ is a limit in $\mathbf{K}$ of vectors in the at most n-particle subspace $\mathbf{K}^{(n)} = \oplus_{j\leq n}\mathbf{K}_j$, where n is the degree of p, whence $\phi_p(f)v$ is an element of this space. Denoting the spectral measure for N as $E(B)$, where B is an arbitrary Borel subset of $\mathbf{R}$, then $\mathbf{K}^{(n)}$ is in the range of $E([0, n])$. Now by Theorem 8.1,

$$\|e^{-tN}\phi_p(f)v\|_q \leq \|\phi_p(f)v\|_2,$$

showing that $\|\phi_p(f)v\|_q \leq e^{tn} \|\phi_p(f)v\|_2$ for sufficiently large t. □

The solution of a nonlinear quantized wave equation at a fixed time cannot be expected to be similar to the free field at this time, and in particular the physical vacuum will certainly be different from the free vacuum for a physically nontrivial theory. Accordingly, what is needed in order to give precise meaning to such an equation is a notion of renormalized powers that is applicable to a general type of vacuum. The next two theorems will show the existence of a unique local concept of powers extending that treated in Section 8.3, to the case of an arbitrary (nonfree) vacuum of appropriate regularity.

THEOREM 8.3. *Let B, C, G, **M**, and **N** be as in Theorem 8.2. Let $(\mathbf{K}, W, \Gamma, v)$ denote the free boson field over **H**, set $U(f) = W(C^{-1}f)$ for $f \in$ **M**, and*

$V(g) = W(iCg)$ for $g \in \mathbf{N}$. Then, if u is any unit vector in $\mathbf{K}_2$ and $u \in \mathbf{K}_p$ with $p > 2$, there exists a unique renormalized power system for this Weyl pair with ground state vector u.

Let $:\phi^n:$ denote the renormalized power system (r.p.s.) for the free vacuum, whose existence is asserted by Theorem 8.2. The putative r.p.s. with the ground state vector u will be denoted as $\psi^n(\cdot)$. By definition, $\psi^0(f) = \int_G f(x)dx = \phi^0(f)$ for $f \in L_1 \cap \mathbf{M}$. Thus the following induction hypothesis (H_r) holds (trivially) for $r = 1$:

For $s < r$, $\psi^s(\cdot)$ exist satisfying the r.p.s. conditions, and have expressions of the form

$$\psi^s(f) = \phi^s(f) - s\psi^{s-1}(fk_1) - \cdots - \binom{s}{j}\psi^{s-j}(fk_j) - \cdots - \psi^0(fk_s). \quad (H_r)$$

It will be shown that (H_r) implies (H_{r+1}), thereby establishing the theorem.

LEMMA 8.3.1. *There exists k_n in $\mathbf{M}$ such that*

$$\langle \phi^n(f)u, u\rangle = \langle f, k_n\rangle, \qquad \text{for } n > 0.$$

PROOF. From Hölder's inequality and the bound

$$\|\phi^n(f)\|_q \leq C_q \|\phi^n(f)\|_2$$

for $q \in [2, \infty)$, it follows that u is in the domain of $\phi^n(f)$, and that $|\langle \phi^n(f)u, u\rangle| \leq C \|\phi^n(f)\|_2$, where C is a constant (dependent on n). By the proof of Theorem 8.2,

$$\|\phi^n(f)\|_2 = (n!)^{1/2} [\int_{G^*} B^{-1(*n)}(y)|\hat{f}(y)|^2 dy]^{1/2}$$

for $n > 0$. Since $B^{-1(*n)}(\cdot)$ is in L_q for any $q \in (1, \infty]$, this implies, in conjunction with Fourier transform theory that for all positive n, $\|\phi^n(f)\|_2 \leq C_r \|f\|_r$ for any $r \in (1, 2]$. By the Riesz representation theorem there exists a $k_n \in L_2(G)$ such that

$$\langle \phi^n(f)u, u\rangle = \langle f, k_n\rangle.$$

Since $|\langle f, k_n\rangle| \leq C \|f\|_r$ for any $r \in (1, 2]$, $k_n \in \mathbf{M}$. □

PROOF OF THEOREM 8.3. Suppose that H_r holds. Define T as the closure of

$$\phi^r(f) - r\psi^{r-1}(fk_1) - \cdots - \binom{r}{j}\psi^{r-j}(fk_j) - \cdots - \psi^0(fk_r).$$

Since all the summands are selfadjoint and affiliated with $\mathbf{R}$, the indicated closure exists. It will next be shown that T satisfies the r.p.s. conditions for

$\psi^r(f)$. To this end, conjugate by $V(g)$ and use the induction hypothesis to obtain the equation

$$V(g)^{-1}TV(g) = \text{closure of}$$
$$\{[\phi^r(f) + \tbinom{r}{1}\phi^{r-1}(fg) + \tbinom{r}{2}\phi^{r-2}(fg^2) + \cdots + \tbinom{r}{r}\phi^0(fg^r)]$$
$$- \tbinom{r}{1}[\psi^{r-1}(fk_1) + (r-1)\psi^{r-2}(fk_1g)$$
$$+ \tbinom{r-1}{2}\psi^{r-3}(fk_1g^2) + \cdots + \psi^0(fk_1g^{r-1})]$$
$$- \tbinom{r}{2}[\psi^{r-2}(fk_2) + (r-2)\psi^{r-3}(fk_2g) + \tbinom{r-2}{2}\psi^{r-4}(fk_2g^2)$$
$$+ \cdots + \psi^0(fk_2g^{r-2})] - \cdots - \psi^0(fk_r)\}.$$

All the summands here are affiliated with $\mathbf{R}$, and the closure indicated exists and is selfadjoint as earlier. Summing by columns instead of by rows shows that $V(g)^{-1}TV(g)$ is the closure of

$$\{[\phi^r(f) - \Sigma_1^r\tbinom{r}{j}\psi^{r-1}(fk_j)]$$
$$+ [\tbinom{r}{1}\phi^{r-1}(fg) - \Sigma_1^{r-1}\tbinom{r}{2}\tbinom{r-j}{1}\psi^{r-1-j}(fk_jg)]$$
$$+ [\tbinom{r}{2}\phi^{r-2}(fg^2) - \Sigma_1^{r-2}\tbinom{r}{2}\tbinom{r-j}{2}\psi^{r-1-j}(fk_jg^2)]$$
$$+ \cdots + \phi^0(fg^r).$$

Applying the induction hypothesis, this last expression is expressible as the closure of

$$T + \tbinom{r}{1}\psi^{r-1}(fg) + \tbinom{r}{2}\psi^{r-2}(fg^2) + \cdots + \psi^0(fg^r).$$

Thus the transformation properties of T relative to conjugation by the $V(g)$ are as required for an r.p.s. It remains to check the vanishing of the expectation value $\langle Tu, u\rangle$, but by the induction hypothesis, $\langle Tu, u\rangle = \langle\phi^r(f)u, u\rangle - \psi^0(fk_r) = 0$, completing the proof. □

The foregoing proof is easily seen to be reversible as regards the two ground states, and indeed both may be distinct from the free vacuum state.

COROLLARY 8.3.1. *With the hypothesis of Theorem 8.3, let u and w be unit vector in* $\mathbf{K}_p$ *for some* $p > 2$, *with corresponding renormalized power systems* ψ^n *and* γ^n. *Then there exist functions* k_j $(j = 1, 2, \ldots)$ *in* $\mathbf{M}$ *such that*

$$\gamma^n(f) = \psi^n(f) - \tbinom{n}{1}\gamma^{n-1}(fk_1) - \tbinom{n}{2}\gamma^{n-2}(fk_2) - \cdots - \gamma^0(fk_n).$$

PROOF. Estimates on $\langle\psi^n(f)u, u\rangle$ similar to those on $\langle\phi^n(f)v, v\rangle$ in the proof of the theorem follow from the recursive expression relating $\psi^n(f)$ and $\phi^n(f)$.

With these estimates the proof applies equally to the case when the r.p.s. for v is replaced by that for u. □

A basic property of the renormalized powers is their locality. From a foundational physical standpoint, and also for the derivation of the finite propagation velocity of the associated nonlinear quantized wave equations, it is important that $\psi^n(f)$ depends essentially only on the $\psi(x)$, with x in the support $S(f)$ of f. For any subset S of G, we denote by $\mathbf{R}(S)$ the W^*-algebra generated by the $U(f)$ for those f in $\mathbf{M}$ such that a neighborhood of $S(f)$ is contained in S.

COROLLARY 8.3.2. *For arbitrary $f \in \mathbf{M}$, and $n = 1, 2, \ldots, \psi^n(f)$ is affiliated with $\mathbf{R}(S)$, for any neighborhood S of the support of f.*

PROOF. In the proof of Theorem 8.2, the approximation g to δ may be chosen to have arbitrarily small support Ω. The approximation $u(g)$ to $\phi^n(f)$ is then affiliated with $\mathbf{R}(S)$, where S is $S(f) + \Omega$ (sum in G), and hence so also is its limit $\phi^n(f)$. The proof of Corollary 8.3.1 applied to the case when $w = v$ shows that the same is true of the $\psi^n(f)$. □

Problems

1. Show that the renormalized powers treated in the preceding section are covariant with respect to translations on G:

$$\Gamma(a)^{-1}\phi^n(f)\Gamma(a) = \phi^n(f_a) \qquad (f_a(x) = f(x-a)).$$

2. Modify the preceding results by using the algebra $L_{[1,\infty)}(G)$ in place of $\mathbf{M}$ as the algebra containing the functions f on which ϕ^n is defined.

3. Show that the map $f \to \phi^n(f)$ is continuous from $\mathbf{M}$ and from $L_{[1,\infty)}(G)$ into the selfadjoint operators in $\mathbf{K}$, where the topology on $L_{[1,\infty)}(G)$ is the sequential topology in which convergence means convergence in every L_p space, $p \in [1, \infty)$.

4. Show that $\phi^n(f)$ may alternatively be defined by the equation

$$\phi^n(f)v = \lim_{g\to\delta} P(n)\left[\int \phi(g_x)^n f(x)dx\right]v,$$

where $P(n)$ denotes the projection of $\mathbf{K}$ onto its n-particle subspace. (This simple expression for renormalized powers does not generalize to interacting fields.)

5. Show that if G is a finite group, and if n is even, then $\phi^n(1)$ is bounded below; but that if $G = S^1$, then $\phi^2(1)$ is unbounded below. By explicit computation show however that $\exp(-\phi^2(1))v$ is in $L_{[1,\infty)}$.

8.5. The semigroup generated by the interaction Hamiltonian

Next we treat the semigroup expressed formally as e^{-tV}, where V is an interaction Hamiltonian of the form $\phi_p(f)$ treated earlier in this chapter. This semigroup is fairly singular, and is hardly tractable at all without the assumption that p is bounded below. It will clarify the origin of the arguments involved to use the suggestive notation $\phi_p(f) = \int :p(\varphi(0,x)): f(x)dx$, while recognizing that φ does not exist as a point function, to which p is directly applicable. As seen in Chapter 7 $\varphi(0,x)$ can be given appropriate rigorous meaning as a densely defined sesquilinear form, but this would not facilitate our immediate object—roughly, to show that the semigroup e^{-tV} degrades the L_p-status of a vector in **K** more slowly than the semigroup e^{-tH} improves it.

The underlying wave equation for which V has the indicated form is formally

$$\Box\varphi + p'(\varphi) = 0.$$

In the case of a classical wave equation, it is only when the energy is bounded below, or substantially when p is bounded below, that solutions generically exist globally. This is physically natural, since the lower bound of the energy is necessary for stability. The same motivation suggests that an effective theory for quantized wave equations will require similar restrictions on p. However, the nonnegativity of p by no means implies the semiboundedness of the operator V, as a consequence of the infinite number of degrees of freedom. This can be seen explicitly in simple cases (cf. Prob. 5, Sec. 8.4).

Thus, e^{-tV} is not a bounded operator even when p is bounded below. But the unbounded semigroup e^{-tV}, $t > 0$, acts appropriately on the L_p scale for the solution of nonlinear quantized wave equations in two space-time dimensions. This is described in

THEOREM 8.4. *Let $G = R^1$ or S^1, $B = (m^2 - \Delta)^{1/2}$ where $m > 0$ in the context of Theorem 8.3, and assume further that $f \in$ **M** is a nonnegative integrable function on G. If p is a nonnegative polynomial vanishing at 0, and $V = \phi_p(f)$, then e^{-V} is in $L_{[1,\infty)}(\mathbf{R}, \nu)$.*

PROOF. Note that p being bounded below means that p has the form $p(x) = a_0x^n + a_1x^{n-1} + \cdots + a_{n-1}x$, where $a_0 > 0$ and n is even. If X is a normal random variable of vanishing mean and variance c^2, then all of its moments, and in consequence all of its renormalized powers, scale with c. Thus, if G is taken in Theorem 8.2 to consist only of its unit element, X may be identified with $c\phi(1)$, and its renormalized powers correspondingly defined. It follows that

$$:p(X): = a_0X^n + a_1'cX^{n-1} + \cdots + a_n'c^n,$$

where the a_j' are independent of c, and subsequently that $:p(X): \geq -kc^n$ for an appropriate constant k. For $g \in \mathbf{M}$, let

$$V_g = \int_G :p(\phi(g_x)):f(x)dx,$$

and let $Z = V - V_g$. Then $\phi(g)$ has variance $\|B^{-1/2}g\|^2$ whence, by the preceding paragraph, $V_g \geq -k\|B^{-1/2}g\|^n \int f(x)dx$.

We now estimate the distribution of V on the negative half-axis. For any $\lambda \in \mathbf{R}$, $Pr[V \leq -\lambda - 1] = Pr[Z + V_g \leq -\lambda - 1]$. If $V_g \geq -\lambda$, then $Z + V_g$ can be $\leq -\lambda - 1$ only if $Z \leq -1$. Setting $P(\lambda) = Pr[V \leq -\lambda - 1]$, this implies that

$$P(\lambda) \leq Pr[|Z| \geq 1] \leq E[|Z|^q]$$

for arbitrary $q > 1$.

We take advantage of the arbitrariness in q as follows. Since Zv is in the spectral manifold $E(n)$ of the number of particles operator $N = \int \lambda\, dE(\lambda)$, it follows from Theorem 8.1 that if $q \leq 2e^{\varepsilon t}$, then $P(\lambda) \leq (e^{nt}\|Z\|_2)^q$. We take $q = e^{-1}(\|Z\|_2)^{-\varepsilon/n}$ (which will be greater than 2 if $\|Z\|_2$ is sufficiently small, which in turn will be the case when g approximates δ sufficiently closely) in order to optimize the inequality. With t correspondingly chosen such that $q = 2e^{\varepsilon t}$, it results that

$$P(\lambda) \leq \exp[-ne^{-1}\varepsilon^{-1}(\|Z\|_2)^{-\varepsilon/n}].$$

In the course of the proof of Theorem 8.2, an expression was obtained for $\|V_g - V\|_2^2$ which leads to the following estimate in the limit $h \to \delta$:

$$\|V_g - V\|_2^2 \leq n!\,\langle|\hat{f}|^2, (B(\cdot)^{-1})^{(*n)} - (\hat{g}B(\cdot)^{-1})^{(*n)}\rangle,$$

where it is assumed that $p(x)$ has the form x^n, the general case being reducible to this case by Minkowski's inequality. Since $f \in \mathbf{M}$ and $f \in L_1$, $\hat{f} \in L_{[2,2+\varepsilon)}(G^*)$ for some $\varepsilon > 0$. Accordingly, if r is sufficiently large,

$$\|V_g - V\|_2^2 \leq c\,\|(B(\cdot)^{-1})^{(*n)} - (\hat{g}B(\cdot)^{-1})^{(*n)}\|_r \qquad (c = c(n,f)).$$

Using the Hausdorff-Young inequality as in the proof of Theorem 8.2 it follows that

$$\|(B(\cdot)^{-1})^{(*n)} - (\hat{g}B(\cdot)^{-1})^{(*n)}\|_r \leq c\,\|B(\cdot)^{-1} - \hat{g}B(\cdot)^{-1}\|_q$$

if $n/q = 1/r + (n - 1)$, and $q > 1$, where the uniform boundedness in $L_q(G^*)$ of the $\hat{g}B(\cdot)^{-1}$ is used, g being restricted by the constraint that $|\hat{g}(y)| \leq 1$. Choosing the conjugate index q' to q so that $q' > n$, p is then given by the equation $r^{-1} = 1 - n/q'$, showing that if q' is sufficiently close to n, r becomes arbitrarily large. In consequence, the preceding inequalities apply, and it results that

$$\|V_g - V\|_2^2 \leq c\,\|B(\cdot)^{-1} - \hat{g}B(\cdot)^{-1}\|_q.$$

Choosing $g = g_s$, where g_s is the inverse Fourier transform of the characteristic function of the interval $(-s, s)$, then $\|B^{-1/2}g_s\|_2$ is asymptotic to $\log s$ as $s \to \infty$. Accordingly, the inequality above on $P(\lambda)$ is applicable if $\lambda = c\,(\log s)^n$. Noting that $\|(\hat{g}_s - 1)B(\cdot)^{-1}\|_q = 0(s^{-1/q'})$, it follows that

$$P(\lambda) \leq \exp[-ne^{-1}\varepsilon^{-1}(\|Z\|_2)^{-\varepsilon/n}] \leq \exp[-a\exp(b\lambda^c)],$$

where a, b, and c are constants depending on ε, n. Thus, as $\lambda \to \infty$, $P(\lambda)$ tends to 0 sufficiently rapidly to imply that e^{-V} is in $L_{[1,\infty)}(\mathbf{R},\nu)$. □

8.6. The pseudo-interacting field

The putative interacting quantized field satisfying the given scalar wave equation $\Box\varphi + p'(\varphi) = 0$ has as its Hamiltonian, in symbolic physical form,

$$H = \int[\tfrac{1}{2}(\nabla\varphi)^2 + \tfrac{1}{2}(\partial_t\varphi)^2 + p(\varphi)]dx,$$

where the integration is over space at any fixed time, e.g., at time 0. But this form of the Hamiltonian appears of little use for rigorous construction, since the interacting field is not known—at time 0, or at any finite time. According to one widely accepted physical model, there is, or is expected to be, an "incoming" field in the infinite past (or, heuristically, "at time $-\infty$" that represents the quantized field before the interaction has commenced, and so is a free field). But this raises new issues, of temporal asymptotics (or "scattering"), etc., rather than resolving the question of how to interpret the symbolic expression for H at a finite time.

What has been shown in the preceding sections is that each of the two terms in H, namely the quadratic or "pseudo-free" term, and the higher order or "pseudo-interacting" term, become analytically somewhat controllable when the free quantized field φ_0 is substituted for the putative interacting one φ. The total pseudo-Hamiltonian H' obtained by adding these terms bears in principle only a specious resemblance to the true physical Hamiltonian, but if the physical interacting field does not deviate very greatly from the free field at finite times, it can reasonably be expected to provide a first approximation to it.

On the other hand, even if the free and interacting fields were unitarily equivalent at finite times, there would be a fundamental distinction between H and H' in the interpretation of the higher-order term $p(\varphi)$, which has no physical reason to be renormalized relative to the free-field vacuum (represented by the lowest eigenvector of H_0) rather than to the putative physical vacuum (represented by the lowest eigenvector of H). Moreover, the appropriate free field is arguably of unknown mass a priori, since the interaction should influence the mass of the putative free incoming field.

The replacement of the symbolic physical Hamiltonian H by the mathematically well defined operator H' is thus *physically* without justification, and indeed replaces the original nonlinear problem by a linear one (which is obtained by substitution of a *known* function φ_0 for the *unknown* function φ in the leading nonlinear term). It is nevertheless mathematically interesting, and it will be seen that, a posteriori, it does lead to a solution for a nonlinear local quantized wave equation. This equation is not the same as the original one, but differs only in lower order terms and provides a reasonable first step in the resolution of the underlying physical nonlinear problem.

The pseudo-interacting field φ (where, for simplicity of notation, we use the same symbol as for the putative interacting one) will be defined symbolically by the equation

$$\varphi(t, x) = e^{-itH'}\varphi_0(0, x)e^{itH'},$$

where φ_0 denotes the free field. There is no special reason to choose the time $t = 0$, rather than some other time, as the starting point, and indeed H' is materially dependent on this initial time t_0, but different choices for t_0 lead to unitarily equivalent H', and ultimately to unitarily equivalent $\varphi(t, x)$. To show actual existence of $\varphi(t, x)$, it is necessary to give appropriate meaning to H' as a selfadjoint operator. Having done this, the mathematical modeling of the physical context is further validated by establishing an appropriate version of the fundamental constraint of causality, in the sense of finite propagation velocity. This is also needed to deal with the case of a noncompact space, which requires a spatial cutoff in the definition of the interaction pseudo-Hamiltonian, with the result that the Hamiltonian is mathematically interpretable as a derivation of a C^*-algebra, but not a priori as a selfadjoint operator. Following this, the unicity of the physical pseudo-vacuum will be shown. Corresponding renormalized powers of the pseudo-interacting field are then established (relative to the physical pseudo-vacuum), and finally it is shown that $\varphi(t, x)$ does indeed satisfy an explicit *local* quantized wave equation relative to this vacuum.

Turning now to the rigorous development, we first recall a general result on the interpretation as a selfadjoint operator of the sum of two given selfadjoint operators, having properties exemplified by the semigroup features shown in Theorems 8.1 and 8.4. The general idea is that a scale of subspaces $\mathbf{K}_p$ of the underlying Hilbert space $\mathbf{K}$ with increasingly stronger norms as p increases is developed, relative to which the respective semigroups generated by given selfadjoint operators A and B have a kind of finite velocity: e^{-tA} improves the status of a given vector at a rate a, while e^{-tB} lowers the status at a rate b. If $a - b \geq 0$, and if A and B are moderately regular in a general way, then a semigroup naturally identifiable with the putative one generated by $A + B$ may be derived, and will improve status at a rate $\geq a - b$. To treat the present pseudo-interacting field, the L_p scale relative to the free vacuum is adequate.

As earlier, for any probability measure space M, the notation $L_{[a,b)}(M)$ will denote the space of all functions that are in $L_p(M)$ for $p \in [a, b)$, in the topology of convergence in each L_p norm; the notation $\mathbf{K}_p$ will also be used on occasion for $L_p(M)$, and $\mathbf{K}_2$ will be denoted simply as $\mathbf{K}$. For any given linear operator T in $\mathbf{K}$, $\|T\|_{p,q}$ will denote $\sup_{u \neq 0}\|Tu\|_q/\|u\|_p$.

We recall the definition of the strong convergence of A_n in $\mathbf{H}$ and note that if the A_n are uniformly bounded below, then $A_n \to A$ if and only if either $\exp(-sA_n) \to \exp(-sA)$ (strongly) for *all* $s > 0$, or for *one* $s > 0$.

THEOREM 8.5. *Let H be a given selfadjoint operator in the space $\mathbf{K} = L_2(M)$, where M is a given probability measure space, with the property that e^{-tH} is a contraction from $\mathbf{K}$ to $\mathbf{K}_{2\exp(\varepsilon t)}$ for some $\varepsilon > 0$. Let F be a real measurable function in $L_{[2,\infty)}(M)$ such that e^{-F} is also in $L_{[2,\infty)}(M)$, and let V denote the operation of multiplication by F, in $\mathbf{K}$. Then $H + V$ is essentially selfadjoint, and if H' denotes its closure, $H + b_n(V) \to H'$ for any sequence of bounded Borel functions $\{b_n\}$ on $\mathbf{R}^1$ that is pointwise convergent to the function $b(x) = x$ and such that $|b_n(x)| \leq |x|$.*

PROOF. We refer to SK, p. 320, Theorem 11.6, for the proof which is based on successive use of the Lie-Trotter and Duhamel formulas. The proof gives useful information about H', including

COROLLARY 8.5.1. *H' is bounded below by $-\frac{1}{2}\varepsilon\log[\int \exp(-2e^{-1}F)]$. Moreover, every entire vector w for H' is in $L_{[2,\infty)}$ and in $\mathbf{D}(H)$; and $H'w = Hw + Vw$.*

The proof of Theorem 8.5 also establishes the regularity of H' as a function of F, for fixed H_0. This may be stated as follows:

COROLLARY 8.5.2. *Let $\mathbf{C}$ denote a collection of functions F satisfying the conditions of Theorem 8.5, and such that for each $p \in [1, \infty)$, $\|F\|_p$ and $\|e^{-F}\|_p$ are bounded as F varies over $\mathbf{C}$. Then H' is continuous as a function of $F \in \mathbf{C}$, in the relative topology on $\mathbf{C}$ as a subset of $L_{[2,\infty)}(M)$. Moreover, there exist positive constants ε' and a (depending on $\mathbf{C}$) such that*

$$\|e^{-tH'}u\|_{2\exp(\varepsilon' t)} \leq e^{at}\|u\|_2 \qquad (u \in L_2(M)).$$

From this regularity, a certain symmetry between H and H' follows, as noted in

COROLLARY 8.5.3. *If $a > 1$, there exists $b \in \mathbf{R}$, such that $H \leq a H'(H, F) + bI$ $(F \in \mathbf{C})$. Moreover, H' is affiliated with the W*-algebra determined by H and V; and H is affiliated with the W*-algebra determined by H' and V.*

We are now in a position to treat the existence and properties of the pseudo-physical vacuum. This is the state represented by the lowest eigenvector, if such exists; in the present context, it does exist and is unique, and also has further convenient properties. The development of these properties involves positivity-preserving features of the semigroups treated earlier, and their application in the context of the theory of positive operators. For any measure space M, an operator T in $L_2(M)$ is called *positivity-preserving* in case $Tf \geq 0$ whenever $f \geq 0$ (in the sense that $f(x) \geq 0$ for all $x \in M$). We call an operator S *inverse compact* in case there exists a scalar c such that $(cI + S)^{-1}$ is compact. The basic result may be formulated as follows:

COROLLARY 8.5.4. *In addition to the hypotheses of Theorem 8.5, suppose that e^{-tH} is positivity-preserving for all $t > 0$. Then $e^{-tH'}$ is also positivity-preserving for all $t > 0$. Moreover, if H is inverse compact, then so is H', and both H and H' have a nonnegative lowest eigenvector. And, if the lowest eigenvector of H is unique, so is that of H'.*

PROOF. In the expression for $e^{-tH'}$ given by the Lie-Trotter formula, each approximating operator is positivity-preserving since the operators $\exp(-tV/n)$ and $\exp(-H/n)$ are positivity-preserving and any product of such operators is again such. It follows that the limiting operator $e^{-tH'}$ is positivity-preserving. The inequality $H \leq a H' + bI$ implies that $a^{-1}(H - bI) \leq H'$, whence, for any sufficiently large constant c, $(a^{-1}(H - bI) + cI)^{-1} \geq (H' + cI)^{-1}$, where the operators on both sides of this inequality exist and are positive. This shows that if H is inverse compact, then so is H'.

We recall now some classic results in the general theory of positivity-preserving operators, where it suffices here to treat the case of a space of the form $\mathbf{H} = L_2(M)$. If T is a compact positivity-preserving operator, there exists a nonnegative function $v \neq 0$ in $\mathbf{H}$ such that $Tv = \|T\|v$. Moreover, T acts "indecomposably" on $\mathbf{H}$, in the sense that T together with the algebra $\mathbf{A}$ of all multiplications by bounded measurable functions act irreducibly on $\mathbf{H}$, if and only if v is unique.

Taking $T = e^{-tH}$ or $e^{-tH'}$ for some $t > 0$, it follows that both H and H' have a nonnegative lowest eigenvector. Moreover, the part of Corollary 8.5.3 referring to affiliation shows that if e^{-tH} is indecomposable for some t, so is $e^{-tH'}$, so that H' has a unique nonnegative lowest eigenvector if and only if H does. □

COROLLARY 8.5.5. *Suppose* $G = S^1$, $B(k) = (m^2 + k^2)^{1/2}$, *and p is an arbitrary real polynomial that is bounded below, in the context of Theorem* 8.2. *Let* $V = \phi_p(1)$, *let* $H_0 = \partial\Gamma(B)$, *and let H denote the closure of* $H_0 + V$. *Then H is inverse compact and has a unique lowest eigenvector of the form Av, where A is a positive self adjoint operator in* $L_2(\mathbf{R}, v)$ *with null space consisting only of* 0.

PROOF. B is easily seen to be inverse compact. In the n-particle subspace, H_0 is bounded below by nm, from which it follows that H_0 is also inverse compact. The remainder of the corollary follows from the preceding general result. The triviality of the null space of A is equivalent to the nonvanishing a.e. of the lowest eigenvector as a vector in $L_2(G)$. □

8.7. Dynamic causality

What is usually called "relativistic causality," which can be regarded as a mathematical formulation of Einstein's principle that the velocity of propagation of a physical effect can not exceed that of light, is the commutativity at a fixed time of physical fields that are conceptually observable. In this form, relativistic causality has been fundamental in this chapter. But Einstein's principle can be given a variety of natural extensions, which are in part exemplified in the finite propagation velocity features of quantized wave equations. As seen in Chapter 6, these closely parallel those of classical linear wave equations. In this section we are concerned with the similar features of quantized nonlinear equations, which are not implied by the parallel features of classical wave equations. In addition, in the nonlinear case, temporal evolution is often not given a priori as a one-parameter unitary group, but rather may be represented as a one-parameter group of automorphisms of a C^*-algebra. Here we extend the considerations of Chapter 6 to the case of such a group, as a step toward the establishment of the finite propagation velocity feature of the solution of a local nonlinear quantized wave equation, treated later.

We recall the definition of concrete C^*-algebra and make the

DEFINITION. A *graduation* on a concrete C^*-algebra $\mathbf{A}$ is a function p from $\mathbf{A}$ to the interval $[0, \infty]$ having the following properties:

i) $[A \in \mathbf{A}: p(A) < \infty]$ is dense in $\mathbf{A}$;
ii) for arbitrary A and B in $\mathbf{A}$, and arbitrary complex number a, $p(A + B)$, $p(AB)$, $p(aA)$, and $p(A^*)$ are all bounded by $\max[p(A), p(B)]$; and
iii) if $A_n \to A$ in the strong operator topology and if $\{p(A_n)\}$ is bounded, then $A \in \mathbf{A}$, and $p(A) \leq \liminf_{n\to\infty} p(A_n)$.

A C^*-algebra with a given graduation is called *graduated*. The subset $[A \in$

$\mathbf{A}: p(A) \leq t]$, which is evidently a C^*-subalgebra, will be denoted as $\mathbf{A}_t$.

A one-parameter group $\alpha(t)$ of automorphisms of a graduated concrete C^*-algebra on the given Hilbert space $\mathbf{K}$ will be called *proper* in case:

a) there exists a constant $a < \infty$ such that $p(\alpha(t)A) \leq p(A) + a|t|$ for all $A \in \mathbf{A}$ and $t \in \mathbf{R}$; and

b) given any positive numbers s and t, there exists a selfadjoint operator $H = H(s, t)$ in $\mathbf{K}$ such that $\alpha(u)(A) = e^{iuH}Ae^{-iuH}$ for all $A \in \mathbf{A}_s$ and $u \in (-t, t)$. A map $(s, t) \rightarrow H(s, t)$ ("a" rather than "the," since $H(s, t)$ is not unique) will be called a *generator* for the automorphism group $\alpha(t)$. The infimum of the constants a for which (a) holds will be called the *velocity* of $\alpha(t)$.

Theorem 8.6 expresses the principle that the composite of finite velocity automorphism groups of a graduated C^*-algebra is again of finite velocity, which velocity cannot exceed the sum of the component velocities.

THEOREM 8.6. *If $\alpha(\cdot)$ and $\beta(\cdot)$ are proper automorphism groups of the graduated (concrete) C^*-algebra $\mathbf{A}$, and if they admit generators $A(s, t)$ and $B(s, t)$ such that $A(s, t) + B(s, t)$ is essentially selfadjoint, then there exists a proper automorphism group $\gamma(\cdot)$ of velocity not greater than the sum of the velocities of $\alpha(\cdot)$ and $\beta(\cdot)$ such that for arbitrary $X \in \mathbf{A}$ with $p(X) < \infty$,*

$$\gamma(t)(X) = \lim_{n\rightarrow\infty} (\alpha(t/n)\beta(t/n))^n(X).$$

PROOF. Setting $C(s, t)$ for the closure of $A(s, t) + B(s, t)$, and setting $\exp(iuA(s, t)/n)\exp(iuB(s, t)/n) = V_n(u)$, then, by Trotter's theorem,

$$e^{iuC(s,t)}Xe^{-iuC(s,t)} = \lim_{n\rightarrow\infty} V_n(u)^n X\, V_n(u)^{-n}$$

for arbitrary $X \in \mathbf{A}$. Now suppose that $X \in \mathbf{A}_r$, $r \leq \infty$, and that $r + (a + b)|u| < t$. Then it is easily verified by induction on $m = 1, 2, \ldots, n$, that

$$V_n(u)^m X V_n(u)^{-m} = (\alpha(u/n)\beta(u/n))^m(X),$$

and is in $\mathbf{A}_{r+(a+b)mu/n}$. Taking $m = n$, it follows that the limit of $(\alpha(u/n)\beta(u/n))^n(X)$ as $n \rightarrow \infty$ exists and equals $e^{iuC(s,t)}Xe^{-iuC(s,t)}$. It follows also that this limit is in $\mathbf{A}_{r+(a+b)|u|}$. The latter expression is thus independent of s and t for sufficiently large s and any fixed r.

Let $\gamma_0(u)$ denote the corresponding linear transformation $X \rightarrow e^{iuC(s,t)}Xe^{-iuC(s,t)}$ defined on the union $\mathbf{A}_+$ of the $\mathbf{A}_r$ with finite r; this is unique and maps $\mathbf{A}_+$ onto itself. If X and Y are arbitrary in $\mathbf{A}_+$, then taking s and t so large that for $Z = XY$, $\gamma_0(u)(Z)$ has the form $e^{iuC(s,t)}Ze^{-iuC(s,t)}$, it follows that $\gamma_0(t)$ is a $*$-automorphism of $\mathbf{A}_+$. From C^*-algebra theory, it follows that $\gamma_0(t)$ extends uniquely to a one-parameter group $\gamma(\cdot)$ of automorphisms of all of $\mathbf{A}$, and it is clear from its construction that $\gamma(\cdot)$ is given by Lie's formula, and that its velocity is bounded by $a + b$. Finally—since $\gamma(u)(X) = e^{iuC(s,t)}Xe^{-iuC(s,t)}$ if $X \in \mathbf{A}_r$, $r + (a + b)|u| < s$, and $|u| < t$—condition b) is satisfied, with generator $C(s - (a + b)t, t)$. □

The domain of dependence and the region of influence properties of classical wave equations, which reflect their hyperbolicity, have analogs in the quantized case. In these, functional dependence is naturally replaced by appropriate affiliation with an operator algebra. In this connection it should be recalled that a given selfadjoint operator A in a separable Hilbert space $\mathbf{H}$ is a function of a second selfadjoint operator B on the space if and only if $A = F(B)$, where F is an ordinary real-valued Borel function. The following result exemplifies the domain of dependence in the case of a quantized equation in a form that permits global dynamics to be in part reduced to local dynamics, and is an essential step in the treatment of nonlinear quantized equations in a noncompact space. The point is that the formal relativistic Hamiltonian, of the form $\phi_p(1)$, is not an operator, since in the noncompact case the function identically 1 on space is not in the requisite L_p spaces. However, appropriately limited functions f_1 and f_2 will lead to interaction Hamiltonians $V_j = \phi_p(f_j)$ that define the same field dynamics, locally in space-time, provided f_1 and f_2 agree in a sufficiently large region. In particular, taking f to be 1 in increasingly large regions, although of compact support, the theory developed earlier in this chapter can be used to define globally on space-time a natural interpretation of the transform of the initial field by the motion generated by the formal Hamiltonian $H_0 + \phi_p(1)$. Thus, in physical terms, a *spatial* cutoff is vastly less singular than a *momentum* cutoff (or convolution of the field involved in the interaction with a smooth function). Indeed, a spatial cutoff does not affect the resulting field in some finite region, which may in fact be taken arbitrarily large. (On the other hand, the *vacuum* is a highly *nonlocal* entity associated with an interacting field, and will be materially sensitive to a spatial cutoff.)

We begin with the precise formulation of $\phi_p(1)$ in the noncompact case, which is as the generator of a one-parameter group of C^*-algebraic automorphisms. These will be unitarily implementable in the free field if the underlying space G is compact, but in general not otherwise.

THEOREM 8.7. *In the context of Theorem* 8.2, *for any open subset Q of G, let $\mathbf{R}_Q$ denote the W^*-algebra generated by the $e^{i\phi(f)}$ and $e^{i\pi(g)}$ for f and g supported by Q. Let $\mathbf{R}$ denote the uniform closure of the union of the $\mathbf{R}_Q$ for Q's having compact closures. If p is any polynomial, there exists a unique one-parameter group $\alpha(t)$ of C^*-automorphisms of $\mathbf{R}$ such that*

$$\alpha(t)(X) = \exp[it\phi_p(f)]X \exp[-it\phi_p(f)]$$

if $X \in \mathbf{R}_Q$ and if f is a continuous function of compact support that is 1 on Q.

PROOF. It suffices to show that the given expression for $\alpha(t)(X)$ is independent of f, within the given constraints on f; the further conclusions then follow directly. To do this means to show that if the f_j ($j = 1, 2$) are continuous functions of compact support on G, that are 1 on Q, then for all $X \in \mathbf{R}_Q$, $\exp[i\phi_p(f_1)]X \exp[-i\phi_p(f_1)] = \exp[i\phi_p(f_2)]X \exp[-i\phi_p(f_2)]$. This is implied for all $X \in \mathbf{R}_Q$ by its validity for any set of generators for $\mathbf{R}_Q$, in particular, for the $e^{i\phi(g)}$ and $e^{i\Pi(h)}$ with g and h supported by Q. This is trivial in the case of the $e^{i\phi(g)}$.

To deal with the $e^{i\Pi(h)}$, note that the $V_j = \phi_p(f_j)$ ($j = 1, 2$) commute strongly (i.e., their spectral projections do so). It follows that it suffices to show that the $e^{i\Pi(h)}$ with h supported by Q commute with the closure of $V_1 - V_2$. The closure is $\phi_p(f_1 - f_2)$, which by its defining properties satisfies the equation

$$e^{-i\Pi(h)}\phi_p(f_1 - f_2)\, e^{i\Pi(h)}$$

$= \text{closure of } \phi_p(f_1 - f_2) + \phi_{p'}((f_1 - f_2)h) + 1/2!\, \phi_{p''}((f_1 - f_2)h^2) + \cdots$.
But $(f_1 - f_2)h = 0$, so that all terms in the last expression vanish except the first. □

EXAMPLE 8.1. Let $G = \mathbf{R}^n$, let $B = (m^2 - \Delta)^{1/2}$ with $m > 0$, and define p on $\mathbf{R}$ as follows: $p(X) =$ infimum of s such that $X \in R_Q$ where Q is a sphere around the origin of radius s (or $+\infty$ if no such s exists). Then $\mathbf{R}$ is graduated, and the one-parameter automorphism group given by the theorem is proper and has velocity 0.

Combining this example with Theorems 8.4, 8.5, and 8.7, the unitary and C^*-algebraic finite propagation velocity features of nonlinear quantized wave equations may be represented as follows:

COROLLARY 8.7.1. *Let $G = \mathbf{R}$, let B denote the usual relativistic scalar single-particle Hamiltonian. Let p be any nonnegative real polynomial. There exists a unique one-parameter automorphism group $\gamma(t)$ of $\mathbf{R}$ such that if $X \in \mathbf{R}_Q$, and $H(f) =$ closure of $H_0 + \phi_p(f)$, then $\gamma(t)(X) = e^{itH(f)}Xe^{-itH(f)}$ for any continuous nonnegative function f of compact support that is 1 on $Q + [-t, t]$.*

Thus the formal integrated interaction Hamiltonian, $\phi_p(1)$, can be given a natural rigorous interpretation, as a derivation of the algebra $\mathbf{R}$, which is the earlier-defined space-finite Weyl algebra. Although this algebra is not representation-independent, it is useful for expressing results in a form that makes manifest their independence from the use of spatial cutoffs. The case of a nonlinear scalar wave equation in two-dimensional Minkowski space is summarized in Corollary 8.7.2. In this connection, an operator on **K** will be said to be compactly supported in case it is in R_Q for some compact Q. $S(f)$ will denote the support of the function f.

COROLLARY 8.7.2. *For any given real polynomial p that is bounded below, there exists a unique one-parameter automorphism group* $\alpha(\cdot)$ *of the space-finite Weyl algebra* **R** *for the corresponding quantized Klein-Gordon equation (of given mass* $m > 0$*), having the following property: for an arbitrary operator X in the dense subalgebra* $\mathbf{R}_+$ *of* **R** *consisting of operators of compact support,* $\alpha(t)(X) = e^{itH(f)}Xe^{-itH(f)}$*, where* $H(f)$ *is the closure of* $H_0 + \phi_p(f)$*, f being any continuous nonnegative function of compact support Q on space that is* 1 *on the region* $Q + [-t, t]$.

8.8. The local quantized equation of motion

At this point it is appropriate to review the results thus far in relation to our original objective: to "solve" the quantized differential equation $\Box\varphi + p'(\varphi) = 0$. The underlying framework is that of Hilbert space, on which φ is a generalized operator-valued function; of canonical commutation relations at fixed times; and other general desiderata such as stability (positive energy), the existence of a vacuum, etc. But no such general constraints serve to define a priori the purely symbolic expression $p'(\varphi)$, and without such a definition the equation itself is purely symbolic. An alternative to consideration of the *equation* is the corresponding *Hamiltonian* formulation. However, this leads to the nonlinear operation $\varphi \to p(\varphi)$, as well as the quadratic terms representing the free component of the total Hamiltonian, and so does not avoid the issue of the meaning of a nonlinear function of a quantum field.

The formulation of the symbolic expression $p(\varphi)$ as the renormalized polynomial $:p(\varphi):$ relative to the putative *physical* vacuum provides a rigorous and natural interpretation of either the equation or its total Hamiltonian. But as indicated above, because of the difficulty in dealing with $:p(\varphi):$ when both φ and the underlying vacuum are unknown, we elected rather to insert the known free field and vacuum into the nonlinear terms, thereby obtaining a tractable *linear* equation, and deferring consideration of the underlying nonlinear issue.

In the case of a classical nonlinear equation, the analogous procedure would be as follows: to simplify the solution of the nonlinear equation $\Box\varphi + p'(\varphi) = 0$, with given data $\varphi(0, x) = f(x)$ and $\partial\varphi(0, x)/\partial t = g(x)$, we first solve the linear equation $\Box\varphi + m^2\varphi = 0$, where $m^2 = p''(0)$, with the same Cauchy data, obtaining a function φ_0. We then solve the equation $\Box\varphi + m^2\varphi +$

$(p'(\varphi_0) - m^2\varphi_0) = 0$, i.e., insert φ_0 in place of φ in the nonlinear term with the same Cauchy data. This equation is of course linear and hyperbolic, and has a global solution (say, for finite energy Cauchy data and appropriate p) enjoying the familiar finite propagation velocity features. Its solution is not a solution to the original nonlinear equation, but a presumptive first approximation to this solution—whose existence remains in doubt—both a priori and at the present stage of classical theory.

Again, in place of the dynamical equations one may use the Hamiltonian formalism, which applies equally in the classical context, via an infinite-dimensional generalization of the Hamiltonian-Jacobi-Lie theory. The solution manifold of the given nonlinear classical equation has a natural invariant symplectic structure, and otherwise plays the role of a phase space for a dynamical system. The Hamiltonian for the given nonlinear equation is the function on this phase space

$$H(\varphi) = \int \{\tfrac{1}{2}[(\nabla\varphi)^2 + (\partial_t\varphi)^2 + m^2\varphi^2] + p(\varphi)\}dx,$$

where the integration is over space at an arbitrary time t; as noted earlier, the result of the integration is time independent. Here also one could modify the difficult nonlinear problem by replacing the source of the nonlinearity, the higher-order term $p(\varphi)$, by $p(\varphi_0)$, where φ_0 is the solution of the free equation that has the same Cauchy data at some fixed time. The resulting Hamiltonian is no longer time independent, and in general agrees with $H(\varphi)$ only at the initial fixed time.

To be sure, as already indicated, the nonlinear (physical, interacting) field may be asymptotic to the linear ("free") field as the time approaches $\pm\infty$. This is the idea behind scattering theory, which is well developed in the classical case, and in which the physically expected temporal asymptotics of the nonlinear equation has been rigorously established in important cases. But this relation between an interacting and a free field—intuitively to the effect that the interaction is diffused after a long time, resulting in a field that is nearly free—does not rationalize the assumption that the two fields are coincident at some *finite* time while the interaction is presumptively taking place. Indeed, the determination of the difference between the free and interacting fields is the primary objective of the theory.

It is remarkable that our results at this point represent considerably more than a first approximation to a putative solution of a nonlinear quantized equation. They do lead to solutions of local nonlinear wave equations—which, however, are not the original equations, but fairly close to them. More specifically, instead of solving the equation $\Box\varphi + p'(\varphi) = 0$, one solves the equation $\Box\varphi + q'(\varphi) = 0$, where q is a polynomial whose leading term coincides with that of p, but has (in general) different lower order terms. Here $q'(\varphi)$ is defined by renormalization with respect to the *physical* vacuum. This is represented by the lowest eigenvector of the total Hamiltonian for the interacting field, whose nonquadratic term takes the form $\int q(\varphi)dx$.

This development could hardly have been anticipated from the analogy with the classical case. A classical solution of the equation $\Box\varphi + m^2\varphi + [(p'(\varphi_0) - m^2\varphi_0] = 0$, where φ_0 is a solution of the linear equation $\Box\varphi + m^2\varphi = 0$ (here $m^2 = p''(0)$, and it is assumed that $p'(0) = 0$), has no reason to satisfy any equation of the form $\Box\varphi + q'(\varphi) = 0$, for *any* polynomial q, or any other smooth function. This difference between the quantized and the classical case can be understood as a consequence of the greater specificity of the quantized case, in which the fixed-time commutation relations constrain the initial data much more strongly than in the classical case, in which the data are arbitrary functions, apart from being in designated function spaces. This is the same feature of the quantized case that enables an appropriate definition of powers of the operator-valued distributions that represent quantized fields, notwithstanding the lack of an effective definition of powers for classical distributions of comparable singularity. Thus in certain respects the quantized case is more coherent, if not acutally simpler, than the formally analogous classical case. Note also that the vacuum plays a fundamental role in the quantized case, but has no analog in basic classical theory.

The main additional theory involved in the derivation of the quantized local nonlinear wave equation cited is functional analytic in character: in consequence it is more succinctly and simply expressed in the following format: let $\mathbf{H}$ be a given complex Hilbert space, let B be a given strictly positive selfadjoint operator, and let $\varkappa$ be a conjugation on $\mathbf{H}$ that commutes with B. We refer to a vector or subspace that is pointwise invariant under $\varkappa$ as *real* and similarly for an operator that commutes with $\varkappa$. Let $(\mathbf{K}, W, \Gamma, v)$ denote the free boson field over $\mathbf{H}$, let $w = \partial W$ and $H_0 = \partial\Gamma(B)$, and set $C = B^{1/2}$.

For any real number a, $\langle\langle \mathbf{D}(B^a)\rangle\rangle$ will denote the completion of the domain of B^a with respect to the inner product

$$\langle x, y\rangle_a = \langle B^a x, B^a y\rangle.$$

We extend $\varkappa$ by continuity to these spaces, denoting the extensions also as $\varkappa$. The subscript $\varkappa$ will denote the real subspace of a given space; e.g., $\mathbf{D}_\varkappa(B^a)$ is the set of all real vectors in $\mathbf{D}(B^a)$.

The next theorem establishes the differential equation satisfied by the transform of the free field by the unitary group generated by a total Hamiltonian of the type earlier considered in this chapter. While the differentiability condition used is adapted to this application, the theorem is more simply expressed in a general form.

THEOREM 8.8. *Let* $\mathbf{R}$ *denote the* W^**-algebra generated by the* $W(x)$, $x \in \mathbf{H}_\varkappa$. *Let* V *denote a selfadjoint element of the Hilbert space* $L_2(\mathbf{R}, v)$ *(which will be* *denoted as* $\mathbf{K}'$), *such that* V *and* $e^{-V} \in L_{[2,\infty)}(\mathbf{R}, \nu)$. *Let* H *denote the closure of* $H_0 + V$. *And for arbitrary* $x \in \mathbf{D}_\kappa(C)$, *let*

$$\phi(x, t) = e^{itH}w(C^{-1}x)e^{-itH}, \qquad \Pi(x, t) = e^{itH}w(iCx)e^{-itH}.$$

Then the following equations hold:

$$(\partial/\partial t)\langle\phi(x, t)u_1, u_2\rangle = \langle\Pi(x, t)u_1, u_2\rangle;$$

$$(\partial^2/\partial t^2)\langle\phi(x, t)u_1, u_2\rangle + \langle\phi(B^2x, t)u_1, u_2\rangle + \langle e^{itH}V'(x)e^{-itH}u_1, u_2\rangle = 0,$$

provided

i) x *is an element of* $\mathbf{D}_\kappa(C^3)$ *such that the map*

$$s \to e^{isw(iCx)}\, V\, e^{-isw(iCx)},$$

from $\mathbf{R}$ *into* $\mathbf{K}'$, *is differentiable at* $s = 0$ *with derivative* $V'(x)$; *and*

ii) the u_j *are analytic vectors for* H.

PROOF. We define $\phi_0(x, t) = \exp(itH_0)w(C^{-1}x)\exp(-itH_0)$, and $\Pi_0(x, t) = \exp(itH_0)w(iCx)\exp(-itH_0)$ for $x \in \mathbf{D}_\kappa(C)$.

LEMMA 8.8.1. *If* $x \in \mathbf{D}_\kappa(C)$, *and if* u_1 *and* u_2 *are arbitrary in* $\mathbf{D}(H_0^{1/2})$, *then* $\langle\phi_0(x, t)u_1, u_2\rangle$ *is a differentiable function of* $t \in \mathbf{R}$ *with derivative* $\langle\Pi_0(x, t)u_1, u_2\rangle$. *If in addition* $x \in \mathbf{D}(C^3)$, *then the latter expression is also differentiable with derivative* $-\langle\phi_0(B^2x, t)u_1, u_2\rangle$.

PROOF. This follows straightforwardly from results in Chapter 1. □

LEMMA 8.8.2. *If* P *and* Q *are selfadjoint operators on* $\mathbf{K}$ *of which* Q *is bounded, and if* T *is an arbitrary bounded linear operator on* $\mathbf{K}$, *then*

$$e^{it(P+Q)}T\, e^{-it(P+Q)} = e^{itP}T\, e^{-itP} + i\int_0^t e^{i(t-s)(P+Q)}[Q, e^{isP}Te^{-isP}]\, e^{-i(t-s)(P+Q)}ds.$$

PROOF. If P is bounded, this is a special case of Duhamel's formula, applied to the perturbation ad(Q) of the operator ad(P) in the Banach space of all bounded linear operators on $\mathbf{K}$. If P is unbounded, let $\{P_n\}$ be a sequence of bounded selfadjoint operators such that $P_n \to P$. Then $P_n + Q \to P + Q$, and a limiting argument employing dominated convergence completes the proof. □

LEMMA 8.8.3. *Let G be an arbitrary bounded linear operator on* $\mathbf{K}$, *and let* u_1 *and* u_2 *be arbitrary analytic vectors for* H. *Then*

$$\langle e^{itH}G\, e^{-itH}u_1, u_2\rangle = \langle \exp(itH_0)\, G\, \exp(-itH_0)u_1, u_2\rangle$$

$$+\ i\int_0^t \langle e^{i(t-s)H}[V, \exp(isH_0)\, G\, \exp(-isH_0)]\, e^{-i(t-s)H}u_1, u_2\rangle ds.$$

PROOF. When V is bounded, this follows from Lemma 8.8.2. For V unbounded, take $V_n = f_n(V)$, where $\{f_n\}$ is any sequence of real valued, bounded, continuous functions on $\mathbf{R}$ such that $|f_n(x)| \le |x|$ and $f_n(x) \to x$ for all x, and replace V by V_n. The only question in passing to the limit as $n \to \infty$ is with the integral on the right. To this end, set $w_j(s) = e^{-i(t-s)H}u_j$. Then, by earlier estimates, $\|w_j(s)\|_p \le c\|e^{\varepsilon H}u_j\|$, where p and the constant c depend only on the $\varepsilon > 0$ for which the u_j are in $\mathbf{D}(e^{\varepsilon H})$, and so are uniform in s. Applying Hölder's inequality to the integrand, and noting that $V_n \to V$ in L_q, $q < \infty$, including the value of q such that $q = p/(p-2)$, the requisite domination for appropriate convergence of the right side follows. □

LEMMA 8.8.4. *For arbitrary analytic vectors* u_1 *and* u_2 *for* H,

$$\langle e^{itH}\Pi(x)\, e^{-itH}u_1, u_2\rangle = \langle \Pi_0(t, x)u_1, u_2\rangle$$

$$+\ i\int_0^t \langle e^{i(t-s)H}[V, \exp(isH_0)\Pi(x)\exp(-isH_0)]\, e^{-i(t-s)H}u_1, u_2\rangle ds.$$

PROOF. This follows by a succession of arguments similar to those used in the proofs of the preceding lemmas, and details are omitted. □

PROOF OF THEOREM, CONTINUED. For arbitrary $\varepsilon > 0$,

$$\varepsilon^{-1}\langle(e^{i(t+\varepsilon)H}\Pi(x)\, e^{-i(t+\varepsilon)H} - e^{itH}\Pi(x)e^{-itH})u_1, u_2\rangle$$
$$= \varepsilon^{-1}\langle(e^{i\varepsilon H}\Pi(x)\, e^{-i\varepsilon H} - \Pi(x))v_1, v_2\rangle,$$

where $v_j = e^{-itH}u_j$. By Lemma 8.8.4, the last expression is

$$\varepsilon^{-1}\langle(\Pi_0(\varepsilon, x) - \Pi_0(0, x))v_1, v_2\rangle$$

$$+\ \varepsilon^{-1}i\int_0^\varepsilon \langle e^{i(\varepsilon-s)H}[V, \exp(isH_0)\Pi(x)\exp(-isH_0)]e^{-i(\varepsilon-s)H}v_1, v_2\rangle ds.$$

By Lemma 8.8.1,

$$\varepsilon^{-1}\langle(\Pi_0(\varepsilon, x) - \Pi_0(0, x))v_1, v_2\rangle \to -\langle\phi_0(B^2x)v_1, v_2\rangle.$$

Since $\langle e^{i(\varepsilon-s)H}[V, \exp(isH_0)\Pi(x)\exp(-isH_0)]\, e^{-i(\varepsilon-s)H}v_1, v_2\rangle$ is a jointly continuous function of ε and s,

$$\varepsilon^{-1}i\int_0^{\varepsilon} \langle e^{i(\varepsilon-s)H}[V, \exp(isH_0)\Pi(x)\exp(-isH_0)]\, e^{-i(\varepsilon-s)H}v_1, v_2\rangle ds$$

$$\to -\langle V'(x)v_1, v_2\rangle.$$

It follows that $(\partial/\partial t)\,\langle\Pi(x,t)u_1, u_2\rangle$ exists and equals

$$\begin{aligned} &-\langle\phi_0(B^2x)v_1, v_2\rangle - \langle V'(x)v_1, v_2\rangle \\ =\ &-\langle\phi(t, B^2x)u_1, u_2\rangle) - \langle e^{itH}V'(x)e^{-itH}u_1, u_2\rangle. \end{aligned}$$

Finally, noting that the Hu_j are again analytic vectors for H, and using the fact that every analytic vector for H is in the domains of H_0 and V, and similar arguments to the foregoing,

$$(\partial/\partial t)\,\langle\phi(t,x)u_1, u_2\rangle = \langle\Pi(t,x)u_1, u_2\rangle. \qquad \square$$

As yet in this section, $\phi_0(t, \mathrm{x})$ and $\phi(t,x)$ were treated only for suitably regular $x \in \mathbf{H}_\times$, so that $\phi_0(t,x)$ and $\phi(t,x)$ are generalized and not strict functions on space. They may be extended to actual point functions, at the cost of using—in place of densely defined operators—generalized operators, formulated as sesquilinear forms on a dense domain in Hilbert space.

The following notational conventions will be used in the treatment here of sesquilinear forms: if B is a given such form with domain $\mathbf{L}$ in the Hilbert space $\mathbf{K}$, the value of B on the given ordered pair of vectors u, u' in $\mathbf{L}$ will be denoted as $\langle Bu, u'\rangle$. If P and Q are operators in $\mathbf{K}$ of which P is bounded while Q and Q^* are defined on $\mathbf{L}$, then $\langle[P,Q]u, u'\rangle$ will denote $\langle Qu, P^*u'\rangle - \langle Pu, Q^*u'\rangle$. We recall that $\mathbf{D}_{-\infty}(B)$ denotes the union of the $\mathbf{D}(B^{-k})$ for $k = 1, 2, \ldots$, with the topology on $\mathbf{D}_{-\infty}(B)$ of convergence in $\langle\mathbf{D}(B^{-k})\rangle$ for some k, this space is denoted as $\langle\mathbf{D}_{-\infty}(B)\rangle$.

For the free equation the extension is straightforward.

Lemma 8.8.5. *The sesquilinear forms $\langle\phi_0(t,x)u_1, u_2\rangle$ and $\langle\Pi_0(t,x)u_1, u_2\rangle$ ($u_j \in \mathbf{D}_\infty(H_0)$) extend continuously (and uniquely so) from the earlier given domain for x to the domain $\langle\mathbf{D}_{-\infty}(B)\rangle$.*

Proof. This is by recursion from the relations

$$\begin{aligned} \langle\Pi_0(t,x)u_1, u_2\rangle &= \langle[iH_0, \phi_0(t,x)]u_1, u_2\rangle \\ \langle\phi_0(t, B_x^2)u_1, u_2\rangle &= \langle[iH_0, \Pi_0(t,x)u_1, u_2\rangle. \end{aligned}$$

Thus $\langle\Pi_0(t,x)u_1, u_2\rangle$, originally defined only for $x \in \mathbf{D}(C)$, is bounded by $c\, \|C^{-1}x\|\, \|(I + H)u_1\|\, \|(I + H)u_2\|$, and so extends in a unique continuous fashion to $\langle \mathbf{D}(C^{-1})\rangle$. Similarly, $\langle \phi_0(t,x)u_1, u_2\rangle$ extends in a unique continuous fashion from $x \in \langle \mathbf{D}(C^{-1})\rangle$ to $x \in \langle \mathbf{D}(C^{-3})\rangle$. The original relations remain valid for these extensions, and the argument just made may be iterated indefinitely, leading to the stated conclusion. □

For the perturbed equation, the argument requires some features of the perturbation theory developed earlier.

LEMMA 8.8.6. *Let* **N** *denote the domain of all entire vectors for* H, *in its natural topology. Then the sesquilinear forms* $\langle \phi(t,x)u_1, u_2\rangle$ $(X \in \langle \mathbf{D}_\times(C^{-1})\rangle)$ *and* $\langle \Pi(t,x)u_1, u_2\rangle$ $(x \in \langle \mathbf{D}_\times(C)\rangle,\ u_1, u_2 \in \mathbf{N})$ *extend continuously (and uniquely so) from the given domains for x to the domains* $\langle \mathbf{D}_\times(C^{-3})\rangle$ *and* $\langle \mathbf{D}_\times(C^{-1})\rangle$; *and*

$$(\partial/\partial t)\ \langle \phi(t,x)u_1, u_2\rangle = \langle \Pi(t,x)u_1, u_2\rangle.$$

PROOF. By the argument used in Chapter 7, for arbitrary $z \in \mathbf{D}(B)$,

$$\|w(z)u\| \leq c\ \|z\|\ \|(I + H_0)^{1/2}u\|,$$

whence $\|w(z)(I + H_0)^{-1/2}u\| \leq c\|z\|\ \|u\|$. Taking adjoints, it follows that $\|(I + H_0)^{-1/2}w(z)u\| \leq c\|z\|\ \|u\|$. Noting that for $u' \in \mathbf{D}_\infty(H_0)$,

$$w(iBz)u' = i\,[H_0, w(z)]u',$$

it follows that

$$(I + H_0)^{-1}w(iBz)\,(I + H_0)^{-1}u' = [iw(z)\,(I + H_0)^{-1} - i\,(I + H_0)^{-1}w(z)]u'.$$

Taking norms on both sides, it results that

$$\|(I + H_0)^{-1}w(iBz)\,(I + H_0)^{-1}\| \leq c\ \|z\|,$$

from which it follows in turn that if $u_1, u_2 \in \mathbf{D}(H)_0)$ and z is arbitrary in **H**, then

$$|\langle w(z)u_1, u_2\rangle| \leq c\ \|B^{-1}z\|\ \|(I + H_0)u_1\|\ \|(I + H_0)u_2\|.$$

Finally, note that if u is entire for H, then by Corollary 8.5.1 $\|H_0u\| \leq c\ \|e^{sH}u\|$ for some $s > 0$. The extension claimed in the lemma follows, and the proof of the final equation is left as an exercise. □

The last equation obtained in Theorem 8.8 is not a local differential equation in time or in space. Thus, the "source" term $\langle e^{itH}V'(x)e^{-itH}u_1, u_2\rangle$ is in no effective sense a function of the Cauchy data for the "interacting" field $\phi(x, t)$ at time t. A truly local differential equation in time is rather of the general form $dy/dt = F(y, t)$, where $F(\cdot, t)$ is a well-defined function from the space Y in which $y(t)$ lies. The equation is thus not yet in a form that is satisfactory from a foundational view, according to which the central problem of quantum field theory is in the solution of a *local* nonlinear quantized partial differential equation. Indeed, besides locality in time, locality in space is also required. That is, the spatial support of $F(y, t)$, where y is any given vector in Y, should be essentially contained in that of y.

As noted earlier, there is no a priori physical reason for the putative physical (interacting) field to resemble the free field at a finite time, apart from such general features as their common satisfaction of the Weyl relations. But it is nevertheless the case that the field $\phi(x, t)$ defined by Theorem 8.8, with V taking as $\phi_p(f)$ as earlier, may satisfy a local nonlinear wave equation, the renormalization of nonlinear terms being with respect to the *physical* vacuum. This is the case in two space-time dimensions, with the usual relativistic energy-momentum relation, i.e., $B = (m^2 - \Delta)^{1/2}$. In higher dimensions, it seems unlikely that if solutions exist to analogous wave equations, the corresponding Weyl system at a fixed time is unitarily equivalent to that of a free field. We state the two-dimensional result only in the case of the space S^1; a slightly more complicated statement applies to the case of $\mathbf{R}$ with a spatial cutoff, in extension of Corollary 8.7.2. Moreover, S^1 is the simplest prototype for the space S^3 that is important for the treatment of conformally invariant fields in four-dimensional space-time.

Before stating the existence theorem for quantized nonlinear fields in two space-time dimensions, it is necessary to state precisely what is meant by the term "solution," since the interpretation of this term is one of the key mathematical issues in the treatment of quantized equations. In order to free the concept of solution from any appearance of dependence on the dimensionality or special structure of space-time, and for potential general application, we treat the case in which space is represented by a complete Riemannian manifold S. We use the canonical measure in S that derives from the assumed Riemannian structure, and denote the usual inner product in $L_2(S)$ as $\langle f, g\rangle = \int f(x)\overline{g(x)}dx$. The Laplacian on S, in its usual selfadjoint formulation, will be denoted as Δ, and the wave operator on $\mathbf{R} \times S$, $\partial_t^2 - \Delta$, will be denoted as $\Box$.

DEFINITION. Let S be a given complete Riemannian manifold. Let p be a given real polynomial. Let h be a given real function on S. Let $c \in \mathbf{R}$ be such that $cI - \Delta > \varepsilon I$ for some $\varepsilon > 0$. A *solution* of the *quantized* nonlinear equation

$$\Box\varphi + c\varphi + hp'(\varphi) = 0 \tag{8.1}$$

is the following mathematical structure:

1) a multiplicative dual couple $(\mathbf{M}, \mathbf{N}, A)$, where each of $\mathbf{M}$ and $\mathbf{N}$ is dense in $L_2(S, \mathbf{R})$, $h \in \mathbf{N}$, and

$$A(f_1 \oplus g_1, f_2 \oplus g_2) = \langle f_2, g_1 \rangle - \langle f_1, g_2 \rangle;$$

2) a Weyl system $(\mathbf{K}, W)$ over $(\mathbf{M}, \mathbf{N}, A)$;
3) a nonnegative selfadjoint operator H in $\mathbf{K}$;
4) a unit vector v in $\mathbf{K}$ such that $Hv = 0$;
5) a renormalized power system for the grounded Weyl pair $(U, V, \mathbf{K}, v)$, where $U(f) = W(C^{-1}f)$ and $V(g) = W(iCg), f \in \mathbf{M}, g \in \mathbf{N}$, and $C = (cI - \Delta)^{1/4}$; and
6) functions $\phi(t, f)$ and $\Pi(t, g)$, from $\mathbf{M} \times \mathbf{R}$ and $\mathbf{N} \times \mathbf{R}$ respectively to selfadjoint operators in $\mathbf{K}$, having the following properties:

   i) $\phi(t, f) = e^{itH} w(C^{-1}f) e^{-itH}, \qquad \Pi(t, g) = e^{itH} w(iCg) e^{-itH}$

   where $f \in \mathbf{M}$, $g \in \mathbf{N}$, $t \in \mathbf{R}$, and $w = \partial W$; and

   ii) for any entire vector u for H and $f \in \mathbf{D}(C^3) \cap \mathbf{M}$,

$$(\partial/\partial t)\, \phi(t, f)u = \Pi(t, f)u;$$

$$[(\partial/\partial t)^2\, \phi(t, f) + \phi(t, C^4 f) + \phi_{p'}(t, hf)]u = 0,$$

   where here $\phi_{p'}(\cdot, t)$ refers to the renormalized power system for $(U(\cdot, t), V(\cdot, t), \mathbf{K}, v)$, which are defined by the equations $U(\cdot, t) = e^{itH} U(\cdot) e^{-itH}$, $V(\cdot, t) = e^{itH} U(\cdot) e^{-itH}$, whose existence follows from the fact that $e^{itH} v = v$.

In order to treat locality of quantized equations, we introduce the following notation: for any open subset Z of S and $t > 0$, let $R(Z, t)$ denote the domain of dependence, at time t, for the classical Klein-Gordon equation on $\mathbf{R} \times S$, of Z. The quantized equation (8.1) is *local* if: a) it is soluble and remains soluble when h is replaced by hk for an arbitrary C_0^∞ function k on S; b) for any neighborhood Z of the support of an arbitrary given vector $f \in \mathbf{M}$, $\int f(x)h(x)k(x) :p(\phi(x)):dx$ is affiliated with the W^*-algebra generated by the $\exp[i\phi(f_1)]$ for those f_1 in $\mathbf{M}$ that are supported by Z; and c) for arbitrary f and g in $\mathbf{M}$ and $\mathbf{N}$ supported by the given open subset Z of S, and $t > 0$, $\phi(t, f)$ and $\Pi(t, g)$ are affiliated with the W^*-algebras generated by the $U(f_1)$ and $V(g_1)$, where f_1 and g_1 are in $\mathbf{M}$ and $\mathbf{N}$ and supported by an arbitrary neighborhood of $R(Z, t)$.

It should be noted that in the definitions there is no reference to the free field or to any other specific field. The representation of the Weyl relations has at no time an a priori restriction. Moreover, the underlying vacuum is not given a priori, either at a finite time, or at a limiting time such as $\pm\infty$, as in the scattering theory. The imposition of such restrictions, with whatever degree of physical plausibility, can only serve to overdetermine the equation, from a general mathematical standpoint, and may well result in the nonexistence of solutions.

In these terms we may describe the results earlier attained on nonlinear equations in two-dimensional space-time as

COROLLARY 8.8.1. *Let $S = \mathbf{R}$ or S^1 and let* **M** *and* **N** *denote the spaces of all real C^∞ functions of compact support on S. Then, for any given nonnegative function h in* **N**, *and for given real polynomial p that is bounded below, there exists a polynomial q having the same leading term as p such that equation* 8.1 *is soluble. The equation and its solution are local and the temporal evolution defined by the equation is causal in the sense of Section* 8.7.

Again there is no reference to the free field, which now appears only as an instrument that is useful in two space-time dimension as a first approximation to the local interacting field described by Corollary 8.8.1. It appears unlikely to serve effectively as such in higher dimensions.

EXAMPLE 8.2. Let $S = S^1$, let $p(x) = \frac{1}{4}gx^4$, and let $\phi_0(x, t)$ denote the free Klein-Gordon quantized field with mass m; let $(\mathbf{K}, W, \Gamma, v)$ denote the free boson field over the Hilbert space **H** of normalizable classical Klein-Gordon wave functions. Let N_0 denote the operation of renormalization with respect to the free vacuum v, and let $V = \int N_0[(\phi(0,x)^4]dx$. Let H_0 denote the Hamiltonian for the free field, and let H denote the closure of $H_0 + V$. Then H has a unique ground state u, within proportionality. Let E denote the expectation functional corresponding to u, and let E_0 denote that corresponding to v. By Corollary 8.3.1, $N_0[(\phi(0,x)^4]$ may be expressed as follows in terms of the $N[(\phi(0,x)^j]$, where we use distribution-type notation:

$$N_0[\phi(0,x)^4] = N[\phi(0,x)^4] + 4k_1N[\phi(0,x)^3] + 6k_2N[\phi(0,x)^2] + 4k_3N[\phi(0,x)] + k_4,$$

where $k_j(x) = E(N_0[\phi(0,x)^j])$.

If T_y denotes the classical transformation $\phi(x, t) \to \phi(x + y, t)$, then $\Gamma(T_y)$ transforms $\phi(0,x)$ into $\phi(0,x+y)$, V therefore commutes with $\Gamma(T_y)$ for arbitrary $y \in S$. It follows in turn that H also commutes with the $\Gamma(T_y)$, whence, by the unicity of u, E is invariant under the indicated action of the $\Gamma(T_y)$ on the expectation value functionals. Accordingly, the functions $k_j(x)$ are invariant under translations on S, and so are constants.

Furthermore, transformation by $\Gamma(-I)$ carries $\phi(0,x)$ and $\Pi(0,x)$ into their negatives, and therefore leaves V invariant. H_0 and, hence, H are similarly invariant, whence E is invariant also under the induced action of $\Gamma(-I)$ on expectation value functions. It follows that $k_j = 0$ if j is odd, so that we have simply $N_0[\phi(0,x)^4] = N[\phi(0,x)^4] + 6k_2N_0[\phi(0,x)^2] + k_4$. From this it follows in turn that $\phi(t,x)$ satisfies the differential equation

$$(\partial/\partial t)^2\phi + (m^2 + g_2)\phi + g\,N(\phi^3) = 0$$

as an operator equation on the dense domain of all entire vectors for H. More specifically, $N(\phi(t,x)^4)$ is defined relative to the r.p.s. at time t indicated earlier, with time-independent ground state vector u. At each time t, N is a local function as regards space, which is explicitly definable in accordance with Section 8.7. Note that $N_0(\phi(t,x)^4)$ is well defined via Section 8.3, and is quite distinct from both $N_0(\phi_0(t,x)^4)$ and $e^{itH}N_0(\phi_0(0,x)^4)e^{-itH}$.

LEXICON. The definition of a solution to a quantized equation given previously embodies the following elements from a physical standpoint: parts (1) and (2) embody the canonical commutation relations at a fixed time; (3) is to the effect that there is a Hamiltonian; (4) is to the effect that the Hamiltonian has a normalizable lowest eigenvector; (5) is the existence of suitably defined powers $:\phi(t,x)^n:$, when renormalized relative to the physical vacuum described in (4); (6i) is to the effect that the temporal evolution of the field ϕ and its first time derivative Π is generated by the Hamiltonian of (3); and (6ii) is the underlying wave equation in an analytically controlled form.

The local character of the equation assures that the nonlinear term that drives the solution is in fact a local function of the unknown field, in space, at any given time. That the equation is local in time is automatic from its form, i.e., the nonlinear term at time t depends only on the field at time t. The locality of the solution implies that the field value at relatively spacelike points commute, irrespective of a possible difference in time. The causality of the equation confirms that one of the essential motivations of local field theory is embodied in a strong form.

The difference between p and q represents "counter terms" that are not negligible, and that cannot simply be discarded. The dependence of the coefficients of q on those of p is smooth but complex, and a nonperturbative class of polynomials that are possible q's is not explicitly known. Both this problem and the treatment of solutions in higher dimensions appear to depend on the development of a class of grounded Weyl systems that is closer to the putative interacting field than the free field.

Problems

1. Let φ denote the free quantized Klein-Gordon field in two-dimensional Minkowski space, and let H_0 denote the Hamiltonian of the field. Let p be an arbitrary real polynomial on $\mathbf{R}$, and demonstrate that the operation $T \to [T, \int_{\mathbf{R}} :p(\varphi(0,x)):dx]$ can be interpreted as a densely defined derivation of the space-finite Weyl algebra. Does $\int_{\mathbf{R}} :p(\varphi(0,x)):dx$ exist as a form on $\langle \mathbf{D}_\infty(H_0)\rangle$?

2. Show that for the free quantized Klein-Gordon field over $\mathbf{R} \times S$, $H_0 + V(g)$, where S is the n-torus, $V(g) = \int_S :\phi(0, g_x)^2: dx$, and H_0 denotes the free Hamiltonian, remains uniformly bounded below as g (assumed nonnegative, even, and of Fourier coefficients bounded by 1) tends to δ only if $n = 1$.

3. In the context of Problem 2, show that the L_2 norm of $\exp(-V)$ tends in $L_2(\mathbf{R}, \nu)$ to ∞ as $g \to \delta$ if $n > 1$.

4. Let $p(x, y)$ denote a real polynomial in the two real variables x and y that is bounded below. Let ϕ and ψ denote two independent free quantized Klein-Gordon fields on $\mathbf{R} \times S^1$, of free Hamiltonians H_0 and H_0', on Hilbert spaces $\mathbf{K}$ and $\mathbf{K}'$. Define $V = \int_{S^1} :p(\phi(0, x), \psi(0, x)): dx$ in extension of the analysis in Chapter 8 and show its essential selfadjointness on $\mathbf{D}_\infty(H_0'')$, where H_0'' is the closure of $H_0 \times I + I \times H_0'$. Show also that $H_0'' + V$ is essentially selfadjoint, and derive the other basic results of Chapter 8 (unique vacuum vector, local differential equations of motion, etc.).

5. Treat the case of a selfinteracting scalar field in Minkowski space of arbitrary dimension, with a "momentum cutoff." This means that in place of the interaction Hamiltonian $V = \int :p(\varphi(0, x)): f(x) dx$, one uses the interaction Hamiltonian $V(g) = \int :p(\varphi(0, g_x)): dx$, where g is a smooth function. (Either introduce a spatial cutoff, to be removed later, or use "periodic boundary conditions," i.e., replace $\mathbf{R}^n$ by n-torus.) Show the existence of a unique physical vacuum for a selfadjoint total Hamiltonian, and renormalized equations of motion that are local in time, but nonlocal in space.

6. With the notation of Theorem 8.7, show that if S is a product of intervals, then $(\mathbf{R}(S))' = \mathbf{R}(S^c)$, where S^c is the complement of S. (Cf. Araki, 1963.)

7. Show that the only vectors in the space $\mathbf{K}$ for the quantized Klein-Gordon field over Minkowski space that are invariant under the induced action of spatial translation are proportional to ν. (This indicates that a spatial-translation invariant interacting field cannot have a normalizable vacuum vector in the free field representation, and is referred to as Haag's theorem.) Show also that there are no trace-class density operators on $\mathbf{K}$ that are invariant under the induced action of spatial translations.

8. Let $\{A_n\}$ and $\{B_n\}$ be two sequences of mutually strongly commutative operators in a Hilbert space $\mathbf{M}$. Show that there exists a Weyl system $(\mathbf{K}, W)$ over the pre-Hilbert sequence space ℓ_2^0 (of all finite sequences $\{a_n\}$ with $\langle\{a_n\}, \{b_n\}\rangle = \Sigma a_n \bar{b}_n$) such that $\Sigma_n(A_n \times P_n + B_n \times Q_n)$ is essentially selfadjoint on $\mathbf{M} \times \mathbf{K}$, where $P_n = \partial W(e_n)$, $Q_n = \partial W(ie_n)$. (Cf. Segal, 1960.)

9. For the nonlinear wave equation $(\partial_t^2 - \Delta + 1)\varphi + p'(\varphi) = 0$ on $M = \mathbf{R} \times S^3$, where p is a given real polynomial, the energy may be expressed as

$$H(\varphi) = (2T)^{-1} \int_{-T}^{T} \int_{S^3} [(\partial_t \varphi)^2 + (\nabla \varphi)^2 + \varphi^2 + p(\varphi)] dx dt$$

(with the usual measure on S^3). Let φ denote the quantized free field for the equation $(\partial_t^2 - \Delta + 1)\varphi = 0$. Show that $H(\varphi)$, when interpreted in accordance with Wick products, is an operator if $T = \pi$, and is in fact essentially selfadjoint on $\mathbf{D}_\infty(H_0)$, where H_0 is the free Hamiltonian. (Hint: show that φ is periodic in time with period 2π, and use Poulsen's theorem. [Cf. Paneitz and Segal, 1983.].)

10. Show that the full Poincaré group, and not merely time evolution and spatial transformations, acts as C^*-automorphisms of the space-finite Weyl algebra, for the quantized scalar field that coincides with the free Klein-Gordon field in two-dimensional Minkowski space at time 0, and $H_0 + \int :p(\varphi(0, x)):dx$, where p is a nonnegative polynomial as symbolic Hamiltonian, interpreted rigorously as in Corollary 8.7.2. (This shows the relativistic covariance of the associated nonlinear field. [Cf. Klein, 1973; Cannon and Jaffe, 1970.])

11. a) In the real wave representation of the free boson field over the Hilbert space $\mathbf{H}$, show that e^{-tN} (where N is the number of particles) is bounded from $L_p(\mathbf{H}_\varkappa)$ to $L_q(\mathbf{H}_\varkappa)$, where $\varkappa$ is the relevant conjugation, if and only if it is a contraction from the former to the latter.

b) Show more specifically that the boundedness holds if and only if $q - 1 \leq (p - 1)e^{2t}$. (Cf. Nelson, 1973b; Gross, 1975.)

12. a) In the complex wave representation of the free boson field over $\mathbf{H}$, show that e^{-tN} is a contraction from $AL_p(\mathbf{H})$ to $AL_q(\mathbf{H})$ if $q \leq pe^{\varepsilon t}$ for a suitable constant ε. Here $AL_p(\mathbf{H})$ denote the space of all antientire functions F on $\mathbf{H}$ for which the norm

$$\|F\|_p = (\sup_M \textstyle\int_M |F(z)|^p dg_{1/2}(z))^{1/p}$$

is finite, where the supremum is taken over all finite-dimensional subspaces of $\mathbf{H}$.

b) Show more specifically that if e^{-tN} is bounded from the one space to the other, then it is a contraction, and the boundedness holds if and only if $q = pe^{2t}$. (The results of Prob. 12 are due to Z. Zhou.)

13. Show that in the context of Theorem 8.8 that if p is an even polynomial, and $V = \int_{S^1} :p(\varphi(0, x)):dx$ then the physical vacuum expectation values of $\int_{S^1} :\varphi(0, x)^n:dx$ vanish for all odd n.

14. Show in the case $p(x) = ax^4 + bx^2$ ($a > 0$) that the renormalized polynomial $q(x) = ax^4 + b'x^2 + c$ for suitable b' and c.

15. In Problem 14, show that c is a C^∞ function of the original mass m and of the constants a and b.

Bibliographical Notes on Chapter 8

The construction of nonlinear quantized fields along the present lines was initiated by Segal (1967), where the nonlinear Weyl relations in rigorous form and the effective independence from spatial cutoffs on the interaction were first treated. The use of the L_p-scale in the real wave representation for the analysis of the free and interaction semigroups was introduced by Segal (1969c).

The semiboundedness of the Hamiltonian for the model on S^1 in which the free field is substituted into the nonlinear expression for the interaction Hamiltonian was first indicated by Nelson (1966). This analysis was based on study of the analytic continuation from real to pure imaginary times, and is correspondingly known as the *euclidean* approach, in distinction from the directly physical (or *Lorentzian*) approach. A starting point was the observation that the one-dimensional harmonic oscillator semigroup e^{-tN} carries $L_2(\mathbf{R}, g)$ into $L_p(\mathbf{R}, g)$ with $p > 2$ if t is sufficiently large. Glimm (1968) showed that e^{-tN} was a contraction from the former to the latter space for sufficiently large t, and Segal (1969c; 1970d) established the analogous result in infinitely many dimensions.

The real-time approach used here may be more adaptable to general types of fields, and follows Segal (1970d, 1971). A summary of the basic results and ideas for the proofs was given by Segal (1967; 1969b, c). Both approaches involve properties of e^{-tV}, where V is the interaction Hamiltonian, which were first established by Nelson (1966). Renormalization with respect to the physical rather than free vacuum, and the establishment of fully local equations of motion for the interacting quantized field, were developed by Segal (1971).

The euclidean approach has been extensively treated, notably by Nelson (1972; 1973a, b). Relevant to both approaches are studies of Gross (1972; 1973; 1974; 1975a, b).

Appendix A. Principal Notations

X	A point of space-time, also denoted (t, x), where t is the time and x is the corresponding point of space.
x	1) A point of space. 2) A vector in a vector space.
$\mathbf{M}_0$	Minkowski space.
$(x_0, x_1, \ldots, x_n)$	The coordinates of the point X in $(n + 1)$-dimensional Minkowski space.
$\mathbf{L}^*$	The dual of the topological vector space $\mathbf{L}$.
$^*\mathbf{L}$	The antidual of the topological vector space $\mathbf{L}$.
$\mathbf{H}$	A complex Hilbert space.
$\langle x, y\rangle$	The inner product of two vectors x and y in a Hilbert space (linear in x and antilinear in y).
$\varkappa$	A conjugation on $\mathbf{H}$.
$\mathbf{H}_\varkappa$	The real-linear subspace of $\mathbf{H}$ consisting of all vectors left invariant by $\varkappa$.
$\mathbf{H}^\#$	$\mathbf{H}$ as a real Hilbert space, i.e., disregarding its complex structure, with the inner product $\langle x, y\rangle' = \mathrm{Re}(\langle x, y\rangle)$.
$\mathbf{K}$	The Hilbert space of the free field over the single-particle space $\mathbf{H}$.
$\varphi(X)$	A classical field, evaluated at the point X.
$\boldsymbol{\varphi}(X)$	A quantized field, evaluated at the point X (boldface letters are used for *quantized* fields).
$\boldsymbol{\phi}(t, f)$	The weighted space integral of the field $\boldsymbol{\varphi}$, at the time t, with weight function f. Formally, $\boldsymbol{\phi}(t, f) = \int_S \boldsymbol{\varphi}(t, x)f(x)\, dx$, where S is space. $\boldsymbol{\phi}(0, f)$ is denoted as $\boldsymbol{\phi}(f)$.
$\boldsymbol{\Phi}(f)$	The weighted space-time integral of the field $\boldsymbol{\varphi}$ with weight function f. Formally, $\boldsymbol{\Phi}(f) = \int_M \boldsymbol{\varphi}(X) f(X) dX$, where M is space-time.
v	The vacuum vector in $\mathbf{K}$.
$\Gamma(U)$	The unitary operator on the free field space $\mathbf{K}$ corresponding to the unitary operator U on the single-particle space $\mathbf{H}$.
$C(z)$	The creation operator in $\mathbf{K}$ for the state vector z in the single-particle space $\mathbf{H}$.
$C(z)^*$	The annihilation operator in $\mathbf{K}$ for the state vector $z \in \mathbf{H}$.
$\phi(z)$	The hermitian field operator $\sqrt{2}^{-1}(C(z) + C(z)^*)$ for the state vector $z \in \mathbf{H}$.
$W(z)$	The Weyl operator $\exp(i\phi(z))$ on $\mathbf{K}$ for the vector z in $\mathbf{H}$.
∂	$-id$, where d is the usual differential. (Example: $\partial\Gamma(A)$ is the

free field Hamiltonian corresponding to the single-particle Hamiltonian A.)

N — The total particle number, $\partial\Gamma(\mathbf{I})$ (unless otherwise indicated).

g_C — The isonormal probability distribution of covariance operator C and mean 0.

$GL(\mathbf{L})$ — The group of all invertible continuous linear transformations on the topological vector space $\mathbf{L}$.

$(\mathbf{L}, F)$ — A topological vector space $\mathbf{L}$ together with a nondegenerate bilinear form F on $\mathbf{L}$.

$Sp(\mathbf{L}, A)$ — The symplectic group over $(\mathbf{L}, A)$, where A is antisymmetric, consisting of all transformations T in $GL(\mathbf{L})$ leaving A invariant.

$O(\mathbf{L}, S)$ — The orthogonal group over $(\mathbf{L}, S)$, where S is symmetric, consisting of all transformations T in $GL(\mathbf{L})$ leaving S invariant.

$\mathbf{B}(\mathbf{H})$ — The algebra of all bounded operators on the Hilbert space $\mathbf{H}$.

$\mathbf{A}'$ — For any subset $\mathbf{A}$ of $\mathbf{B}(\mathbf{H})$, the set of all bounded operators in $\mathbf{B}(\mathbf{H})$ that commute with every operator in $\mathbf{A}$.

T^* — The adjoint of the operator T.

$L_p(M, \mathbf{L})$ — For any measure space M and Banach space $\mathbf{L}$ (both assumed separable), the space of all strongly measurable functions f from M to $\mathbf{L}$ for which the norm

$$\|f\|_p = (\textstyle\int_M \|f(x)\|^p dx)^{1/p} < \infty.$$

$L_p(\mathbf{A}, v)$ — For any abelian W^*-algebra $\mathbf{A}$ and vector v, the space of all closed operators T affiliated with $\mathbf{A}$ such that $v \in \mathbf{D}(|T|^{p/2})$, with the norm $\|T\|_p = \langle |T|^{p/2}v, |T|^{p/2}v\rangle^{1/p}$, where $|T|$ is the self-adjoint component of T in its polar decomposition.

$L_p(\mathbf{H}', g_C)$ — If $\mathbf{H}'$ is a real Hilbert space, the space $L_p(\mathbf{H}', g_C)$, where usually $C = I$, defined by completion of the real polynomials over $\mathbf{H}'$.

$L_\mathbf{I}$ — For any real interval $\mathbf{I} \subset [1, \infty]$, $\bigcap_{p \in \mathbf{I}} L_p$ with the topology of convergence in each L_p.

$L_{p,r}(M)$ — The Sobolev space of all functions in L_p together with their first r derivatives, on the manifold M.

$\mathbf{D}(T)$ — The domain of the operator T.

$\langle \mathbf{D}(T) \rangle$ — The domain of the operator T in a Hilbert space, as a pre-Hilbert space with respect to the inner product $\langle x, y\rangle_T = \langle Tx, Ty\rangle$.

$\langle\langle \mathbf{D}(T) \rangle\rangle$ — The completion of $\langle \mathbf{D}(T) \rangle$ (or $\langle \mathbf{D}(T) \rangle$ if already complete).

$\mathbf{D}_\infty(T)$ — The set of all vectors in the domain of T^n for all $n = 1, 2, \cdots$.

$\langle \mathbf{D}_\infty(T) \rangle$ — The set $\mathbf{D}_\infty(T)$ in the sequential topology of convergence in every pre-Hilbert space $\langle \mathbf{D}(T^n) \rangle$.

$\mathbf{D}_{-\infty}(T)$	The antidual of $\mathbf{D}_{\infty}(T)$.
$\langle \mathbf{D}_{-\infty}(T) \rangle$	$\mathbf{D}_{-\infty}(T)$ in the sequential topology of convergence in some $\langle \mathbf{D}(T^m) \rangle$, where m may be negative (for invertible T).
$\oplus$	The direct sum (of vector spaces or of Hilbert spaces).
$\otimes$	The tensor, or direct, product (of vector spaces, operators).
$\bar{f}$	Complex conjugate of the function f.
$f^{(*r)}$	r-fold convolution of the function f with itself.
$C_0^\infty(M)$	The space of all C^∞ functions of compact support on the manifold M.
Δ	The (nonpositive) Laplacian on the Riemannian manifold M.
$\Box$	The wave operator $\partial_t^2 - \Delta$ on $\mathbf{R} \times M$, M as under Δ.
$\varnothing$	Empty set.
$\mathbf{W}$	Wiener space, normally the set of all real continuous functions on $[0, 1]$ vanishing at 0.
$\{x: P\}$	The set of all elements x satisfying condition P.
$\mathbf{R}$	The set of all real numbers.
$\mathbf{C}$	The set of all complex numbers.

Appendix B. Universal Fields and the Quantization of Wave Equations

There are four main points in the subsumption of all types of boson or fermion fields under the universal fields over an abstract Hilbert space.

1) There is a mutual correspondence between covariant wave equations and irreducible unitary representations of an underlying symmetry group G.

2) For each type of field there is a covariant mapping T from the space of test functions appropriate to the field into a dense subset of Hilbert space $\mathbf{H}$.

3) T carries the commutator (resp. anticommutator) into the imaginary (resp. real) part of the inner product in $\mathbf{H}$.

4) In consequence, the quantization of a given wave equation is effectively equivalent to the quantization of a given unitary group representation in Hilbert space (or some variant thereof, such as a symplectic or orthogonal representation). This latter *quantization* is simply a specialization of the quantization for the full unitary group on Hilbert space, and so is universal, i.e., the same for all wave equations.

To exemplify these points, we consider the case in which G is the Poincaré (i.e., "inhomogeneous Lorentz") group, acting on Minkowski space $\mathbf{M}_0$. Mathematically, this is the pseudo-euclidean group for the four-dimensional real vector space with fundamental quadratic form

$$x_0^2 - x_1^2 - x_2^2 - x_3^2,$$

where the x_j are suitable coordinates, of which x_0 is called the "time." A unitary representation of G is a continuous mapping $g \to U(g)$ from G to unitary operators on a complex Hilbert space $\mathbf{H}$ that preserves the group operations: $U(ab) = U(a)U(b)$ for arbitrary a, b in G. The simplest nontrivial irreducible unitary representation U of G that has *positive energy* (meaning that the selfadjoint generator of the one-parameter unitary group $U(x_{0 \to} x_{0} + t)$ is positive) is that associated with the Klein-Gordon equation (or physically speaking, neutral scalar fields). This is the equation

$$(*) \qquad \Box\varphi + m^2\varphi = 0 \quad (m > 0).$$

This is a slight variant of the classical wave equation that is readily solved. For example, if $f(x)$ and $g(x)$ are arbitrary C_0^∞ functions of the space variable $x = (x_1, x_2, x_3)$, there is a unique solution φ of equation (*) with the Cauchy data f and g at time 0: $\varphi(0, x) = f(x)$, $\partial_t\varphi(0, x) = g(x)$. This solution φ is C^∞

and has compact support in space at all times. The set $\mathbf{H}_0$ of all such solutions is G-invariant, where G operates as follows:

$$U(g): \varphi(X) = \varphi(g^{-1}(X)) \qquad (X = (x_0, x_1, x_2, x_3).$$

An important aspect of this situation is that the space $\mathbf{H}_0$ admits continuous symplectic and quadratic forms that are invariant under G and unique (within multiplication by a constant). The symplectic form A is local, taking the form of an integral over space.

$$A(\varphi, \psi) = \int [\psi(\partial_t\varphi) - (\partial_t\psi)\varphi]dx.$$

The orthogonal form is nonlocal, taking the following form:

$$S(\varphi, \varphi) = \int [(C^{-1}(\partial_t\varphi))^2 + (C\varphi)^2]dx,$$

where $C = (m^2 - \Delta)^{1/4}$. The form A is closely related to the commutator form for the quantized Klein-Gordon equation, and the form S to the two-point function. Together they determine a unique complex Hilbert space $\mathbf{H}$, by appropriate introduction of the action of the complex unit i, characterized as a transformation on $\mathbf{H}_0$ (which maps $\mathbf{H}_0$ outside of itself but into its completion with respect to the form S) such that $i^2 = -I$, and $S(\varphi, \psi) = A(i\varphi, \psi)$. (In terms of the Fourier transform of φ, i simply multiplies the positive-frequency component by the complex number i and the negative-frequency component by the complex number $-i$; but the Fourier transform approach doesn't work in general space-time and appears less directly physical.)

The upshot is a unitary representation U of G in the completion $\mathbf{H}$ of $\mathbf{H}_0$; $\mathbf{H}$ is the space of normalizable solutions of the Klein-Gordon equation. Conversely, as shown in the classical work of Wigner (1939) (cf. Mackey, 1963), every unitary positive-energy representation of the Poincaré group arises in a similar way, from some invariant wave equation.

Turning now to (2), a familiar formulation of the notion of quantum field is as an operator-valued distribution on space-time. In this approach, corresponding to any C_0^∞ function h on space-time, there should be an operator $\phi(h)$ on the quantum field Hilbert space $\mathbf{K}$, satisfying the underlying wave equation and appropriate commutation relations. In the universal field over the Hilbert space $\mathbf{H}$, there is available an essentially unique mapping $\phi_0(x)$ that is defined for all vectors x in $\mathbf{H}$. To derive a mapping ϕ in the present particular case from the universal mapping ϕ_0, what is needed is an appropriate mapping T from C_0^∞ over space-time into $\mathbf{H}$; the definition $\phi(h) = \phi_0(Th)$ then applies. To preserve relativistic invariance, the mapping T must be covariant, i.e., intertwine the respective actions of G: $TU_0(g) = U(g)T$, where $U_0(g)$ denotes the action of g on $C_0^\infty(\mathbf{M}_0)$, which happens to be formally identical to the action of $U(g)$, i.e., $U_0(g)$: $h(X) \rightarrow h(g^{-1}(X))$, where X is arbitrary in $\mathbf{M}_0$.

This mapping T is not entirely unique, but different choices for T lead to quantizations that are unitarily equivalent, and thereby physically indistinguishable. The technically most succinct procedure is to use Fourier analysis, and restrict the Fourier transform of h to the "mass hyperboloid" $K^2 = m^2$, obtaining the Fourier transform of a function φ in $\mathbf{H}$, which can now be defined as Th. More generally applicable is a partial differential equation approach, using the fundamental singular functions associated with the wave equation. A simple local formulation uses the commutator function (more exactly, distribution) D, defined as the solution of equation (*) with the Cauchy data $\varphi(0, x) = 0$, $\partial_t\varphi(0, x) = \delta(x)$. The mapping T

$$h(X) \rightarrow \int D(X - Y)h(Y)d^4Y,$$

is then a covariant map from $C_0^\infty(\mathbf{M}_0)$ into a dense subset of $\mathbf{H}$. Since the quantized Klein-Gordon field satisfies the Klein-Gordon equation, nothing material to this field is lost in this procedure, and some redundancy in the labeling of field variables is eliminated, by discarding components of the test functions that are "off the mass shell," and so give vanishing contributions to the field operators.

In particular, the commutator $[\phi(h), \phi(k)] = i \int h(X)k(Y)D(X - Y)dXdY$ may equally be expressed as $iA(Th, Tk)$ in terms of the imaginary part of the inner product in $\mathbf{H}$. The two-point function of the quantized field is similarly expressible by the real part of the inner product. The underlying differential equation for the quantized field φ may be expressed as

$$\phi((\Box + m^2)h) = 0$$

after multiplication by the arbitrary function h in $C_0^\infty(\mathbf{M}_0)$ followed by integration. This follows from the fact that

$$T(\Box + m^2)h = (\Box + m^2)Th = 0$$

since $\Box$ commutes with convolutions and Th satisfies the Klein-Gordon equation.

Thus the quantization of the Klein-Gordon equation is derivable from a type of quantization of the associated unitary group representation. But the latter quantization is just the restriction to the Poincaré group of the corresponding quantization of the full unitary group $U(\mathbf{H})$ on the Hilbert space $\mathbf{H}$. This quantization carries an arbitrary unitary operator V on $\mathbf{H}$ into a unitary operator $\Gamma(V)$ on the quantized field Hilbert space $\mathbf{K}$. The quantization for the Klein-Gordon case is obtained simply by substituting $U(g)$ for V. Since all Hilbert spaces of a given dimension are unitarily equivalent, there is essentially just one universal quantization, for each type of statistics, Bose or Fermi.

It is a considerable clarification and economy to reduce the quantization of wave equations in this way to the treatment of the universal boson and fermion fields over a Hilbert space. This reduction applies irrespective of the covariance group, the underlying space-time, or the transformation properties of the underlying fields, as long as there is, as is typical, a canonical invariant complex inner product on the solution manifold of the wave equation in question. For groups such as the Poincaré or conformal groups, the solution manifold is normally irreducible, and the essential unicity of the mapping T from the space of test functions to the solution manifold of the wave equation follows by group-theoretical considerations. But even when the group is only a one-parameter temporal evolution group, essential unicity may be deduced from stability, or positive-energy, considerations. Thus, for the (real) equation $\Box\varphi + V(x)\varphi = 0$, where $V(x)$ is a given bounded nonnegative function on space, there is a unique temporally invariant positive-energy (complex) Hilbert space structure on the solution manifold, and quantization is again reducible to the universal boson field treated in Chapter 1. This follows from a variant of the universal theory in which the infinite-dimensional symplectic group takes over the role of the unitary group on a Hilbert space.

For detailed treatment of the quantization of specific wave equations, see Chapter 6 and also some of the earlier lexicons and problems. For nonlinear functions of specific quantum fields, including generalized Wick products, see Chapters 7 and 8.

Glossary

ADJOINT. If T is a densely defined operator in the Hilbert space $\mathbf{H}$, its adjoint T^* is defined as follows: a vector y is in the domain $\mathbf{D}$ of T^* if and only if there exists a vector y' in $\mathbf{H}$ such that $\langle Tx, y\rangle = \langle x, y'\rangle$ for all vectors x in the domain of T. For any such vector y, T^*y is defined as y'.

This defines T^* uniquely, but there is nothing to keep T^* from having a domain that is vacuous, apart from the vector 0. But, in case T^* is densely defined, its adjoint T^{**} is an extension (q.v.) of T. The adjoint of an operator is always closed (q.v.), and T^{**} (when it exists) is the minimal closed extension of T, meaning that any closed operator that extends T also extends T^{**}. Conversely, if T is densely defined and has a closed extension, then T^* is densely defined.

Note that just because $\langle Tx, y\rangle = \langle x, Sy\rangle$ for two densely defined operators S and T, for x in the domain of T and y in the domain of S, S doesn't need to be the adjoint of T, even if closed, but only extended by T^*.

ADJUNCTION OPERATION. An operation, usually denoted*, on a complex associative algebra $\mathbf{R}$, with the following properties: $(A+B)^* = A^* + B^*$, $(AB)^* = B^*A^*$, $(cA)^* = \bar{c}\,A^*$, and $A^{**} = A$, for arbitrary A, B in $\mathbf{R}$ and complex number c.

AFFILIATION. An unbounded or partially defined operator A is said to be affiliated with a W^*-algebra $\mathbf{R}$ in case it commutes with all unitary operators in the commutor $\mathbf{R}'$ (q.v.) of $\mathbf{R}$, When A is selfadjoint or normal, this is equivalent to the condition that every spectral projection of A be in $\mathbf{R}$. When $\mathbf{R}$ is abelian, any densely defined operator that is affiliated with it is automatically essentially normal (i.e., has normal closure); in particular, if hermitian it is automatically essentially selfadjoint.

ANALYTIC VECTOR. See REGULAR VECTOR.

BANACH ALGEBRA. This is a Banach space $\mathbf{B}$ that is also an algebra, with the property that $\|xy\| \leq \|x\|\,\|y\|$ for arbitrary x and y in $\mathbf{B}$.

BANACH SPACE. This is a vector space $\mathbf{L}$, together with a norm $\|x\|$ for vectors x in $\mathbf{L}$, that is "complete," in the sense that if $\{x_n\}$ is a (Cauchy) sequence such that $\|x_m - x_n\| \to 0$ as $m, n \to \infty$, then there exists a vector x in $\mathbf{L}$ such that $\|x - x_n\| \to 0$ as $n \to \infty$. A Banach space is a particular type of topological vector space in which convergence of a sequence $\{x_n\}$ to a vector x is defined to mean that $\|x - x_n\| \to 0$. The L_p-spaces are Banach spaces.

BOREL FUNCTION. Let B be the smallest class of functions on $\mathbf{R}^n$ that contains all polynomials and has the property that if $f_n \in B$ and $f_n(x) \to f(x)$ for all x, then $f \in B$. A function in B is called a Borel function.

BOREL SET. A Borel subset of a topological space X is one that is in the smallest σ-ring (q.v.) containing all open subsets of X. Its characteristic function is Borel.

C^*-ALGEBRA. A *concrete* C^*-algebra is an algebra of bounded operators on a complex Hilbert space **H** having the property that it is closed in the uniform topology and under the adjunction operation *. Unlike W^*-algebras, C^*-algebras can be characterized purely algebraically, leading to the concept of an *abstract* C^*-algebra. This is a complex Banach algebra, having an adjunction operation * such that $\|A\| = \|A^*\|$ and $\|AA^*\| = \|A\|\,\|A^*\|$. Any abstract C^*-algebra is isomorphic to some concrete C^*-algebra, which, however, is in general not at all unique *spatially*, in that if two concrete C^*-algebras are algebraically *-isomorphic (meaning that the *'s correspond as well as the usual algebraic operations), they are in general *not* at all *unitarily equivalent* (although the norms $\|A\|$ automatically correspond). Whereas the fruitful equivalence relation for W^*-algebras is primarily that of unitary equivalence (and so is spatial), that for C^*-algebras is primarily *-algebraic isomorphism. The selfadjoint elements of a C^*-algebra turn out to be a natural and effective model for the conceptual observables of a physical system, in such a way that *-algebraically isomorphic C^*-algebras correspond to physically equivalent systems. In particular, the concept of *state* of a physical system is conveniently expressible in terms of C^*-algebra, and is also important in the mathematical theory of C^*-algebras. (Cf. Segal, 1963, chap. 1.)

CAUCHY PROBLEM. The Cauchy problem (also known as the initial value problem) is the solution of an evolutionary differential equation, e.g., the abstract equation $u'(t) = Au(t) + K(u(t))$ where A is linear and K is nonlinear, given the solution at a fixed time, as $u(t_0) = u_0$. *Duhamel's principle* provides a corresponding integral equation that incorporates the initial condition and provides a slightly more tractable and physically appropriate problem than the literal differential equation.

CLIFFORD ALGEBRA. If Q is a nondegenerate quadratic form on a linear vector space **L**, the Clifford algebra **C** over $(\mathbf{L}, Q)$ is the algebra generated by **L** and a unit e ($e^2 = e$ and e commutes with all vectors in **L**) subject only to the relations $x^2 = Q(x, x)e$, for arbitrary x in **L**. In principle, the field of scalars applicable to **L** is arbitrary, but only the case of *real* scalars in **L**, together, however, with *complex* coefficients for the algebra **C**, is used here. If **L** has a finite basis $e_1, e_2, \ldots, e_n$, then **C** has dimension 2^n and is spanned by e together with the $e_{i_1} e_{i_2} \cdots e_{i_r}$ with $i_1 < i_2 < \cdots < i_r$. When n is even, **C** is isomorphic to the algebra of all $m \times m$ complex matrices, where $m = 2^{n/2}$. When Q is positive definite, there is an adjunction operation * on **C** that is uniquely determined by the condition that $x^* = x$ for all x in **L**. The definition of Clifford algebra does not require that **L** be finite-dimensional, and if **L** is infinite-dimensional and Q is positive-definite, the corresponding Clifford algebra is dense in the trace-endowed infinite-dimensional analog of a complete matrix algebra discovered by von Neumann, known as the *approximately finite* II_1 factor.

CLOSED OPERATOR. An operator T in a Banach space **H** is closed in case whenever $\{x_n\}$ is a convergent sequence of vectors in the domain **D** of T such that the sequence $\{Tx_n\}$ is also convergent, then the limit x of $\{x_n\}$ is also in **D**, and Tx is the limit of $\{Tx_n\}$.

CLOSURE. A linear operator T in a Banach space **B** has a closure $\overline{T}$ provided it has a closed *extension* (q.v.). Equivalently, it has a closure if and only if the closure $\overline{G(T)}$ of its graph $G(T)$ is single-valued, in which case $\overline{G(T)}$ is the graph of $\overline{T}$. The graph of T is defined as the subset of the topological direct sum $\mathbf{B} \oplus \mathbf{B}$ consisting of all vectors of the form $x \oplus Tx$ with x in the domain of T.

COMMUTOR. The commutor (also known as commutant) of a set S of continuous linear operators on a topological vector space **L** is the set (denoted S') of all such operators on **L** that commute with every operator in S.

COMPACT OPERATOR. An operator on a Banach space is compact if it carries the unit ball into a set whose closure is compact. In an infinite-dimensional separable Hilbert space **H**, a selfadjoint operator T is compact if and only if **H** has an orthonormal basis $\{e_n\}$ such that $Te_n = \lambda_n e_n$, where $\lambda_n \to 0$ as $n \to \infty$.

COMPLEX STRUCTURE. A complex structure in a real topological vector space **L** is a continuous linear transformation J on **L** such that $J^2 = -I$, where I denotes the identity on **L**. If $(\mathbf{L}, A)$ is a symplectic space (q.v.), J is called *symplectic* in case $A(Jx, Jy) = A(x, y)$ for all $x, y \in \mathbf{L}$; and is called *positive* in case $A(Jx, x) \geq 0$ for all $x \in \mathbf{L}$. A positive symplectic complex structure in **L** gives rise to a complex pre-Hilbert structure in **L** in which $(a + ib)x$ is defined as $ax + bJx$ for arbitrary real a and b, and $\langle x, y\rangle$ is defined as $A(Jx, y) + iA(x, y)$. Similarly in the case of an orthogonal space.

CONFORMAL GROUP. Every one-to-one transformation of Minkowski space $\mathbf{M}_0$ onto itself that preserves causality is in the Ponicaré group apart from a scale transformation when the space dimension $n > 1$. However, there exist local transformations in a neighborhood of any point in $\mathbf{M}_0$ that are one-to-one and preserve causality inside the neighborhood, but typically cannot be extended to all of $\mathbf{M}_0$ (without mapping $\mathbf{M}_0$ outside of itself). These are the *conformal* transformations. They are also definable as the transformations preserving the conformal structure associated with the Lorentz (or pseudo-Riemannian) quadratic differential form $dx_0^2 - dx_1^2 - \cdots - dx_n^2$. The conformal group is (locally isomorphic to) the group of all causal transformations on the "Einstein Universe," in which $\mathbf{M}_0$ is covariantly imbedded. The wave, Maxwell, and neutrino equations are covariant relative to the conformal group, but equations involving nonvanishing mass, such as the Klein-Gordon and Dirac electron equations, are not.

CONJUGATION. A conjugation on a complex Hilbert space $\mathbf{H}$ is an involutory antilinear isometry. Thus if $\varkappa$ is a conjugation on $\mathbf{H}$, $\varkappa^2 = I$, $\varkappa(ax) = \bar{a}\,\varkappa(x)$ for all complex numbers a and x in $\mathbf{H}$, and $\langle \varkappa x, \varkappa y\rangle = \langle y, x\rangle$. The elements of $\mathbf{H}$ that are invariant under $\varkappa$ form a real Hilbert space $\mathbf{H}_\varkappa$ relative to $\langle\cdot,\cdot\rangle$ and every element of $\mathbf{H}$ has the form $z = x + iy$ for unique x and y in $\mathbf{H}_\varkappa$. When $\mathbf{H}$ has a more structured form such as $L_2(M)$, M being a given measure space, there is a natural special conjugation, namely complex conjugation, but in general there is no unicity about a conjugation on a Hilbert space. Any two conjugations $\varkappa$ and $\varkappa'$ are conjugate via a unitary operator: $\varkappa' = U^*\varkappa U$ for some unitary U.

CYCLIC VECTOR. A cyclic vector for a set S of operators on a space $\mathbf{L}$ is a vector z in $\mathbf{L}$ such that the set of all vectors Tz, where T is in the algebra generated by S, is dense in $\mathbf{L}$. If S is an abelian selfadjoint algebra of bounded operators on a Hilbert space that has a cyclic vector, then $S' = S''$, i.e., the algebra S' is maximal abelian; conversely, in a separable Hilbert space every maximal abelian selfadjoint algebra has a cyclic vector. The algebra is then said to have simple spectrum. In particular, if S is the W^*-algebra generated by the spectral projections of a given selfadjoint operator A, then A is said to have simple spectrum. This notion generalizes to Hilbert space that of a selfadjoint matrix all of whose eigenvalues are distinct. See also SEPARATING VECTOR.

DIAGONALIZABLE OPERATOR. In heuristic usage, this is an operator that is appropriately conjugate to one that is in diagonal form. As used in lexicons here, it is an operator on Hilbert space that is unitarily equivalent to the operation of multiplication by a measurable function, acting in L_2 over a measure space. Such an operator T is densely defined and satisfies the equation $TT^* =$

T^*T, and conversely such a operator (called "normal" in mathematical literature) is diagonalizable.

DIFFERENTIABLE VECTOR. See REGULAR VECTOR.

DIFFERENTIAL. If T is an operator (not necessarily linear) from a topological vector space **L** to another such space **M**, its differential $(dT)_x$ at the point x of **L** is the linear mapping in **L** to **M** defined by the equation $(dT)_x y = \lim_{t\to 0} t^{-1}[T(x + ty) - T(x)]$.

DIRECT SUM. A Hilbert space **H** is the direct sum of subspaces $\mathbf{H}_j$ if the $\mathbf{H}_j$ are mutually orthogonal and every vector in **H** is the sum of its components x_j in the $\mathbf{H}_j$ (or, equivalently, the $\mathbf{H}_j$ span **H**). One writes $\mathbf{H} = \oplus_j \mathbf{H}_j$.

DIRECTED SYSTEM. This is a generalization of the positive integers used to label generalized sequences, called nets (q.v.). Specifically, it is a partially ordered system $(S, <)$, with the property that any two elements a and b in S have an upper bound c in S. A typical example is the set of all neighborhoods of a point in a topological space, ordered by reverse inclusion. Little will be lost if the reader thinks of the term as either the positive integers or this example.

DISTRIBUTION. A predistribution on a topological linear space **L** is a linear map D from the dual $\mathbf{L}^*$ of **L** to random variables on a probability measure space P. A distribution (also known as generalized random process) is an equivalence class of predistributions relative to the following equivalence relation: the predistributions D and D' on **L** to random variables on P and P' are equivalent if and only if for arbitrary finite subsets $f_1, f_2, \ldots, f_n$ in $\mathbf{L}^*$, the joint probability distribution of $D(f_1), \ldots, D(f_n)$ is the same as that of $D'(f_1), \ldots, D'(f_n)$. When **L** is finite-dimensional, this notion of distribution is effectively coincident with the usual one of a probability distribution in the space **L**, but this is not the case when **L** is infinite-dimensional. Thus the *isonormal distribution* (q.v.) in a real Hilbert space **H** is a distribution in the present sense that cannot be represented by a countably additive probability distribution of the usual type. The theory extends to "noncommutative" distributions, in which the values of $D(f)$ are effectively selfadjoint operators. See RANDOM VARIABLE and SPECTRAL THEORY.

DUAL SPACE. The dual of a topological vector space **L** is the vector space denoted $\mathbf{L}^*$ consisting of all continuous linear functionals on **L**. The *antidual*, denoted $^*\mathbf{L}$, is the space of all continuous conjugate (or anti-) linear functionals, which functionals f differ from linear functionals in the property that $f(ax) = \bar{a} f(x)$ for arbitrary scalars a and vectors x.

DUHAMEL'S PRINCIPLE. In a succinct abstract form, this classic principle for obtaining the solution of an inhomogeneous linear equation from the solution of the corresponding linear one may be stated as follows. For $t \geq 0$, let $S(t)$ be a bounded operator on the Banach space $\mathbf{B}$, and suppose that $S(0) = I$, $S(t + t') = S(t)S(t')$ for arbitrary t and t', and that the map $t \rightarrow S(t)x$ is continuous for all fixed $x \in \mathbf{B}$. (Such a function $S(t)$ is called a one-parameter semigroup.) Let $f(t)$ be a continuous function from $[0, \infty]$ to $\mathbf{B}$, and let u_0 be arbitrary in $\mathbf{B}$. Then there is a unique (slightly generalized) solution $u(t)$ of the equation $u' = Au + f$, where A is the infinitesimal generator of S, given explicitly as

$$u(t) = S(t)u_0 + \int_0^t S(t - s)f(s)ds.$$

A special case is a formula for $e^{t(A+B)}$. If A and B are bounded linear operators on $\mathbf{B}$, then

$$e^{t(A+B)} = e^{tA} + \int_0^t e^{(t-s)A}\, B\, e^{s(A+B)}\, ds,$$

for arbitrary real t. The formula also applies for $t \geq 0$ in case A is unbounded but generates a continuous semigroup. The tactical use of Duhamel's formula is typically different from but complementary to that of the Lie-Trotter formula (q.v.).

EXTENSION. An extension of an operator A in a Hilbert space $\mathbf{H}$ is an operator B whose domain includes that of A and agrees with A on their common domain. This is expressed symbolically as $A \subset B$.

FATOU'S LEMMA. If $\{f_n\}$ is an arbitrary sequence of nonnegative measurable functions on a measure space, then

$$\int \liminf_{n\rightarrow\infty} f_n \leq \liminf_{n\rightarrow\infty} \int f_n$$

FIELD. Physical usage regarding this term cannot be made entirely precise, but essentially, or normally, it refers to a section of a vector bundle over space-time M. The treatment of this notion is beyond the scope of this book (cf., Choquet-Bruhat et. al. 1982). Group-covariant, or homogeneous, vector bundles are determined by (or "induced from") a representation of the subgroup leaving fixed an arbitrary point of M, known as the "isotropy" subgroup (different points lead to equivalent results). Scalar, spinor, and vector fields on $\mathbf{M}_0$ are induced from the following representations of the Lorentz group $\mathbf{L}$, which is the isotropy subgroup of the Poincaré group as it acts on $\mathbf{M}_0$: taking $\mathbf{L}$ in the simply connected form $SL(2, \mathbf{C})$, the representations are (i) for scalar fields, $L \to 1$ (on a one-dimensional space); (ii) for spinor field, $L \to L \oplus L^{*-1}$ (on a four-dimensional space); and (iii) for vector fields, on the space of 2×2 complex hermitian matrices H, L maps into the transformation $H \to LHL^*$. Vector fields on $\mathbf{M}_0$ may be identified with differential 1-forms in such a way that their Poincaré transformation properties are the same.

GENERAL LINEAR GROUP. If $\mathbf{L}$ is a topological vector space, the general linear group $GL(\mathbf{L})$ over $\mathbf{L}$ is defined as the group of all continuous linear transformations T that have continuous inverses. If L is a Hilbert space, every element T of $GL(\mathbf{L})$ has a *polar decomposition* $T = US$ in which U is unitary and S is selfadjoint.

GROUP, TOPOLOGICAL AND LIE. A topological group G is a group having in addition a topology such that xy^{-1} is a continuous function of x and y, as they range (independently) over G. A Lie group is one in which a neighborhood of the unit element can be given euclidean coordinates in such a way that group composition is defined by analytic functions. Any Lie group has a *Lie algebra*, also known as an infinitesimal Lie group, which is much easier to deal with from an algebraic standpoint, but whose relation to the group involves analytic problems (see REPRESENTATION OF A GROUP).

The simplest Lie groups are $\mathbf{R}^n$, with addition of vectors as the group operation. More representative are the pseudo-orthogonal groups $O(p, q)$, consisting of all homogeneous linear transformations on a vector space of dimension $p + q$ that leave invariant the quadratic form $x_1^2 + \cdots + x_p^2 - x_{p+1}^2 - \cdots - x_{p+q}^2$, where the x_j are the coordinates. The group $O(3, 1)$ is the usual Lorentz group, and $O(n, 1)$ is its analog for $(n + 1)$-dimensional Minkowski space $\mathbf{M}_0$. Transformations in this group, as they act on $\mathbf{M}_0$, either preserve or reverse causality, in the sense that if X is in the future of Y, then gX is in the future (resp. past) of gY. If $n > 1$, the most general one-to-one transformation from $\mathbf{M}_0$ onto itself that is causal in this sense is the product of one in $SO(n, 1)$ with a vector translation $x_j \to x_j + a_j$, where the a_j are constants, and a scale transformation, $X \to \lambda X$, where λ is a positive constant. All these transformations together form the scaling-extended Poincaré group. If the scale transformations are omitted, the result is one version of the Poincaré group. This version includes the *discrete symmetries* of time reversal: $x_0 \to -x_0$ and $x_j \to x_j$ for $j > 0$; and of space reversal: $x_0 \to x_0$ and $x_j \to -x_j$ for $j > 0$. When the discrete symmetries are excluded by limiting the group to its connected component, the result is the *connected Poincaré group*, generally called simply the Poincaré group here and denoted $\mathbf{P}$.

HAUSDORFF-YOUNG INEQUALITY. This specifies an L_p space to which the convolution of two or more functions belongs in terms of the L_p spaces to which the factors belong. If G denotes $\mathbf{R}^n$, or any locally compact abelian group (with Lebesgue measure replaced by the invariant measure); if f_j is in $L_{p_j}(G)$, for $j = 1,\ldots,m$; and if $p^{-1} = p_1^{-1} + \cdots + p_m^{-1} - 1$, where the p_j and p are all in the range $[1,\infty]$, then the convolution $f = f_1 * \cdots * f_m$ is in L_p (G) and

$$\|f\|_p \leq \|f_1\|_{p_1} \cdots \|f_m\|_{p_m}.$$

HERMITE POLYNOMIALS. Normalization conventions vary regarding the definition of these polynomials $H_n(x)$, which have long figured in the expression of particle-wave duality in the free boson field. The classic definition is

$$H_n(x) = (-1)^n \exp(x^2)\,(d^n/dx^n)\exp(-x^2).$$

The $H_n(x)$ satisfy simple recursion relations that can be regarded as an expression of the actions of creation and annihilation operators on an n-particle state:

$$(d/dx)H_n(x) = 2n\,H_{n-1}(x); \qquad H_{n+1}(x) = 2x\,H_n(x) - 2nH_{n-1}(x)$$

They also represent the eigenfunctions of the harmonic oscillator in a suitable representation; thus $H_n(x)$ is the unique polynomial solution $H_n(x)$, within a constant factor, of the differential equation $u'' - 2xu' + 2nu = 0$. The addition formula

$$2^{n/2}H_n(x + y) = \sum_{k=0}^{n} H_{n-k}(\sqrt{2}x)\,H_k(\sqrt{2}x)\binom{n}{k}$$

can be regarded as a version of the binomial theorem in combination with the Wiener transform (q.v., Ch. 1). In classical terms, the representation of $H_n(x)$ as the Wiener transform of x^n, within a constant factor, corresponds to the formula

$$H_n(x) = \pi^{-1/2}\,2^n \int_0^\infty (x + iy)^n \exp(-y^2)dy.$$

The $H_n(x)$ form an orthogonal basis for $L_2(\mathbf{R}, g_{1/2})$, where in general g_c denotes the Gaussian measure of variance c on $\mathbf{R}$.

HERMITIAN OPERATOR. A hermitian (or symmetric) operator T is one whose domain $\mathbf{D}$ is dense and that has the property $\langle Tx, y\rangle = \langle x, Ty\rangle$ for all x and y in $\mathbf{D}$. This is by no means sufficient to insure that T is diagonalizable (unless T is bounded). In general, a hermitian operator has no selfadjoint extension, and even when it does, the extensions, when they exist, may have materially different spectra. In distinction to a selfadjoint operator T, which is characterized by the equality $T = T^*$, a hermitian operator is characterized by the inclusion $T \subset T^*$, meaning that T^* is an extension of T. But there is no assurance that T^* is itself hermitian; only the minimal closed extension T^{**} is surely hermitian.

HILBERT SPACE. The unmodified term Hilbert space always means *complex* Hilbert space here. Real Hilbert spaces are designated as *real*. The dimension of a Hilbert space, complex or real, may be finite, countable, or uncountable. A Hilbert space is separable if and only if it is finite or countably-dimensional.

HILBERT-SCHMIDT OPERATOR. The Hilbert-Schmidt norm $\|T\|_2$ on a Hilbert space **H** is defined by the equation $\|T\|_2^2 = \Sigma_{j,k}|\langle Te_j, e_k\rangle|^2$, where the e_j form an orthonormal basis in **H**. A Hilbert-Schmidt operator is one whose Hilbert-Schmidt norm is finite. The norm is independent of the particular basis, and may also be described as the trace $tr(T^*T)$. The totality of all Hilbert-Schmidt operators on a Hilbert space is itself a Hilbert space relative to the inner product $\langle A, B\rangle = tr(B^*A)$. In particular, a Hilbert-Schmidt operator is compact, and if selfadjoint, $\|T\|_2^2 = \Sigma_j \lambda_j^2$, where the λ_j are the eigenvalues of T.

HÖLDER'S INEQUALITY. If f and g are arbitrary measurable functions on a measure space M, and if $1 \leq p, q \leq \infty$ and $p^{-1} + q^{-1} = 1$, then, $\int|fg| \leq \|f\|_p \|g\|_q$. The case $p = q = 2$ is the Cauchy-Schwarz inequality.

INVERTIBLE OPERATOR. A continuous operator on a topological vector space is invertible if it has a continuous inverse (defined on the entire space). Partially defined inverses are not used in this book.

IRREDUCIBLE. A set of bounded operators on a Banach space is called irreducible if it leaves no nontrivial closed linear subspace invariant. For a selfadjoint set S on a Hilbert space, irreducibility is equivalent to either of the following conditions: (1) the commutor S' consists only of scalars, or (2) the W^*-algebra generated by S consists of all bounded linear operators. A *-representation ϱ of a C^*-algebra **A** is called irreducible if $\varrho(\mathbf{A})$ is irreducible.

ISONORMAL DISTRIBUTION. This is the isotropic, centered, Gaussian distribution on a Hilbert space. Thus, on a real Hilbert space this distribution D is characterized by the properties that $D(x)$ is Gaussian of mean 0 and variance $c \|x\|^2$, and that $D(x)$ and $D(y)$ are stochastically independent if x and y are orthogonal. The variant for a complex Hilbert space **H** involves in addition invariance under the phase transformations $z \to e^{i\theta}z$ ($z \in \mathbf{H}$, θ an arbitrary real number).

KLEIN-GORDON EQUATION. This is the equation $\Box\varphi + m^2\varphi = 0$ on Minkowski space $\mathbf{M}_0$ where $\Box$ denotes the wave operator $\partial_t^2 - \Delta$, Δ denoting the

Laplacian on space, and m is a constant, usually > 0, and interpreted physically as the mass of the quanta of the field represented by the equation. If f and g are given functions in C_0^∞ on space, then there is a unique solution φ such that $\varphi(t_0, x) = f(x)$, $\partial_t\varphi(t_0, x) = g(x)$, where t_0 is an arbitrary given time. This solution remains in C_0^∞ on space at all times, and its support at time t is contained in the sum of its support at time 0 with the ball in space of radius $|t|$.

There is a Poincaré- (or relativistically) invariant symplectic and local form A defined on the indicated class $\mathbf{H}_0$ of smooth solutions by the equation (where φ is assumed to be real-valued)

$$A(\varphi, \varphi') = \int[(\partial_t\varphi)\varphi' - \varphi(\partial_t\varphi')]dx,$$

the integration being over space at an arbitrary fixed time. The symmetric energy form, which is not relativistically invariant, is also expressible in local form:

$$E(\varphi, \varphi') = \int[(\partial_t\varphi)(\partial_t(\varphi') + m^2\varphi\varphi' + (\nabla\varphi) \times (\nabla\varphi')]dx.$$

The space $\mathbf{H}_0$ is dense in a complex Hilbert space $\mathbf{H}$ of generalized solutions, on which the Poincaré group acts in a unitary irreducible fashion, which is stable in the sense that the (quantum) energy operator, defined as the infinitesimal generator of time displacement in this representation, is positive. The complex unit i acts on this space of real functions as the Hilbert transform with respect to time. The imaginary part of the inner product in $\mathbf{H}$ is given by the form A. The real part of the inner product has an inherently nonlocal form:

$$\langle\varphi, \varphi\rangle = \|C\varphi(t_0, \cdot)\|_2^2 + \|C^{-1}\partial_t\varphi(t_0, \cdot)\|_2^2,$$

where C denotes the operator in L_2 over space, $(m^2 - \Delta)^{1/4}$, and $\|\cdot\|_2$ denotes the norm in this L_2.

In terms of Fourier transforms (or in "momentum space"), this Hilbert space $\mathbf{H}$ becomes an ordinary complex L_2 space, over the positive frequency branch of the "mass hyperboloid" M_m: $K^2 = m^2$, where $K = (k_0, k_1, \ldots, k_n)$ is a dual vector to the space-time position X and $K^2 = k_0^2 - k_1^2 - \cdots - k_n^2$. There is a Lorentz-invariant measure $d\mu(k) = |k_0|^{-1}dk_1\cdots dk_n$ on M_m, unique within normalization, and an arbitrary function f in $L_2(M_m; k_0 > 0)$ corresponding to the vector φ in $\mathbf{H}$:

$$\varphi(X) = \mathrm{Re}\,[\textstyle\int_{M_m} e^{i\langle X,K\rangle} f(k)\,d\mu(k)].$$

LEBESGUE INTEGRATION. A function that is measurable on a measure space $(R, \mathbf{R}, \varrho)$ (q.v.) is one that is in the smallest class of functions that is closed under pointwise convergence of sequences and contains all finite linear combinations of characteristic functions of sets in $\mathbf{R}$. Any nonnegative measurable function f has an integral $\int_R f(x)\,d\varrho\,(x)$, or simply $\int f$, which may be ∞, characterized by the following properties: (i) if $f(x) = C_A(x)$, where C_A denotes the characteristic function of the set A, then $\int f = \varrho(A)$; and (ii) if $f_1(x) \leq f_2(x) \leq \cdots$ is an increasing sequence of nonnegative measurable functions, whose pointwise limit is f, then $\lim_n \int f_n = \int f$.

An integrable function is a measurable function f for which $\int |f| < \infty$. The dominated convergence theorem states that if $\{f_n\}$ is a sequence of measurable functions such that $|f_n| \leq g$ for some fixed integrable function g, and $f_n(x) \to f(x)$ for all x, then f is integrable and $\lim_n \int f_n = \int f$. What happens on a set of measure zero does not affect the Lebesgue integral or convergence theorem for sequences. See also FATOU'S LEMMA and L_p-SPACES.

LIE ALGEBRA. A Lie algebra is a vector space **L** together with a bracket operation $[X, Y]$ for any two elements X and Y of **L** whose values are again in **L** that satisfies the relations

$$[X, Y] = -[Y, X], \qquad [X, [Y, Z]] + [Z, [X, Y]] + [Y, [Z, X]] = 0,$$

and

$$[aX + Y, Z] = a[X, Z] + [Y, Z],$$

for arbitrary X, Y, and Z in **L**, and an arbitrary scalar a. Every Lie group G has associated with it a unique Lie algebra **G**, which may be represented as the space of all vector fields on G that are invariant under left translation, $x \to ax$, on G, as translations act naturally on vector fields. The bracket operation is then the usual commutator of vector fields: $[X, Y] = XY - YX$. There is a unique one-parameter unitary subgroup $g(t)$ of G that is generated by any given element X of **G**, denoted as e^{tX} or $\exp(tX)$, and characterized by the equation

$$(Xf)(p) = (d/dt)f(g(t)^{-1}p)|_{t=0}$$

for all C^∞ functions f on G, of which p is an arbitrary point.

LIE-TROTTER FORMULA. In its simplest form, the Lie-Trotter formula is a (noncommutative) generalization of Riemann integration, in which one forms the limit of a product of values of a matrix-valued function instead of the limit of a sum of values of a numerically-valued function. It states, e.g., that if A and B are any two finite-dimensional square matrices, then $e^{A+B} = \lim_{n\to\infty} (e^{A/n}\, e^{B/n})^n$. The same formula applies more generally when A and B are operators in a Hilbert space, provided A and B are selfadjoint, and $A + B$ is essentially selfadjoint (e^{A+B} being then defined as e^C where C is the closure of $A + B$). At a formal level, this formula underlies the path-integral approach of Feynman and the Feynman-Kac formula; in this book the formula is used directly, rather than indirectly via path integrals.

L_p-SPACES. For any measure space M, the space $L_p(M)$, where $1 \leq p < \infty$, is defined to consist of all measurable functions f such that $\int |f|^p < \infty$. The L_p-norm of f, usually denoted as $\|f\|_p$, is defined as $[\int |f|^p]^{1/p}$. Examination of the limiting behavior as $p \to \infty$ leads to the definition of $L_\infty(M)$ as the space of all bounded measurable functions, with $\|f\|_\infty$ defined as the "essential" supremum of $|f|$, or the supremum when null sets (i.e., sets of measure zero) are disregarded. Strictly speaking, the vectors in $L_p(M)$ are not functions, but equivalence classes of functions that agree except on a null set. The space $L_2(M)$ forms a Hilbert space with the inner product $\langle f, g\rangle = \int f\bar{g}$, and all the L_p spaces are Banach spaces. See also HÖLDER'S INEQUALITY.

MAXIMAL ABELIAN ALGEBRA. A maximal abelian algebra **A** within the algebra **B** of all bounded operators on a Hilbert space is just what its name implies—one such that every operator in **B** that commutes with every operator in **A** is in **A**. If M is a finite measure space, its multiplication algebra, defined as all operators on $L_2(M)$ of the form $f(x) \to k(x)f(x)$, where k is a bounded measurable function, is maximal abelian in this sense. Conversely, every maximal abelian algebra in **B** that is selfadjoint is unitarily equivalent to a direct sum of such multiplication algebras, or if the underlying Hilbert space is separable, to one such algebra.

MEASURE SPACE. A measure space consists of a set R, a σ-ring of subsets $\mathbf{R}$ of R, and a function ϱ from $\mathbf{R}$ to $[0, \infty]$ having the property that if the A_j ($j = 1, 2, \cdots$) are disjoint sets in $\mathbf{R}$ of union A, then $\varrho(A) = \Sigma_j \varrho(A_j)$. A *finitely-additive measure space* is the same except that the last condition is assumed only for finite unions, and so is not necessarily a measure space in the usual sense here adopted. Gaussian measure in Hilbert space is only finitely-additive, but Wiener measure on Wiener space is countably additive. For brevity we often denote the measure space $(R, \mathbf{R}, \varrho)$ as (R, ϱ) when $\mathbf{R}$ is given by the context or is immaterial. A measure space is called *finite* in case $R \in \mathbf{R}$, and $\varrho(R) < \infty$; if, moreover, $\varrho(R) = 1$, it is called a *probability measure space*.

MINKOWSKI SPACE. This space will be denoted as $\mathbf{M}_0$, and its points described by coordinates as $(x_0, x_1, \ldots, x_n)$, where x_0 denotes the time, and the x_j for $j > 0$ denote the space coordinates. For any vectors X and Y in $\mathbf{M}_0$, $X \cdot Y$ denotes $x_0y_0 - x_1y_1 - \cdots - x_ny_n$, and X^2 denotes $X \cdot X$. The future (past) of the origin 0 consists of all points X such that $X^2 > 0$ and $x_0 > 0$ ($x_0 < 0$). The future of any other point Y is defined as the vector sum of Y with the future of 0 (similarly for the past).

NET. This is a generalization of sequence, in which a directed system (q.v.) is used to label the points, rather than the positive integers. The definition of convergent net is analogous to that of convergent sequences, and a set in a topological space is closed if it contains the limits of all convergent nets of its elements.

OPERATOR. This term will usually mean linear operator in a vector space. Thus if T is a linear operator in the space **L**, having the domain **D**, then **D** is a linear subspace, i.e., a subset of **L** that is closed under addition, and under multiplication by scalars. In addition, T has to have the property that $T(ax + y) = aT(x) + T(y)$ for all vectors x and y in **D** and all scalars a. In a complex vector space, one must on occasion consider transformations T that satisfy the preceding condition only for real scalars a; these are called *real-linear*, to distinguish them from the transformations T that satisfy the equation for all complex numbers a, which are then *complex-linear*. An operator (or function) *on* a space will be understood to be defined on the entire space, while an operator *in* a space will be understood to have a domain that may be a proper subset of the entire space.

OPERATOR TOPOLOGY. For bounded operators on a Hilbert space **H**, there are three main topologies: the uniform, the strong operator, and the weak operator topologies. To describe these, it suffices to indicate corresponding neighborhoods N of 0, since the neighborhoods of an arbitrary operator B are obtained simply by vector addition of B to the neighborhoods of 0. For the uniform topology, a neighborhood N of 0 consists of all operators B such that $\|B\| < \varepsilon$, where ε is an arbitrary positive number. For the strong operator topology, N depends on an arbitrary finite set of vectors $x_1, \ldots, x_n$ in **H** and arbitrary positive number ε, and consists of all operators B such that $\|Bx_j\| < \varepsilon$ for all j. For the weak operator topology, N depends also on an arbitrary set of n vectors $y_1, \ldots, y_n$ and consists of all operators B such that $|\langle Bx_j, y_j\rangle| < \varepsilon$ for all j.

Other topologies occur in the literature, but only the foregoing three are needed in this book, and principally the strong operator topology. The qualification "operator" is useful on occasion because, when it is omitted, "strong" may appear to indicate the uniform topology by virtue of another usage, and "weak" may be confused with the "w^*-" topology on the space of all bounded operators on **H** (as the dual of the space of trace-class operators).

In the space of all selfadjoint operators in **H**, both bounded and unbounded, the only topology used is an extension of the strong operator topology, which is most conveniently given in terms of the convergence of a sequence. A sequence $\{A_n\}$ of selfadjoint operators in **H** is said to converge to the selfadjoint operator A in **H**, in the strong operator topology, in case any of a number of equivalent conditions holds, including the following: a) $e^{itA_n} \to e^{itA}$ for all real

t (in the strong operator topology as earlier defined); b) $E_n(\lambda) \to E(\lambda)$ at every point λ of continuity of the spectral resolution $E(\lambda)$ of A, where $E_n(\lambda)$ is the spectral resolution of A_n. For this to take place it is sufficient but not necessary that $A_n x \to Ax$ for all x in a dense subspace **D** of **H** on which A is essentially selfadjoint.

ORTHOGONAL SPACE/GROUP. An orthogonal linear space is a topological vector space **L** together with a continuous nondegenerate symmetric bilinear form S on **L**. The orthogonal group, denoted $O(\mathbf{L}, S)$, is the subgroup of the general linear group $GL(\mathbf{L})$ (q.v.) consisting of transformations that leave S invariant. A complex Hilbert space **H** determines a real orthogonal space by taking $S(x, y) = \mathrm{Re}(\langle x, y \rangle)$, and taking **L** to be **H** with only real scalars, thus doubling its dimension and creating a real Hilbert space $\mathbf{H}^\#$. $O(\mathbf{H}^\#, S)$ will be called the orthogonal group over **H** and denoted $O(\mathbf{H})$. Transformations in $O(\mathbf{H})$ correspond to homogeneous canonical linear transformations of the free fermion field whose single-particle space is **H**.

POSITIVITY-PRESERVING OPERATOR. A classical theory of Perron and Frobenius treats matrices whose entries are nonnegative. Thus, if A is a square matrix with positive entries, $\|A\|$ is an eigenvalue and there exists a corresponding eigenvector all of whose components are nonnegative. The theory has been extended to infinite-dimensional spaces by Krein, Andô, Gross, and others. In particular, if T is a bounded operator on $L_2(M)$ for some probability measure space M, that is positive in the sense that $f(x) \geq 0$ for all x implies that $(Tf)(X) \geq 0$ for all x, then $\|T\|$ will be an eigenvalue with a nonnegative eigenfunction, provided that T is a compact operator, or alternatively if T maps $L_2(M)$ into $L_p(M)$ for some $p > 2$. The theory also gives conditions for the eigenspace of $\|T\|$ to be one-dimensional, analogues of which apply to the unicity of ground states of quantum field Hamiltonians. This type of "positive" operator is not to be confused with the usual notion applicable to selfadjoint operators in Hilbert spaces, i.e., positivity of the spectrum. A selfadjoint operator may be positive in either sense without being positive in the other, but the context usually indicates the relevant sense.

PRE-HILBERT SPACE. This is a space that is a Hilbert space except that it is not necessarily complete. Equivalently, it is a vector space with an inner product $\langle x, y \rangle$ having the properties of being linear in x for fixed y, hermitian: $\langle y, x \rangle = \overline{\langle x, y \rangle}$, and such that $\langle x, x \rangle > 0$ for all $x \neq 0$. When completed by an analog to the Cantor process, it forms a full Hilbert space.

RANDOM VARIABLE. In modern mathematical usage, a random variable is a measurable function on a probability measure space (or more exactly, an equivalence class of such functions, when functions that agree except on a set of probability zero identified). Earlier (before 1933, when Kolmogorov's book laying foundations for probability theory was published) random variables were treated in an informal semi-axiomatic way, which can now be made precise. Random variables form an associative algebra on which is given a partially-defined positive normalized linear functional E, the expectation or integral, and defined on the subalgebra **B** of all bounded elements. Starting from **B** and E as restricted to **B**, a probability measure space on which the random variables become measurable functions whose integrals correspond to the respective expectations may be derived. The individual points of this measure space are not well defined, having in general only a theoretical character, corresponding to their not being conceptually definable from the results of measurements. Probability theory, the conventional theory of random variables, can in part be extended to the context in which the underlying algebra of bounded quantities is noncommutative, provided the expectation functional E is *central*: $E(ab) = E(ba)$ for arbitrary a, b in **B**.

REGULAR (DIFFERENTIABLE, ANALYTIC, ETC.) VECTOR. In treating a one-parameter unitary group $U(t)$ on Hilbert space **H**, or its selfadjoint generator A, it is often necessary to develop subspaces of **H** consisting of vectors that are especially regular in relation to the given group or operator. A *differentiable vector* x in **H** is one such that $U(t)x$ is a differentiable function of t, with values in **H**; an *analytic vector* x is one such that $U(t)x$ is a (real-) analytic function of t in some neighborhood of $t = 0$, with values in **H**. This means that the series $\Sigma_n (itA)^n x/n!$ is defined and convergent for sufficiently small t. More generally, if the latter series is convergent for a given operator A, whether selfadjoint or not, x is called an analytic vector for A. An *entire* vector is one such that the foregoing series is convergent not only in a neighborhood of $t = 0$, but for all values of t. The spectral theorem shows that every selfadjoint operator has a dense set of analytic vectors, and a converse is true: a hermitian operator having a dense set of analytic vectors is essentially selfadjoint.

REPRESENTATION OF A GROUP. A continuous unitary representation of a topological group G on a Hilbert space **H** is a mapping $g \to U(g)$ from G to unitary operators on **H** that is strongly continuous, and a representation: $U(gg') = U(g)U(g')$ for arbitrary g, g' in G and $U(e) = I$, e being the unit of G. Stone's theorem associates with any such representation of a Lie group a corresponding infinitesimal representation dU of its Lie algebra **G**, defined by the property that for arbitrary X in **G**, $-idU(X)$ is the selfadjoint generator of the one-parameter unitary group $U(e^{tX})$. The domains of the $dU(X)$ vary with X, but their common part **D** is dense in **H**, $dU(X)|\mathbf{D}$ is essentially selfadjoint, and the closure of $[dU(X), dU(Y)]$ acts on any w in **D** to give $dU([X, Y])w$ for arbitrary X, Y in **G**.

RIESZ INTERPOLATION THEOREM. This is the original L_p interpolation theorem, from which many variants and extensions derive (proved for real spaces by Riesz, and extended to the complex spaces by Thorin). It states the following: if T is a linear operator defined on the set S of all simple functions on a measure space M (i.e., measurable functions having only finitely many values) to the measurable functions on M, such that $\|Tf\|_{q_j} \leq m_j\|f\|_{p_j}$ $(j = 0, 1)$, where $1 \leq p_0, q_0, p_1, q_1 \leq \infty$, then $\|Tf\|_q \leq m_0^t m_1^{1-t}\|f\|_p$ for all $f \in S$, where $0 \leq t \leq 1$, and

$$(p^{-1}, q^{-1}) = t(p_0^{-1}, q_0^{-1}) + (1-t)(p_1^{-1}, q_1^{-1}).$$

SCALE OF SPACES. Subspaces $\mathbf{M}_s$ of a topological vector space $\mathbf{M}$ that depend on a real parameter s, and have an intrinsic topology, with the properties that $\mathbf{M}_s \subset \mathbf{M}_t$ if $s < t$—and that the inclusion injection of $\mathbf{M}_s$ into $\mathbf{M}_t$, in their intrinsic topologies, is continuous—form a scale of spaces. The parameter range may be discrete as well as continuous. The $L_p(M)$ spaces for a finite measure space M, with $1 \leq p < \infty$ provide a useful example. Another example is that of the domains $\mathbf{D}(A^n)$, where A is a strictly positive selfadjoint operator in Hilbert space, with the intrinsic topology defined by the norm $\|x\|_n = \|A^n x\|$, where n may range over the positive integers, or all integers, etc., as required.

SCHWARZ REFLECTION PRINCIPLE. This treats the analytic continuation of a complex analytic function, and in its simplest form states that if $f(z)$ is continuous on the closed upper half-plane $\{\text{Im}(z) \geq 0\}$ and analytic in the interior, with real values on the boundary, then f extends uniquely to an analytic function on the entire plane.

SELFADJOINT OPERATOR. This is a densely defined operator T in a Hilbert space such that $T^* = T$. Equivalently, it is an operator that is unitarily equivalent to the operation of multiplication by a real measurable function, in an L_2-space. A densely defined operator T is said to be *essentially selfadjoint* in case it has a unique selfadjoint extension. This is the case if and only if $T^* = T^{**}$, in which case T^* or T^{**} is the selfadjoint extension. See also HERMITIAN OPERATOR.

SEPARABLE. A Hilbert space is separable if it has a countable dense subset, or, equivalently, if it is at most countably dimensional. A measure space M is separable if $L_2(M)$ is a separable Hilbert space. Although most of the results in the book do not assume separability of the Hilbert space involved, some peripheral results cited do assume separability in order to avoid the elaboration that would otherwise be required. Inseparable spaces have no essential role in physical quantum field theory, and the reader will lose little by assuming that the Hilbert spaces involved are separable.

SEPARATING VECTOR. A separating vector in a space **L** for a set S of operators on **L** is a vector z such that if T and T' are in S and $Tz = T'z$, then $T = T'$. If S is a selfadjoint algebra of bounded operators on a Hilbert space, then a given vector z is separating for S if and only if it is cyclic (q.v.) for the commutor S'.

SEQUENTIAL TOPOLOGY. This is a topology that is defined by the specification of convergent sequences. The closed sets are those containing the limits of all convergent sequences of its elements. The resulting ensemble of closed sets satisfies the usual axioms for a topological space, and a function on a sequentially topologized space to another such space is continuous if and only if it carries convergent sequences into convergent sequences.

σ-RING. A σ-ring of sets is a collection **R** of sets that is closed under the difference operation and under finite and countable unions and intersections. The Borel sets in $\mathbf{R}^n$ form a σ-ring.

SOBOLEV INEQUALITY. This states basically the following: let $L_{p,r}$ denote the space of all functions on $\mathbf{R}^n$ that are in L_p together with their first r derivatives, in the topology of convergence in L_p for all derivatives up to and including those of order r. Then—only if $p \geq 1$, $r \geq s \geq 0$, $1 \leq q < \infty$, and $q^{-1} = p^{-1} - (r - s)n^{-1}$—any function in $L_{p,r}$ is also contained in $L_{q,s}$, and the inclusion mapping is continuous. The inequality fails if $q = \infty$, but a variant holds, which says that

$$\|f\|_\infty \leq c\,\|f\|_{p,r}, \text{ if } r > n/p.$$

SPECTRAL THEORY. This is the infinite-dimensional generalization of the theory of diagonalization of selfadjoint, unitary, and other normal matrices, in finite-dimensional vector spaces. The diagonalization of commuting *sets* of normal operators is also important, and similar to the finite-dimensional case. The simplest way to describe the theory is to say that the only essential difference from the finite-dimensional case is the replacement of a measure space consisting of a finite number of points (substantially a set of basis vectors in the finite-dimensional vector space case) and corresponding sum by a Lebesgue-type measure space and corresponding integral. This reflects in part the phenomenon of the continuous spectrum.

A natural generalization of a diagonalized matrix is the operation on $L_2(M)$, where M is a given measure space (q.v.), of multiplication by a given measurable function, say $f(x) \to k(x)f(x)$, where $f(x)$ is arbitrary in $L_2(M)$ and $k(x)$ is the given measurable function. This operator T, like a diagonal matrix, is normal, i.e., $TT^* = T^*T$, where T^* is the adjoint of T. In one of its simplest forms, the spectral theorem says simply that the converse is true—the most general densely defined normal operator in a complex Hilbert space is unitarily equivalent to a multiplication operator of the type indicated.

In the finite-dimensional case, the function $k(x)$ is just the jth eigenvalue λ_j ($j = 1, 2, \ldots, n$-dimension of the space), and so is always bounded; but in the infinite-dimensional cases, $k(x)$ may be unbounded. When this happens, T does not operate on the whole space, but its domain is restricted to consist of all functions $f(x)$ such that $k(x)f(x)$ is again in $L_2(M)$. With the appropriate definition of the adjoint for unbounded operators, the spectral theorem is valid for both bounded and unbounded operators, provided they are densely defined (i.e., defined on a dense subset of $\mathbf{H}$; see ADJOINT).

The classic way of expressing the spectral resolution of a given selfadjoint operator T is in the form of an increasing family E_λ of projections in $\mathbf{H}$, λ ranging over the real numbers, such that $T = \int \lambda dE_\lambda$, meaning more specifically that if x and y are any vectors in $\mathbf{H}$, x being in the domain of T, then $\langle Tx, y\rangle = \int \lambda d\langle E_\lambda x, y\rangle$. In the case of the operation of multiplication by the real function $k(x)$, E_λ is just the operation of multiplication by the characteristic function of the subset of M on which $k(x) < \lambda$ (within an inessential question of normalization; one could equally use the subsets on which $k(x) \leq \lambda$). In the case of normal operators that are not selfadjoint, it is similar except that λ ranges over the complex numbers.

More generally, if B is a given Borel set of complex numbers, the spectral projection $E(B)$ for T is the operation of multiplication by the characteristic function of the subset of M for which $k(x)$ lies in B. This spectral resolution function $E(B)$ is a countably-additive function of B; i.e., if $B_1, B_2, \cdots$ is a sequence of disjoint Borel sets whose union is B, then $E(B)$ is the sum of the orthogonal projections $E(B_j)$. A set of normal operators are said to be *simultaneously diagonalizable* in case they are conjugate to multiplication operators via transformation by one and the same unitary transformation. For this to be the case it is necessary and sufficient that their spectral projections $E(B)$ be mutually commutative. If one or more of the operators is unbounded, it is by no means sufficient that the operators commute on a dense subspace $\mathbf{D}$ in $\mathbf{H}$, i.e., $ABx = BAx$, where A and B are the two operators for all x in $\mathbf{D}$. We say the operators are *strongly commutative* when their spectral projections are mutually commutative.

If β is a Borel function of a complex variable, and T is a normal operator consisting of (or unitarily equivalent to) multiplication by k, then $\beta(T)$ is defined as the operation of multiplication by $\beta(k(x))$ (or the transform of this by the corresponding unitary operator). All bounded Borel functions of a set of strongly commutative normal operators commute with each other, and not just their spectral projections (the case in which β is the characteristic function of a Borel set).

The main point of commutative spectral theory is that many issues are reduced to questions in measure theory, and are thereby often relatively straightforward to deal with.

STATE. A state of a C^*-algebra **A** with identity I is a linear functional E on **A** that is positive, in the sense that $E(A^*A) \geq 0$ for arbitrary A in **A**, and normalized, meaning that $E(I) = 1$. For example, if **A** is the algebra of all bounded operators on a Hilbert space **H**, and if x is a unit vector in **H**, then $E(A) = \langle Ax, x\rangle$ is a state of **A**.

The set of all states forms a convex set, i.e., if E and E' are states, so is $aE + bE'$, for arbitrary a, b such that $a, b \geq 0$ and $a + b = 1$. A "pure" state is an *extreme point* of this set, this being defined as a point that is not a nontrivial convex combination of two other states. The preceding example is a pure state, and all pure states are of this form in case **H** is finite-dimensional; but when **H** is infinite-dimensional, there are others, contrary to heuristic folklore. (These may in part be interpreted as arising from nonnormalizable vectors representing eigenvectors for the continuous spectrum of a Hamiltonian.) If D is a trace class nonnegative operator of unit trace in **H**, then $E(A) = tr(DA)$ is a state, "mixed" rather than pure unless D is a one-dimensional projection. D is called the *density operator* (or *density matrix*) of the state.

In physics terminology, the above notion of state was originally called an "expectation value form in a state," or words to that effect. The term "state" is sometimes used for the vector x above; but x is not uniquely determined by E, and is now known as a "state vector." "State" as used here includes all that is conceptually truly measurable for a state as the term is used in physical literature, so only a metaphysical nuance is lost by shortening the term "expectation value form in a state" in this way.

STATE-REPRESENTATION DUALITY. There is a mutual correspondence between states of a C^*-algebra **A** and representations of **A** on Hilbert space, in which pure states correspond to irreducible representations. Given a state E, an inner product in **A** is defined by the equation $\langle A, B\rangle = E(B^*A)$, leading to a Hilbert space $\mathbf{H}_E$ after completion and factoring by (i.e., essentially deletion of) vectors of zero norm. For any element B in **A**, there is a corresponding operator $\varrho(B)$ on $\mathbf{H}_E$, defined essentially by the equation $\varrho(B)A = BA$. The mapping $B \rightarrow \varrho(B)$ is a $*$-representation in the sense that ϱ preserves the $*$, and addition and multiplication. The subset $\varrho(A)I$ is dense in $\mathbf{H}_E$, so that the vector in $\mathbf{H}_E$ corresponding to I is cyclic (q.v.). The state E can be recovered from ϱ and this cyclic vector by the equation $E(A) = \langle\varrho(A)I, I\rangle$. The state E is pure (q.v.) if and only if the representation ϱ is irreducible. The association of a $*$-representation on Hilbert space with a given state is exemplified by the reconstruction of a quantum field from the vacuum expectation values of products of field operators (as in Wightman, 1964), and is also known as the GNS construction.

STEIN INTERPOLATION THEOREM. This is an extension of the Riesz interpolation theorem in which the operator as well as the L_p space is varied. A simplified form adequate for present purpose proceeds as follows. *Hypotheses*: (1) M is a finite measure space; (2) z is a complex variable ranging over the strip $0 \leq \mathrm{Re}(z) \leq 1$; (3) $z \to T_z$ is a map from this strip to operators defined on the class S of simple functions on M, having values that are measurable functions on M; (4) For arbitrary simple functions f and g, $\langle T_z f, g\rangle$ is an analytic function of z in the interior strip and bounded on the boundary; and (5) For arbitrary f in S, $\|T_{j+iy}f\|_{q_j} \leq m_j \|f\|_{p_j}$ $(j = 0, 1)$, where the m_j are constants, y is arbitrary in $\mathbf{R}$, and $1 \leq p_0, q_0, p_1, q_1 \leq \infty$. *Conclusion*: If $(1/p, 1/q) = t(1/p_0, 1/q_0) + (1 - t)(1/p_1, 1/q_1)$, where $0 \leq t \leq 1$, then for all f in S,

$$\|T_t f\|_q \leq m_0^t m_1^{1-t} \|f\|_p.$$

STOCHASTIC INDEPENDENCE. The random variables X in a set S are said to be stochastically independent in case for any finite subcollection $X_1, \ldots, X_n$ the probability of the joint event $X_1 < a_1, \ldots, X_n < a_n$ is the product of the individual probabilities that $X_j < a_j$, the a_j being arbitrary real numbers. The concept can be extended to noncommutative probability algebras, defined as under RANDOM VARIABLE, with the additional constraint that $E(XY) = E(YX)$ for arbitrary random variables X and Y. In this case one requires that for arbitrary bounded continuous functions f_j, $E(f_1(X_1)\cdots f_n(X_n)) = E(f_1(X_1))\cdots E(f_n(X_n))$. The canonical Q's in either a boson or fermion free field are stochastically independent in this sense, relative to the vacuum as expectation functional.

STONE'S THEOREM. If $U(t)$ is a continuous one-parameter unitary group on a Hilbert space $\mathbf{H}$, then there exists a unique selfadjoint operator A in $\mathbf{H}$ such that $U(t) = e^{itA}$. Here a one-parameter unitary group $U(t)$ is defined as a unitary operator-valued function of the real variable t having the properties that $U(t + t') = U(t)U(t')$ for arbitrary real t and t', and $U(0) = I$ (the identity operator). Continuity is in the sense of the strong operator topology, or the weak operator topology, which happen to be the same within the space of all unitary operators.

The theorem also shows that A can be deduced from the group $U(t)$ as follows:

$$Ax = \lim_{t\to\infty} (it)^{-1} (U(t) - I)x$$

for all vectors in the domain of A; and if the limit on the right exists, then x will be in the domain of A. From the spectral theorem it follows also that for a given vector x to be in the domain of A, it suffices that $(it)^{-1}(U(t) - I)x$ be a bounded function of t as $t \to 0$.

STONE–VON NEUMANN THEOREM. Established in a beautiful paper by von Neumann (1931), this theorem underlies the essential equivalence of the Heisenberg and Schrödinger formulations of quantum mechanics. In the mathematical form given by von Neumann, it concerns two continuous unitary representations U and V on a Hilbert space of the additive group of a real n-dimensional vector space $\mathbf{L}$, that satisfy the (Weyl) relations

$$U(x)V(y) = e^{i\langle x,y\rangle} V(y)U(x) \qquad (x \text{ and } y \text{ arbitrary in } \mathbf{L}),$$

where $\langle x, y\rangle$ denotes the usual scalar product in $\mathbf{L}$ when identified with the space $\mathbf{R}^n$ of real n-tuples. The theorem states that there exists a set of mutually orthogonal subspaces $\mathbf{H}_j$ in $\mathbf{H}$, of which $\mathbf{H}$ is the direct sum, each of which is invariant under all $U(x)$ and $V(y)$, and irreducible under the set of all these operators, and in addition is such that if $U_j(x)$ and $V_j(y)$ denote the representations of $\mathbf{R}^n$ obtained by restricting each $U(x)$ and $V(y)$ to $\mathbf{H}_j$, then there exists a unitary transformation T_j from $\mathbf{H}_j$ onto $L_2(\mathbf{R}^n)$ such that

$$U_0(x) = T_j U_j(x) T_j^{-1}, \qquad V_0(x) = T_j V_j(x) T_j^{-1}$$

attain the following form (which is essentially the Schrödinger representation of the Heisenberg commutation relations): for an arbitrary function $f(u)$ in $L_2(\mathbf{R}^n)$, $U_0(x)$ sends $f(u)$ into $e^{i\langle u,x\rangle}f(u)$, while $V_0(x)$ sends $f(u)$ into $f(u + y)$. This conclusion is false (in general) if $\mathbf{L}$ is replaced by an infinite-dimensional space.

STONE-WEIERSTRASS THEOREM FOR PROBABILITY SPACE. This gives a condition for an algebra $\mathbf{A}$ of bounded measurable functions on a probability space M to be dense in $L_p(M)$, for $1 < p < \infty$. $\mathbf{A}$ must be *measure-theoretically separating*, in the sense that if $\mathbf{R}$ is the σ-ring of the measure space, for any two sets S and S' in $\mathbf{R}$ that differ by more than a null set, there exists a function f in $\mathbf{A}$ such that $\int_S f \neq \int_{S'} f$.

STRICTLY POSITIVE. A selfadjoint operator T in a Hilbert space (or a function f) is strictly positive if there is a positive constant ε such that $T \geq \varepsilon I$ (or $f \geq \varepsilon$).

SYMPLECTIC SPACE/GROUP. A (linear) symplectic space is a topological vector space $\mathbf{L}$ together with a nondegenerate antisymmetric bilinear form A on $\mathbf{L}$. The symplectic group $Sp(\mathbf{L}, A)$ is the subgroup of the general linear group (q.v.) $GL(\mathbf{L})$ that leaves A invariant. A complex Hilbert space **H** determines a real symplectic space by taking $A(x, y) = \text{Im}(\langle x, y\rangle)$, and taking **L** to be **H** with only real scalars, thus doubling its dimension and creating a real Hilbert space $\mathbf{H}^\#$. The symplectic group $Sp(\mathbf{H}^\#, A)$ is also called the symplectic group over **H**, and denoted $Sp(\mathbf{H})$. Transformations in $Sp(\mathbf{H})$ correspond to homogeneous canonical linear transformations of the free boson field whose single-particle space is **H**.

TENSOR PRODUCT. The tensor (or in von Neumann's term, "direct") product $\mathbf{H}\otimes\mathbf{H}'$ of two Hilbert spaces **H** and **H′** is the completion of their algebraic tensor product (which consists of all finite combinations of the $u\otimes u'$ with u in **H** and u' in **H′**), with respect to the inner product having the property that $\langle u\otimes u', v\otimes v'\rangle = \langle u, v\rangle \langle u', v'\rangle$, if v and v' are similarly arbitrary in **H** and **H′**. If $\mathbf{H} = L_2(M)$ and $\mathbf{H}' = L_2(M')$, then there is a natural unitary equivalence of $\mathbf{H}\otimes\mathbf{H}'$ and $L_2(M \times M')$. Tensor products of any finite number of Hilbert spaces may be defined in a similar way. If T_j is a bounded linear operator on $\mathbf{H}_j$, the algebraic tensor product $T_1 \times \cdots \times T_n$ is bounded on the algebraic tensor product $\mathbf{H}_1\otimes\cdots\otimes\mathbf{H}_n$, and the topological (or Hilbert space) tensor product of the T_j is defined as the unique continuous extension to the (Hilbert space) tensor product of the $\mathbf{H}_j$ of their algebraic tensor product, and denoted as $T_1\otimes\cdots\otimes T_n$.

TOPOLOGICAL VECTOR SPACE. This is a vector space **L** over the real or complex numbers that has also a topology relative to which the usual linear operations are continuous (i.e., $ax + y$ is a continuous function of the scalar a and the arbitrary vectors x and y in **L**). This is also known as a linear topological space. If it is locally convex, meaning that it has a complete system of neighborhoods, each of which is convex, then for every vector $x \neq 0$ in **L** there exists a functional f in the dual $\mathbf{L}^*$ to **L** such that $f(x) \neq 0$.

TRACE-CLASS OPERATOR. The following conditions on an operator T on Hilbert space are all equivalent, and define the notion of trace-class operator: 1) $T = AB$, where A and B are Hilbert-Schmidt (q.v.); 2) T is a linear combination of selfadjoint operators, each of which has eigenfunctions forming an orthogonal basis, and corresponding eigenvalues λ_j such that $\Sigma_j|\lambda_j|$ is finite; and 3) The traces $tr(TS)$ are bounded as S varies over the bounded operators of finite rank of unit bound (the traces in question are definable in finite-dimensional terms). The supremum over these over these S of $|tr(TS)|$ forms a norm $\|T\|_1$ relative to which the space of all trace-class operators forms a Banach space on which the trace is everywhere defined as $\lim_n tr(TP_n)$ if $\{P_n\}$ is a sequence of projections convergent to I.

TYPE. According to the theory of W^*-algebras (q.v.), every such algebra is a direct sum of three types of components, known as types I, II, III. All bounded operators, all abelian algebras, and all algebras obtainable from these by formation of commutors, and direct sums or products, form the type I algebras. The basic type II algebras, known as type II_1 factors (factor = W^*-algebra **A** that has only scalars in common with its commutor), are distinguished by their possession of a unique continuous trace, or complex-valued linear functional T that $T(AB) = T(BA)$ and $T(I) = 1$. If P is a projection in such an algebra, $T(P)$ has values in the range $[0, 1]$, and all values in this range occur. The universal free fermion field is closely connected with the simplest type II_1 factor (called "approximately finite" by von Neumann). Type III algebras have no numerically-valued trace, and are not involved here.

W^*-ALGEBRA. This is an algebra of bounded operators on a Hilbert space containing the identity operator I, closed in the weak operator topology, and closed under the adjunction operation, originally called a "ring" by Murray and von Neumann, and known also as a "von Neumann algebra." A key property of W^*-algebras is that any such algebra **A** is identical with its second commutor, denoted **A**$''$, i.e., the commutor of its commutor **A**$'$. The W^*-algebra $W^*(A)$ generated by a bounded selfadjoint operator A (i.e., the smallest W^*-algebra containing it) consists of all bounded Borel functions of A (if the underlying Hilbert space is separable), whereas the C^*-algebra $C^*(A)$ generated by A consists of all continuous functions of A. In particular, the spectral projections of a bounded selfadjoint operator are contained $W^*(A)$, but are not in general contained in $C^*(A)$. The same is true for a finite set of commuting selfadjoint operators, and a generalized function of noncommuting bounded operators can be defined as one that is in the W^*-algebra they generate. For unbounded operators, see AFFILIATION.

Bibliography

Andô, T. 1954. Positive linear operators in semi-ordered linear spaces. J. Fac. Sci. Hokkaido Univ. **13**:214–28.

Anzai, H. 1949. A remark on spectral measure of the flow of Brownian motion. Osaka Math. J. **1**:95–97.

Araki, H. 1963. A lattice of von Neumann algebras associated with the quantum theory of a free bose field. J. Math. Phys. **4**:1343–62.

———. 1988. Bogolioubov automorphisms and Fock representations of canonical anticommutation relations. In *Operator algebras and mathematical physics*, ed. P.E.T. Jorgensen and P. S. Muhly. Amer. Math. Soc., Providence.

Baez, J. 1989. Wick products of the free bose field. J. Funct. Anal. **86**:211–25.

Bargmann, V. 1961. On a Hilbert space of analytic functions and an associated integral transform, Part I. Comm. Pure Appl. Math., 187–214.

Berman, S. J. 1974. Wave equations with finite velocity of propagation. Trans. Amer. Math. Soc. **188**:127–48.

Bjorken, J. D., and Drell, S. D. 1965. *Relativistic quantum fields*. New York: McGraw-Hill.

Blattner, R. J. 1958. Automorphic group representations. Pacific J. Math. **8**:665–77.

Bogolioubov, N. N., Logunov, A. A., and Todorov, I. T. 1975. *Introduction to axiomatic quantum field theory*. New York: Benjamin.

Bogolioubov, N. N., and Shirkov, D. V. 1959. *Introduction to the theory of quantized fields*. New York: Interscience.

Bohr, N., and Rosenfeld, L. 1933. Zur Frage der Messbarkeit der Elektromagnetischen Feldgrössen. Mat.-Fys. Medd. Danske Vid. Selsk. **12** (no. 8).

Bongaarts, P.J.M. 1970. The electron-positron field, coupled to external electro-magnetic potentials, as an elementary C^* algebra theory. Ann. Phys. **56**:108–39.

Broadbridge, P. 1983. Existence theorems for Segal quantization via spectral theory in Krein space. J. Austral. Math. Soc., Ser. B **24**:439–60.

Cameron, R. H. 1945. Some examples of Fourier-Wiener transforms of analytic functionals. Duke Math. J. **12**:485–88.

Cameron, R. H., and Martin, W. T. 1944. Transformation of Wiener integrals under translations. Ann. of Math. **45**:386–96.

Cameron, R. H., and Martin, W. T. 1945a. Transformation of Wiener integrals under a general class of linear transformations. Trans. Amer. Math. Soc. **58**:184–219.

———. 1945b. Fourier-Wiener transforms of analytic functionals. Duke Math. J. **12**:489–507.

———. 1947. Fourier-Wiener transforms of functionals belonging to L_2 over the space C. Duke Math. J. **14**:99–107.

Cannon, J. J., and Jaffe, A. M. 1970. Lorentz covariance of the $\lambda(\varphi^4)_2$ quantum field theory. Comm. Math. Phys. **17**:261–321.

Chaiken, J. M. 1967. Finite-particle representations and states of the canonical commutation relations. Ann. Phys. **42**:23–80.

———. 1968. Number operators for representations of the canonical commutation relations. Comm. Math. Phys. **8**:164-84.

Cheng, T.-P., and Li, L.-F. 1984. *Gauge theory of elementary particle physics*. New York: Oxford Univ. Press.

Choquet-Bruhat, Y., Dewitt-Morette, C. and Dillard-Bleick, M. 1982. *Analysis, manifolds, and physics*. 2d ed. New York: North-Holland.

Cook, J. M. 1953. The mathematics of second quantization, Trans. Amer. Math. Soc. **74**:222–45.

Dirac, P.A.M. 1927. The quantum theory of the emission and absorption of radiation. Proc. Roy. Soc. London Ser. A**114**:243–65.

———. 1949. La seconde quantification. Ann. Inst. H. Poincaré:15–47.

———. 1958. *The Principles of Quantum Mechanics*. 4th ed. New York: Oxford University Press.

———. 1974. *Spinors in Hilbert space*. New York: Plenum.

Dixmier, J. 1958. Sur la relation $i(PQ - QP) = 1$. Compositio Math. **13**:263–69.

———. 1981. *Von Neumann algebras*. New York: North-Holland.

Doob, J. L. 1965. *Stochastic Processes*. New York: Wiley.

Faris, W. 1972. Invariant cones and the uniqueness of the ground state for fermion systems. J. Math. Phys. **13**:1285–89.

Feldman, J. 1966. Nonexistence of quasi-invariant measures on infinite-dimensional Hilbert spaces. Trans. Amer. Math. Soc. **17**:142–46.

Fock, V. 1928. Verallgemeinerung und Lösung der Diracschen statistischen Gleichung. Zeits. f. Phys. **49**:339–57.

Friedrichs, K. O. 1953. *Mathematical aspects of quantum theory of fields*. New York: Interscience.

Gaffney, M. P. 1955. Hilbert space methods in the theory of harmonic integrals. Trans. Amer. Soc. **78**:426–44.

Gårding, L., and Wightman, A. S. 1954. Representations of the commutation relations. Proc. Nat. Acad. Sci. USA. **40**:622–26.

Gelfand, I. M., and Neumark, M. 1943. On the imbedding of normed rings into the ring of operators in Hilbert space. Mat. Sbornik N. S. **12**:197–213.

Glimm, J. 1968. Boson fields with nonlinear self-interaction in two-dimensions. Comm. Math. Phys. **8**:12–25.

Goodman, R. W. 1967. On localization and domains of uniqueness. Trans. Amer. Math. Soc. **127**:98–106.

———. 1969. Analytic and entire vectors for representations of Lie groups. Trans. Amer. Math. Soc. **143**:55–76.

Goodman, R. W., and Segal, I. E. 1965. Anti-locality of certain Lorentz-invariant operators. J. Math. Mech. **14**:629–38.

Gross, L. 1972. Existence and uniqueness of physical ground states. J. Funct. Anal. **10**:52–109.

———. 1973. The relativistic polaron without cutoff. Comm. Math. Phys. **31**:25–73.

———. 1974. Analytic vectors for representations of the canonical commutation relations and nondegeneracy of ground states. J. Funct. Anal. **17**:104–11.

———. 1975a. Logarithmic Sobolev inequalities. Amer. J. Math. **97**:1061–83.

———. 1975b. Hypercontractivity and logarithmic Sobolev inequalities for the Clifford Dirichlet form. Duke Math. J. **42**:383–96.

Haag, R. 1955. On quantum field theories. Mat.-Fys. Medd. Danske Vid. Selsk. **29** (no. 12).

Heisenberg, W. 1930. *The physical principles of the quantum theory*. Univ. of Chicago Press.

Heisenberg, W., and Pauli, W. 1929. Zur Quantendynamik der Wellenfelder. Z. Physik **56**:168–90.

Hille, E. 1926. A class of reciprocal functions. Ann. of Math. **27**:427–64.

Jost, R. 1965. *The general theory of quantized fields*. Amer. Math. Soc. Providence.

Kac, M. 1939. On a characterization of the normal distribution. Amer. J. Math. **61**:726–28.

Kakutani, S. 1948. On equivalence of infinite product measures. Ann. of Math. **49**:214–24.

———. 1950. Determination of the spectrum of the flow of Brownian motion. Proc. Nat. Acad. Sci. **36**:319–23.

Källén, G. 1964. *Elementary particle physics*. Reading, Mass.: Addison-Wesley.

Klein, A. 1973. Quadratic expressions in a free bose field. Trans. Amer. Math. Soc. **181**:439–56.

Kolmogoroff, A. 1933. Grundbegriffe der Wahrscheinlichkeitsrechnung. Ergebnisse der Mathematik, Vol. 2.

Koopman, B. O. 1931. Hamiltonian systems and transformations in Hilbert space. Proc. Nat. Acad. Sci. **17**:315–18.

Leray, J. 1981. *Lagrangian analysis and quantum mechanics*. M.I.T. Press.

Mackey, G. W. 1952. Induced representations of locally compact groups. Ann. of Math. (2) **55**:101–39.

———. 1963. Group representations in Hilbert space. Pages 113–30 in Segal, 1963.

Mandl, F. 1959. *Introduction to quantum field theory*. New York: Interscience.

Nelson, E. 1959. Analytic vectors. Ann. of Math. **70**:572–615.

———. 1966. A quartic interaction in two dimensions. In *Mathematical Theory of Elementary Particles*, ed. R. W. Goodman and I. E. Segal. M.I.T. Press.

———. 1972. Time-ordered operator products of sharp-time quadratic forms. J. Funct. Anal. **11**:211–19.

———. 1973a. Construction of quantum fields from Markov fields. J. Funct. Anal. **12**:97–112.

———. 1973b. The free Markov field. J. Funct. Anal. **12**:211–27.

Nelson, E., and Stinespring, W. F. 1959. Representation of elliptic operators in an enveloping algebra. Amer. J. Math. **81**:547–60.

Newton, T. D., and Wigner, E. P. 1949. Localized states for elementary systems. Rev. Mod. Phys. **21**:400–406.

Niederman, D. 1981. *Spinors in Hilbert space and the infinite orthogonal group*. Ph.D. thesis, M.I.T.

Paneitz, S. M., and Segal, I. E. 1983. Self-adjointness of the Fourier expansion of quantized interaction field Lagrangians. Proc. Nat. Acad. Sci. USA **80**:4595–98.

Pauli, W. 1980. *General principles of quantum mechanics*. Berlin: Springer-Verlag.

Poulsen, N. S. 1972. On C^∞-vectors and intertwining bilinear forms for representations of Lie groups. J. Funct. Anal. **9**:87-120.

Reeh, H., and Schlieder, S. 1961. Bemerkungen zur Unitäräquivalenz von Lorentzinvarianten Feldern. Nuovo Cim. **22**:1051–68.

Robbins, S. 1978. A uniform approach to field quantization. J. Funct. Anal. **29**:23–36.

Satake, I. 1971. Factors of automorphy and Fock representations. Advances Math. **7**:83–110.

Segal, I. E. 1947a. Irreducible representations of operator algebras. Bull. Amer. Soc. **61**:73–88.

———. 1947b. Postulates for general quantum mechanics. Ann. of Math. **48**:930–48.

———. 1953. A non-commutative extension of abstract integration. Ann. of Math. **57**:407–57; **58**:595.

———. 1954. Abstract probability spaces and a theorem of Kolmogoroff. Amer. J. Math. **76**:721–32.

———. 1956a. Tensor algebras over Hilbert spaces *I*. Trans. Amer. Math. Soc. **18**:106–34.
———. 1956b. Tensor algebras over Hilbert spaces *II*. Ann. of Math. **63**: 160–75.
———. 1957a. The structure of a class of representations of the unitary group on a Hilbert space. Proc. Amer. Math. Soc. **8**:197–203.
———. 1957b. Ergodic subgroups of the orthogonal subgroup on a real Hilbert space. Ann. of Math. **66**:297–303.
———. 1958a. Direct formulation of causality requirements on the *S*-operator. Phys. Rev. **109**:2191–98.
———. 1958b. Distributions in Hilbert space and canonical systems of operators. Trans. Amer. Math. Soc. **88**:12–41.
———. 1959. Foundations of the theory of dynamical systems of infinitely many degrees of freedom *I*. Math.-fys. Medd. K. Danske Vidensk. Selsk. **31**:1–38.
———. 1960a. Quasi-finiteness of the interaction Hamiltonian of certain quantum fields. Ann. of Math. **72**:594–602.
———. 1961. A generating functional for the states of a linear boson field. Canadian J. Math. **13**:1–18.
———. 1962. Mathematical characterization of the physical vacuum for linear Bose-Einstein fields. Illinois J. Math. **6**:500–23.
———. 1963. *Mathematical problems of relativistic physics*. Proceedings of Summer Seminar on Applied Mathematics, Boulder Colo. 1960. With Appendix by G. W. Mackey. Amer. Math. Soc., Providence.
———. 1964a. Quantum fields and analysis in the solution manifolds of differential equations. In *Proc. Conf. Analysis in Function Space*. M.I.T. Press.
———. 1964b. Interpretation et solution d'équations non linéaires quantifées. C. R. Acad. Sci. Paris **259**:301–3.
———. 1967. Notes towards the construction of nonlinear relativistic quantum fields *I*; The Hamiltonian in two space-time dimensions as the generator of a C^*-automorphism group. Proc. Nat. Acad. Sci. **57**:1178–83.
———. 1968. Quantized differential forms. Topology **7**:147–71.
———. 1969a. Nonlinear functions of weak processes *I*. Funct. Anal. **4**:404–56.
———. 1969b. Notes towards the construction of nonlinear relativistic quantum fields *II*: The basic nonlinear functions in general space-times. Bull. Amer. Math. Soc. **75**:1383–89.
———. 1969c. Notes towards the construction of nonlinear relativistic quantum fields *III*. Properties of the C^*-dynamics for a certain class of interactions. Bull. Amer. Math. Soc. **75**:1390–95.
———. 1970a. Transformations in Wiener space and squares of quantum

fields. Advances in Math. **4**:91–108.

———. 1970b. Local nonlinear functions of quantum fields. From Proc. Conf. in honor of M. H. Stone, May 1968. In *Functional Analysis*, ed. F. E. Browder, 188–210. Berlin: Springer-Verlag.

———. 1970c. Nonlinear functions of weak processes *II*. J. Funct. Anal. **6**:29–75.

———. 1970d. Construction of nonlinear quantum processes *I*. Ann. of Math. **92**:462–81.

———. 1970e. Local non-commutative analysis. In *Problems in Analysis*, ed. R. C. Gunning. Princeton Univ. Press, 111–30.

———. 1971. Construction of nonlinear quantum processes *II*. Inventiones Math. **14**:211–42.

———. 1978. The complex-wave representation of the free boson field. Suppl. Studies, Vol. 3., 321–44. Advances Math. New York: Academic Press.

———. 1981. Quantization of symplectic transformations. In *Mathematical Analysis and Applications*, ed. L. Nachbin, Part B, 749–58. New York: Academic Press.

———. 1989. Algebraic characterization of the vacuum for quantized fields transforming indecomposably. Pacific J. Math. **137**:387–403.

Segal, I. E., and Kunze, R. A. 1978. *Integrals and Operators*. Berlin: Springer-Verlag.

Shale, D. 1962. Linear symmetries of free boson fields. Trans. Amer. Math. Soc. **103**:149–67.

Shale, D., and Stinespring, W. F. 1964. States of the Clifford algebra. Ann. of Math. (2) **80**:365–81.

———. 1965. Spinor representations of the infinite orthogonal groups. J. Math. Mech. **14**:315–22.

Stein, E. M. 1956. Interpolation of linear operators. Trans. Amer. Math. Soc. **83**:482–92.

Stone, M. H. 1930. Linear transformations in Hilbert space *III*. Operational methods and group theory. Proc. Nat. Acad. Sci. **16**:172–75.

Streater, R. F., and Wightman, A. S. 1964. *PCT, spin and statistics, and all that*. New York: Benjamin.

Titchmarsh, E. C. 1952. *The Theory of Functions*. London: Oxford Univ. Press.

van Hove, L. 1952. Les difficultés de divergences pour un modele particulier de champ quantifié. Physica **18**:145–59.

Vergne, M. 1977. Groupe symplectique et seconde quantification. C. R. Acad. Sci. Paris Ser. A-B **285**:A191–94.

von Neumann, J. 1931. Die eindeutigkeit der Schrödingerschen Operatoren. Math. Ann. **106**:570–78.

———. 1935. Quantum mechanics of infinite systems. From mimeographed notes of seminar conducted by W. Pauli on the "Theory of positron," 1935–36. Institute for Advanced Study.

———. 1936. On an algebraic generalization of the quantum-mechanical formalism. Mat. Sbornik (N. S.) **1**:415–84.

———. 1938. On infinite direct products. Compositio Math. **6**:1–77.

———. 1955. *Mathematical foundations of quantum mechanics*. Translated from the German. Princeton Univ. Press.

Weinless, M. 1969. Existence and uniqueness of the vacuum for linear quantized fields. J. Funct. Anal. **4**:350–79.

Wick, G. C. 1950. The evaluation of the collision matrix. Phys. Rev. (2) **80**:268–72; MR **12**:380.

Wiener, N. 1938. The homogeneous chaos. Amer. J. Math. **60**:897–936.

Wightman, A. S. 1956. Quantum field theory in terms of vacuum expectation values. *Phys. Rev.* (2) **101**:860–66.

Wightman, A. S., and Gårding, L. 1964. Fields as operator-valued distributions in relativistic quantum theory. Ark. Fys. **28**:120–84.

Wigner, E. P. 1939. On unitary representations of the inhomogeneous Lorentz group. Ann. of Math. (2) **40**:149–204.

Index

Milton Keynes UK
Ingram Content Group UK Ltd.
UKHW022026100524
442551UK00006B/343

9 780691 605128